Total Synthesis of Steroids

This is Volume 30 of
ORGANIC CHEMISTRY
A series of monographs
Editors: ALFRED T. BLOMQUIST and HARRY WASSERMAN

A complete list of the books in this series appears at the end of the volume.

Total Synthesis of Steroids

ROBERT T. BLICKENSTAFF

Medical Research Laboratory
Veterans Administration Hospital
and
Department of Biochemistry
Indiana University School of Medicine
Indianapolis, Indiana

ANIL C. GHOSH

Sheehan Institute and Sharps Associates (SISA)
Cambridge, Massachusetts

GORDON C. WOLF

Medical Research Laboratory
Veterans Administration Hospital
Indianapolis, Indiana

East Texas Baptist College Library
Marshall, Texas

ACADEMIC PRESS   New York and London   1974

A Subsidiary of Harcourt Brace Jovanovich, Publishers

COPYRIGHT © 1974, BY ACADEMIC PRESS, INC.
ALL RIGHTS RESERVED.
NO PART OF THIS PUBLICATION MAY BE REPRODUCED OR
TRANSMITTED IN ANY FORM OR BY ANY MEANS, ELECTRONIC
OR MECHANICAL, INCLUDING PHOTOCOPY, RECORDING, OR ANY
INFORMATION STORAGE AND RETRIEVAL SYSTEM, WITHOUT
PERMISSION IN WRITING FROM THE PUBLISHER. RIGHTS FOR
PRIVILEGED USE BY THE UNITED STATES GOVERNMENT FOR
GOVERNMENTAL PURPOSES ARE RESERVED.

ACADEMIC PRESS, INC.
111 Fifth Avenue, New York, New York 10003

United Kingdom Edition published by
ACADEMIC PRESS, INC. (LONDON) LTD.
24/28 Oval Road. London NW1

Library of Congress Cataloging in Publication Data

Blickenstaff, Robert T
    Total synthesis of steroids.

    Includes bibliographical references.
    1.   Steroids.   2.   Chemistry, Organic–Synthesis.
I.   Ghosh, Anil Chandra, joint author.   II.   Wolf, Gordon
C., joint author.   III.   Title.
QD426.B55   547′.73   73-20486
ISBN 0–12–105950–2

PRINTED IN THE UNITED STATES OF AMERICA

To our parents,
and to Ruthie, Sumitra, and Mary

Contents

Preface   xi

Abbreviations   xiii

1. Designing Total Syntheses

1.1	Introduction	1
1.2	Defining the Problem	3
1.3	Methods and Reactions	10
1.4	Resolutions	23
	References	29

2. Biogenetic-like Steroid Synthesis

2.1	Cyclization of Terminal Epoxides	37
2.2	Total Synthesis from Nonepoxide Precursors	41
	References	46

3. AB → ABC → ABCD

3.1	Equilenin	49
3.2	Estrone	58
3.3	Bisdehydrodoisynolic Acid	63
3.4	18,19-Bisnorprogesterone and 19-Norpregnanes	65
3.5	Miscellaneous Syntheses	68
3.6	Heterocyclic Steroids	71
	References	79

4. AB+D → ABD → ABCD

4.1	Introduction	85
4.2	The Torgov Synthesis	86
4.3	Alkylation	103
4.4	Michael Addition	105
4.5	Enamines	108
4.6	Acylation	110
4.7	Aldol-Type Condensations	111
4.8	Organometallic Coupling	115
4.9	Miscellaneous Methods	118
	References	118

5. AB+D → ABCD

5.1	Introduction	126
5.2	Carbocyclics	127
5.3	Thiasteroids	131
	References	131

6. A+C → AC → ABC → ABCD

6.1	Synthesis from p-Anisylcyclohexanes	134
6.2	Synthesis from C-5,C-8 Bridged Intermediates	137
	References	140

7. A+CD → ACD → ABCD

7.1	Condensation of Ring-A Anionic Fragments with CD Ketones	142
7.2	Alkylation of an Anionic CD Fragment	146
7.3	The Wittig Reaction in Total Synthesis	151
7.4	CD Enamine Intermediates	152
7.5	Thiasteroids	154
7.6	Miscellaneous Routes	156
	References	158

8. A+D → AD → ABCD

8.1	The Smith-Hughes Approach	161
8.2	8-Azasteroids	166
8.3	8,13-Diazasteroids	169
8.4	Miscellaneous Routes	170
	References	173

9. BC → ABC → ABCD

9.1	Synthesis of Epiandrosterone	176
9.2	Other Methods	180
9.3	Synthesis According to Wilds	181
9.4	Synthesis of Cortisone	182
9.5	Synthesis of Aldosterone	187
9.6	3β-Hydroxy-5α-pregnan-20-one, Latifoline, and Conessine	194
9.7	Ring-C Aromatic Steroids	196
	References	196

10. BC → BCD → ABCD

10.1	Introduction	201
10.2	*Trans*-Benzohydrindane Derivatives	202
10.3	19-Nortestosterone	203
10.4	Adrenocortical Steroids	206
10.5	Estrogens	206
10.6	10-Allyl Steroids	207
10.7	19-Norprogesterone and 11-Hydroxysteroids	208
10.8	Trienes	211
10.9	7α-Methyl Steroids	213
10.10	Miscellaneous Steroidal Derivatives	213
10.11	Isosteroids	218
10.12	11-Oxygenated Steroids and Conessine	222
10.13	Miscellaneous Synthetic Routes	226
	References	228

11. B + D → BD → BCD → ABCD

	Discussion	234
	References	238

12. CD → BCD → ABCD

12.1	CD Intermediates	239
12.2	Hydrochrysene Approach	243
12.3	Modification of the Hydrochrysene Synthesis	254
12.4	C-Nor-D-homosteroids	256
12.5	Woodward Synthesis	260
12.6	Asymmetric Induction	266
12.7	Hoffman-LaRoche Approach	273
12.8	Miscellaneous Syntheses	275
	References	279

Appendix. **Supplementary References** 286

Author Index 293

Subject Index 317

Preface

Few ventures in organic chemistry have presented a challenge as formidable as the synthesis of naturally occurring steroids from common chemicals. These endeavors have occupied the attention of some of the keenest minds in the field, and their successes provide one of the more exciting chapters in this century's accomplishments. Major advancements in synthetic organic chemistry and in conformational theory have resulted from these endeavors. In gathering together an account of this work, it was difficult to avoid a strictly historical approach, but for the most part we have tried to approach the subject as one planning to synthesize a steroid and needing to know what has already been done.

This book is for the organic chemist who desires an overall view of steroid total synthesis, including the general approaches, special problems, stereochemical complexities, expansion or contraction of rings, and insertion of hetero atoms. Not all conceivable steroids have been synthesized, but a large enough number have so that it is difficult to imagine one for which some of the problems to be encountered have not already been solved. We hope the experience detailed between these covers will aid the chemist planning a new steroid synthesis.

Shortly after beginning the book, we became aware of Akhrem and Titov's highly competent work '"Steroid Total Synthesis" in English translation. Their bibliography is complete through 1965 (with many additional references into 1968), and we made frequent use of it in developing our account of the earlier work. We instituted a computor search of *Chemical Abstracts* through Aerospace Research Applications Center, Indiana University Foundation, for the period January 1, 1968, through December 31, 1973. References received too late for inclusion in the text can be found in the Appendix, pp. 286–292. We are indebted to Arthur

Nagel for loaning us his thesis prior to publication of his research, and to G. Saucy and N. Cohen for sending a prepublication copy of their chapter in "Steroids, MTP International Review of Science," Organic Chemistry Series, Vol. 8, 1973.

We wish to thank S. Chatterjee, F. Johnson, R. K. Razdan, J. C. Sheehan, and D. M. S. Wheeler for their critical review of portions of the text. We, of course, accept full responsibility for any errors or omissions. One of us (ACG) would like to express his grateful thanks to H. G. Pars for his constant encouragement. We are endebted to Rita Muncie for drawing the structures and to Rita Stanley, Anne Clarke, and Jeanne Fernald for typing the manuscript.

ROBERT T. BLICKENSTAFF
ANIL C. GHOSH
GORDON C. WOLF

Abbreviations

The following abbreviations have been used throughout the Schemes.

Ac	Acetyl
Bz	Benzyl
DDQ	2,3-Dichloro-4,5-dicyanobenzoquinone
DHP	Dihydropyran or dihydropyranyl
DME	Dimethoxyethane
DMF	N,N-Dimethylformamide
DMSO	Dimethyl sulfoxide
DNP	Dinitrophenylhydrazone
DVK	Divinyl ketone
EDTA	Ethylenediaminetetraacetic acid
EG	Ethylene glycol
Et	Ethyl
EVK	Ethyl vinyl ketone
i-Am	Isoamyl
i-Bu	Isobutyl
IPA	Isopropenyl acetate
i-Pr	Isopropyl
LAH	Lithium aluminum hydride
MCPBA	m-Chloroperbenzoic acid
Me	Methyl
Ms	Methanesulfonyl
MVK	Methyl vinyl ketone
NBA	N-Bromoacetamide
NBS	N-Bromosuccinimide
n-Bu	n-Butyl
Ph	Phenyl
PPA	Polyphosphoric acid
Py	Pyridine
t-Am	tert-Amyl
t-Bu	tert-Butyl
THF	Tetrahydrofuran
THP	Tetrahydropyranyl
Ts	p-Toluenesulfonyl

1
Designing Total Syntheses

1.1 Introduction

The total synthesis of steroids, in one or more of its various aspects, has been, with increasing frequency, the subject of reviews over the period of the past decade. Most of these authors themselves have been prolific contributors to the wealth of literature in which the often epic events were described. In two of the earlier of these reviews, Torgov presented a masterful summary of types of reactions featured in successful syntheses, specific problems faced, and industrial importance of total synthesis [1]. Velluz, Mathieu, and Nominé reviewed total syntheses relative to the production of contraceptive steroids [2], while Johnson and colleagues described their hydrochrysene approach to total synthesis [3]. Mathieu presented syntheses from useful 4,9,11-trienic intermediates [4], and Smith presented routes to equilin and 11-oxygenated steroids [5] at a symposium on Drug Research in Montreal. Akhrem and Titov published the Russian edition of their outstanding book on total synthesis in 1967, and followed that with an English translation by Hazzard [6]. The review of steroidal estrogens by Morand and Lyall contains an extensive section on total synthesis [7]. Gogte's review of heterocyclic steroids dealt primarily with total synthesis [8], a subject also treated by Huisman the same year [9]. Pappo presented new developments important to industrial synthesis of steroids [10], and Stork compared annelation methods [11]. Huisman's review of heterocyclic steroids has been brought up to date as of 1971 [12]. Ninomiya's review of azasteroids is devoted in part to their total synthesis [13]. The total synthesis of carbocyclic steroids has been thoroughly reviewed recently by Saucy and Cohen [14]. Naturally, all of these, in addition to other reviews, were of inestimable help in the writing of the present book.

With the culmination, in the early 1930's, of several decades of work on the determination of structures of naturally occurring steroids, it was inevitable that organic chemists would turn their attention toward total synthesis. The steroids with which most of these structural studies were carried out, the sterols and bile acids [15], were far too complex in structure to be synthesized at that time. But with elucidation of the structure of equilenin, one of the simpler sex hormones, in 1935 [16], an obtainable goal was recognized by the synthetic chemists. Within just four years that bastion fell [17]. Attacks were mounted on other steroids, and in 1948 the stereochemically much more complex steroid estrone was conquered [18]. A spate of syntheses then followed, some of which are chronicled in Table 1-1 [17–28]. On examination of these structures, one is immediately impressed with their stereochemical complexity, and it becomes understandable that the solution of these formidable synthetic problems would have been infinitely more difficult without the timely introduction of the concepts of conformational analysis [29]. During the 1950's much effort was devoted to total synthesis of steroids, but during the early 1960's the pace slowed. Toward the end of the 1960's and into the 1970's the pace has quickened, primarily in the area of synthesis of unnatural steroids.

In any discussion of total synthesis it is necessary to confront the questions as to what constitutes total synthesis and how it differs from other types of synthesis. Some steroids have been synthesized utilizing only starting materials and reactants which are not derived in any sense from any known steroids; such syntheses are total. Other syntheses employ relay compounds obtained by degradation of naturally occurring steroids. Once such a relay compound has been totally synthesized, then in theory it should make no difference whether the relay compound derived by synthesis or by degradation is used to complete the synthesis of a given steroid. If a synthesis employed a relay compound obtained by degradation, the synthesis is said to be total in a formal sense. An example is the synthesis of epiandrosterone by Cornforth and Robinson, wherein they and their colleagues synthesized the Köster-Logemann ketone (see Chapter 9); this ketone, the relay compound, was then

1.2 Defining the Problem

obtained more readily by oxidation of cholesterol to provide material to complete the synthesis [20]. Similarly, once any steroid has been totally synthesized, the construction of a new side chain on it to obtain a different class of steroid represents a formal total synthesis of that class member. The conversion of epiandrosterone to tigogenin and neotigogenin [27] is an example of a formal total synthesis, inasmuch as epiandrosterone had been synthesided previously [20]. Similarly, other formal total syntheses include solasodine [30] and hecogenin [31] from diosgenin, tomatidine from neotigogenin [32], conessine from pregnenolone [33], lanosterol and agnosterol from cholesterol [22], digitoxigenin from 3β-hydroxy-5β-androstan-17-one [28], veratramine [34], bufalin and resibufogenin [35].* The details of most of these achievements, notable as they are, will not be presented here as they are beyond the scope of this book.

It was to be expected also that in the beginning attention would focus on synthesis of the naturally occurring steroids as final verification of the structure determination. The potential for commercial production has also fueled the race to synthesize these medically important compounds. The heaviest emphasis in recent years, however, has been toward syntheseis of steroids that do not exist in nature. When one considers that compounds with either larger or smaller rings, compounds with both possible geometrical configurations at every asymmetric center (though not all combinations of geometries), as well as compounds with heteroatoms replacing nearly every carbon have been synthesized, the conclusion is almost inescapable that *any desired steroid* can be synthesized. Be that as it may, the chemist planning a total synthesis faces a plethora of problems, some of which may resemble problems previously solved in the syntheses to be described in this book.

1.2 Defining the Problem

1.2.1 Structure

Steroids are compounds containing four or more fused rings having the same layout as the chrysene ring system. Any of the rings may be larger or smaller than six-membered, but more often than not the first three rings are six-membered and the fourth five-membered. Heteroatoms may appear anywhere, but nearly all of the naturally occurring

---

* In a trivial sense, every derivative or substitution product of (say) cortisol has been synthesized in a formally total manner.

TABLE 1-1[a]
First Total Syntheses of Selected Naturally Occurring Steroids

Steroid	Year	Authors [Reference]
Equilenin	1939	Bachmann, Cole, and Wilds [17]
Estrone	1948	Anner and Miescher [18]
Cortisone	1951	Woodward, Sondheimer, Taub, Heusler, McLamore [19]
Epiandrosterone	1951	Cardwell, Cornforth, Duff, Holtermann, and Robinson [20]
3-Ketoetio-5β-cholanoic Acid	1953	Wilds, Ralls, Tyner, Daniels, Kraychy, and Harnik [21]

1.2 Defining the Problem

Lanosterol	1954	Woodward, Patchett, Barton, Ives, and Kelly [22]
Testosterone	1955	Johnson, Bannister, Pappo, and Pike [23]
Aldosterone	1955	Schmidlin, Anner, Billeter, Vischer, and Wettstein [24]
Cholecalciferol [a]	1957	Inhoffen, Schutz, Rossberg, Berges, Nordsick, Plenio, Horoldt, Irmscher, Hirscheld, Stache, and Kreutzer [25]
Dihydroconessine	1958	Corey and Hertler [26]

TABLE 1-1 (continued)

Steroid		Year	Authors [Reference]
Tigogenin		1959	Danieli, Mazur, and Sondheimer [27]
Digitoxigenin		1962	Danieli, Mazur, and Sondheimer [28]

[a] The vitamin D group is included in this book even though it does not fit the definition of steroids given in Section 1.2.1.

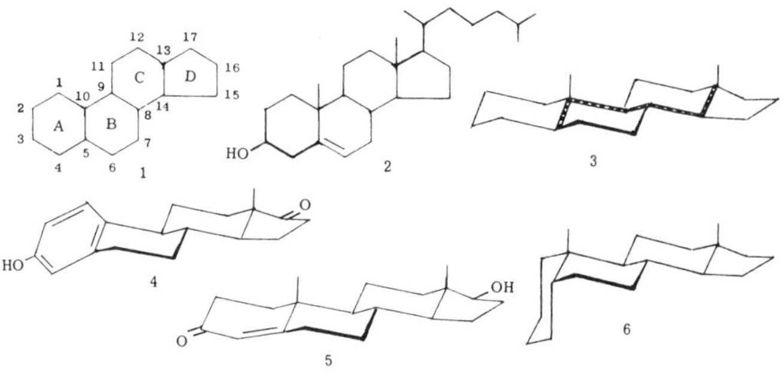

1.2 Defining the Problem

Scheme 1-1

steroids are carbocyclic. The stereochemical complexities become immediately apparent when one realizes that even in the unsubstituted hydrocarbon gonane* (**1**, Scheme 1-1), there are six asymmetric centers (carbons 5, 8, 9, 10, 13, and 14) which give rise to 64 isomers. Many steroids have more; and, as is often pointed out, cholesterol (**2**) has eight asymmetric centers, 256 stereoisomers. It is obvious, then, to avoid unmanageable separation problems, a total synthesis of such a compound must employ steps that are highly stereoselective. That is, reactions which create new asymmetric centers should produce the desired geometry in 80–95% yield. The term "desired geometry" initially referred to that of the naturally occurring steroids, but in future syntheses any conceivable combination of geometries might be the "desired" one.

The *trans-anti-trans-anti-trans* backbone of 5α-androstane (**3**, backbone shown in heavy dotted line) produces a structure that is both stable and relatively flat. The three *trans* terms refer to AB, BC, and CD ring fusions, while the *anti* terms refer to the 9,10- and 8,14- geometries. The 5α indicates that the 5-hydrogen is on the side of the ring opposite to the angular methyl group. This stereochemistry is found in such naturally occurring steroids as 3β-cholestanol, androsterone, and many others. Relatively flat structures are found also in unsaturated compounds like estrone (**4**) and testosterone (**5**), but steroids with a 5β-hydrogen (**6**) such as bile acids, have a decided bend in them.

As illustrated in structure **3**, the saturated A ring exists in a chair conformation [15, page 7] (unless it is forced into a boat conformation

---

* In those instances throughout this book where systematic names are used, we have followed the Revised Tentative Rules published in *Steroids* **13**, 277 (1969). Readers interested in a system of steroid nomenclature for computer use are referred to "Tentative Rules for the SSS Designation of Steroids," J. Seyle, S. Szabo, P. Kourounakis, and Y. Taché, Institut de Médecine et de Chirurgie Expérimentales, University of Montreal, Montreal, Quebec.

by substitution that destabilizes the chair). Because of the rigidity of this whole system, substituents on the A, B, and C rings are conformationally unique. That is 1α, 2β, 3α, 4β, 5α, 6β, 7α, 8β, 9α, 10β, 11β, 12α, 13β and 14α are axial substituents and the rest are equatorial. When ring D is five-membered 17β- and 15α-substituents are quasi-equatorial,, while 17α- and 15β-substituents are quasi-axial. As conformation influences chemical reactivity, advantages can be taken of these conformational differences to perform selective reactions, for example, selective oxidation of an axial hydroxyl in the presence of an equatorial one. Additional complex features will become clear as individual syntheses are discussed.

1.2.2 Synthetic Routes and Organization

It would seem logical that to construct a steroid molecule the synthetic organic chemist will in most instances start with a one- or two-ring moiety and attach chains to it capable of being cyclized. In that way the molecule is built up until the four rings are attached to each other in the required chrysene-type orientation. Stereochemical details are created or corrected, and functional groups are elaborated. Clearly, the ultimate stereochemical and functional group requirements influence the choice of reagents, which are chosen not only to add the necessary number of carbons, but also for their potential to produce functionality required later. As a first approximation, however, we shall ignore these subtleties and concentrate on constructing the carbon skeleton.

The synthesis of a steroid may be viewed in terms of the sequence in which rings are added, either by building up or as preformed rings. In the following arrangement of synthetic plans, the conversion of AB to ABC (AB → ABC) means that ring C is built onto a starting AB-ring system, while the combination of two rings (A + D → AD) signifies that two preformed rings were joined.

The first sequence begins with no preformed rings, but involves, rather, the cyclization of a polyolefin, and is designated → ABCD. It is illustrated by the cyclization of all *trans*-5,13,17-trimethyloctadeca-5,9,13,17-tetraenal ethyleneketal (7, Scheme 1-2) to the tetracyclic compound 8 in a single step [36]. A related synthesis (A → ABCD) employs a polyolefin containing the potential A ring and its conversion to 9 [37]; both are discussed in Chapter 2.

A sequence of great historical significance is AB → ABC → ABCD, 10 → 11 → 12, because it was used to synthesize equilenin, the first steroid to be totally synthesized [17]; Chapter 3 contains this type of synthesis.

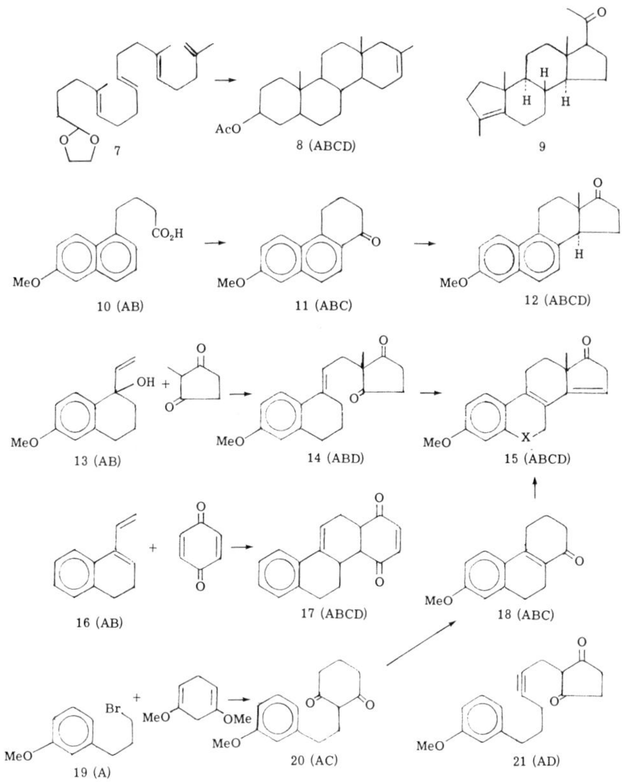

1.2 Defining the Problem

Scheme 1-2

The Torgov synthesis is an outstanding example of an AB + D → ABD → ABCD synthesis, **13** → **14** → **15** (X = CH$_2$) [38]. There are numerous other examples of this general scheme, varying mainly in how the C ring is constructed, in Chapter 4.

An AB + D → ABCD synthesis is illustrated by the Diels-Alder addition of benzoquinone to **16**, forming ring C in the process [39]; it is described in Chapter 5.

Alkylation of the dimethyl dienol ether of cyclohexane-1,3-dione with the bromide **19**, cyclization of **20** and construction of the D ring on **18** giving **15** (X = CH$_2$) illustrates an A + C → AC → ABC → ABCD synthesis [40] (Chapter 6). Coupling the A ring to CD portion **22** (Scheme 1-3), followed by cyclization of **23**, is an A + CD → ACD → ABCD synthesis [41] of the type described in Chapter 7. Another synthesis (A + D → AD → ABCD) begins with alkylation of ring D to give the AD intermediate **21** (Scheme 1-2), followed by ring closure to a tetracyclic steroid [42] (Chapter 8).

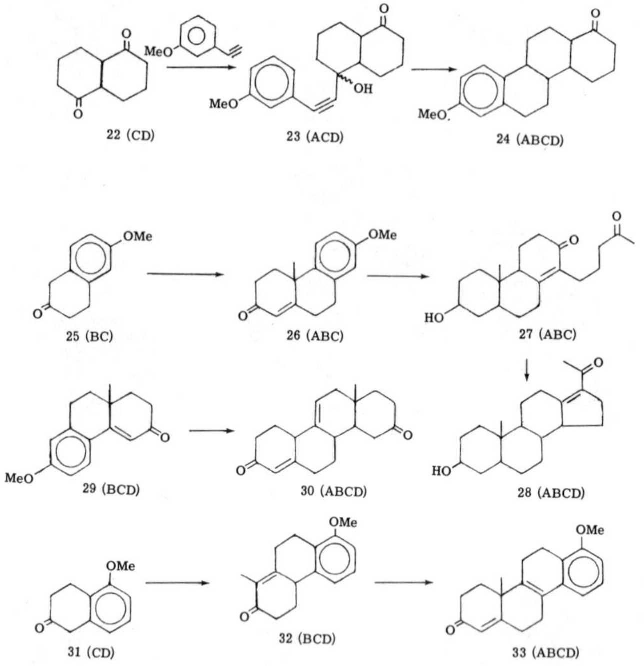

Scheme 1-3

BC intermediates can be converted into steroids by building on first the A ring then the D, or vice versa. The first of these (BC → ABC → ABCD), discussed in Chapter 9, is illustrated by the conversion of **25** (Scheme 1-3) to the pregnane derivative **28** [43]. Annelation of **25** to **29** followed by a second annelation to **30** constitutes the reverse route (BC → BCD → ABCD) [44] (Chapter 10).

Reactions coupling B to D (analogous to AB + D in Chapter 4) constitute the first step in the B + D → BD → ABCD synthesis described [45] in Chapter 11). Finally, in Chapter 12 the CD fragment **31** undergoes two annelation steps to give **33** (CD → BCD → ABCD) [46]. It is obvious that in many of these sequences, use of the appropriate heterocyclic compound at one of the stages in the synthesis would give rise to a heterosteroid. For example, the AB + D → ABD → ABCD synthesis has been used to prepare **15** (Scheme 1-2), where X = O [47], S [48], NH [49], and $SiMe_2$ [50].

1.3 Methods and Reactions

All the methods customarily used by organic chemists for extending chains or for attaching chains to rings have been employed in steroid

1.3 Methods and Reactions

total synthesis. In the following list of reactions, the examples cited are for illustrative purposes only, as the reader will encounter them repeatedly in subsequent chapters.

1.3.1 Michael Reaction

It is difficult to find a more popular reactant for total steroid synthesis than methyl vinyl ketone; the addition to its conjugated double bond of the anion of an appropriate cyclic ketone has ben utilized for construction of each of the four steroid rings (in separate syntheses). This Michael addition is the first half of the Robinson annelation procedure, the second half of which consists of intramolecular aldol-type condensation and dehydration [51]. Thus, in Sarett's synthesis of cortisone, the *cis*-BC intermediate **34** (Scheme 1-4) added stereospecifically to methyl vinyl ketone; the strong base closed the A ring of **35**, giving the ABC intermediate **36** in 39% yield [52]. In many instances it was found advantageous to employ the Mannich base methiodide **37** [53] or 4-chloro-

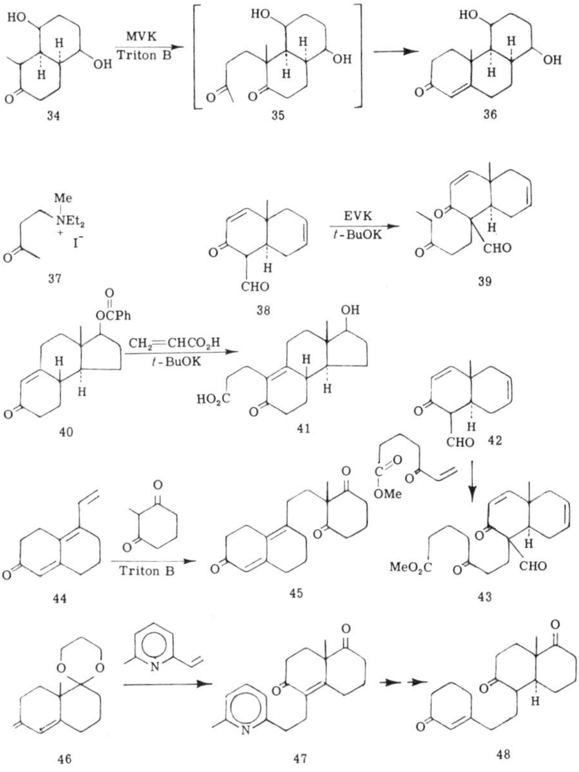

Scheme 1-4

2-butanone, both of which gave rise to methyl vinyl ketone in the reaction medium [54]. When an additional carbon was needed (a carbon which would ultimately become an angular methyl group), ethyl vinyl ketone was employed, as in the conversion of **38** to **39** in the Woodward synthesis of cortisone [19] (in the subsequent ring closure step the 8-formyl group was eliminated).

The attempted Michael addition of **40** to methyl vinyl ketone gave an abnormal product containing an aromatic ring fused to carbons 9, 10, and 11; however, it (**40**) underwent normal Michael addition to acrylic acid giving **41** in 50% yield [54]. Steroidal precursors have been synthesized by Michael addition of 2-methyl-1,3-cyclopentanedione to methyl 2-chloracrylate [55]. Methyl 5-oxo-6-heptenoate was used to furnish parts of both A and B rings by way of its Michael reaction with **42** [56]. The A and B rings were added intact in the reaction of **44** with 2-methyl-1,3-cyclohexanedione to give **45** [57]. A particularly intriguing annelation procedure is illustrated by the Michael addition of **46** to 6-vinyl-2-picoline, which furnished the equivalent of two moles of methyl vinyl ketone. The product **47** was reduced to the corresponding dihydropyridine and hydrolyzed to the seco steroid **48** [58]. A number of annelation procedures developed by Stork will be discussed in the section on alkylation (1.3.3).

1.3.2 Torgov Reaction

The Torgov reaction, like the Michael reaction, involves the addition of an anion to an unsaturated system, but is acid-catalyzed rather than base-catalyzed. Torgov and Ananchenko and their colleagues, as well as many others, have explored this reaction in depth, and have found it particularly efficacious for synthesis of aromatic A-ring steroids. Thus, the allylic alcohol **49** (Scheme 1-5, X = OH) reacted with 2-methyl-1,3-cyclopentanedione to give the ABD intermediate **50** in a synthesis of estrone [59]. Analogs of **49** with X equal to *N*-pyrrolidyl, *N*-piperidyl, and *N*-morpholino, as well as the isothiouronium salt **51** have also been used similarly.

1.3.3 Akylation

Large moieties like **49** and **51** can be attached to rings by simple alkylation procedures also. Thus, the dienol ether of 1-3-cyclohexanedione was alkylated with the bromide **52** in the presence of potassium amide; hydrolysis gave **53** in 86% yield [40]. In a familiar type of chain lengthening procedure, malonic ester was alkylated with the bromide **54** (the product **55** was further acylated to provide the carbons of ring

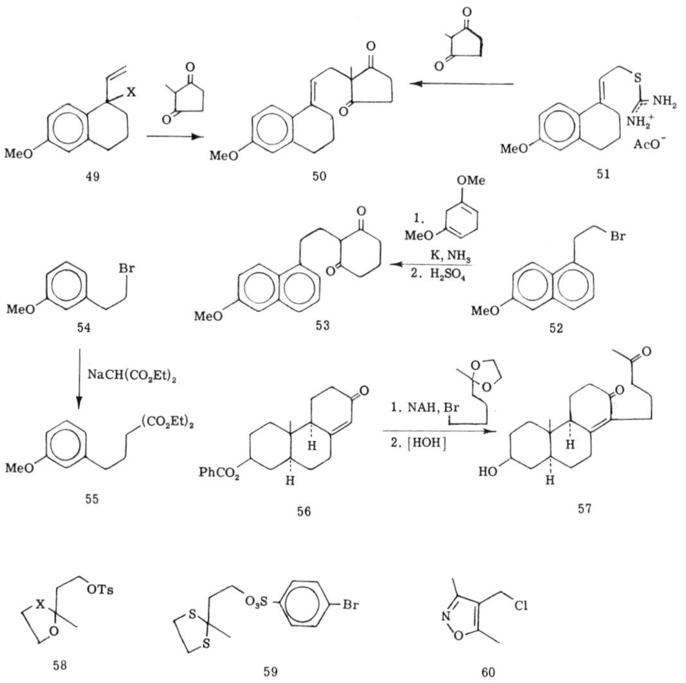

1.3 Methods and Reactions

Scheme 1-5

C before ring B was closed) [60]. The tricyclic ketone **56** was alkylated with the ethylene ketal of 5-bromo-2-pentanone in a synthesis of steroids with the conessine side chain [43]. This is one of the general annelation methods developed by Stork and colleagues [11,61]; its major advantage is the ease with which the protective group is removed, though it suffers from relative low reactivity of the bromide group [62]. In the total synthesis of some tetracyclic precursors of pentacyclic triterpenes, ApSimon and colleagues found the tosylate **58** (X = O) to be more reactive than the corresponding chloride or bromide [63]. Additional annelating agents include the hemithioketal tosylate **58** (X = S), the dithioketal brosylate **59** [64], and most particularly the isoxazole **60** [65]; improvements in the isoxazole annelation method have been reported [66].*

In the synthesis of 11-oxosteroids, **61** (Scheme 1-6), was alkylated with 3-benzyloxy-1-butyl bromide; the benzyloxy group is a potential car-

---

* Cyclization of α-haloketals is a new annelation procedure subject to remarkable steric control: G. Stork, J. O. Gardner, R. K. Boeckman, Jr., and K. A. Parker, *J. Am. Chem. Soc.* **95**, 2014 (1973); G. Stork and R. K. Boeckman, Jr., *J. Am. Chem. Soc.* **95**, 2016 (1973).

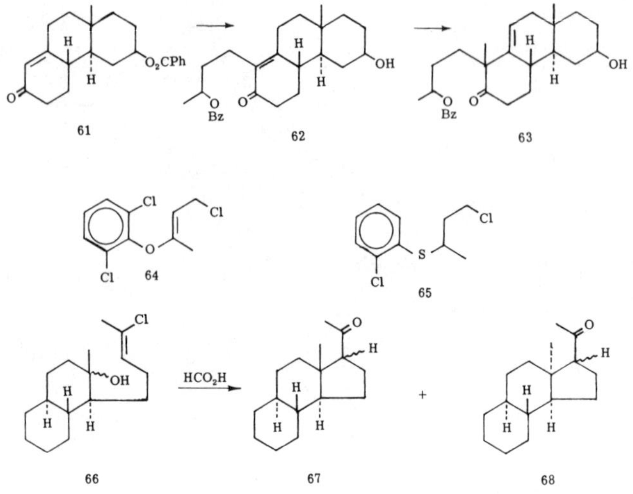

Scheme 1-6

bonyl, but in this case the protective group is more cumbersome to remove than are ketal groups. A second alkylation step employed methyl iodide to provide the angular methyl group, producing **63** and its C-10 epimer in about equal amounts [44]. The dichlorophenyl ether **64** and the chlorophenyl thioether **65** also have been developed as annelating agents [11].

A new annelation procedure shows promise for extension to steroids, its main drawback being that it lacks stereospecificity. $\Delta^{1(9)}$-Octalone was alkylated with 2-chloro-5-iodo-2-pentene, the double bond was reduced with lithium and ammonia, and the ketone group reacted with methyl Grignard reagent to give **66**. Formolysis of **66** at reflux temperature gave a 3:2 mixture of **67** and **68** (95% yield) [67].* Annelations via substituted pyridines were discussed in section 1.3.1. Three additional annelation methods have been developed which have potential utility for steroid synthesis. *Trans*-hydrindane systems resulted ultimately from the addition of 2,4-biscarbomethoxy-2,5-dihydrothiophene to 2-ethoxybutadiene [68], and from hydroboration and carbonylation of 1-vinylcyclohexene [69]. Monocyclics and bicyclics were annelated by alkylation of appropriate epoxides with diphenylsulfonium cyclopropylide [70].†

Angular methyl groups at both C-10 and C-13 have been incorporated

---

* See also: E. P. Woo, K. T. Mak, and H. N. C. Wong, *Tetrahedron Letters*, 585 (1973).

† See also: B. M. Trost and M. J. Bogdanowicz, *J. Am. Chem. Soc.* **95**, 289 and 2038 (1973); M. J. Bogdanowicz, T. Ambelang, and B. M. Trost, *Tetrahedron Letters*, 923 (1973).

1.3 Methods and Reactions

into the steroid molecule by methylation reactions in a variety of syntheses. The degree of stereospecificity (absent in the methylation of **62**) varies from compound to compound, and depends primarily on the stereochemistry of the material being methylated. Some of these influences are seen in the results of methylation of the isomeric 17a-ketones **69–72** (Scheme 1-7) [71–73]. These compounds were first condensed with aromatic aldehyde in order to protect the more reactive 17-position. The carbanions of the resulting arylidine derivatives were not accessible to methyl iodide equally on the α- and β-sides. Compounds **69** and **70** gave 4:1 and 3:1 α:β-methyl products, respectively, β-approach apparently being inhibited by the 8β-, 11β-, and 15β-hydrogens. Attack on the β-side, however, is preferred for **71** (11:1) and exclusive for **72**, α-approach being inhibited by the C-7 methylene group.

The rationalization of methylation results is not always so straightforward, however; in addition to accessibility, relative stabilities of the

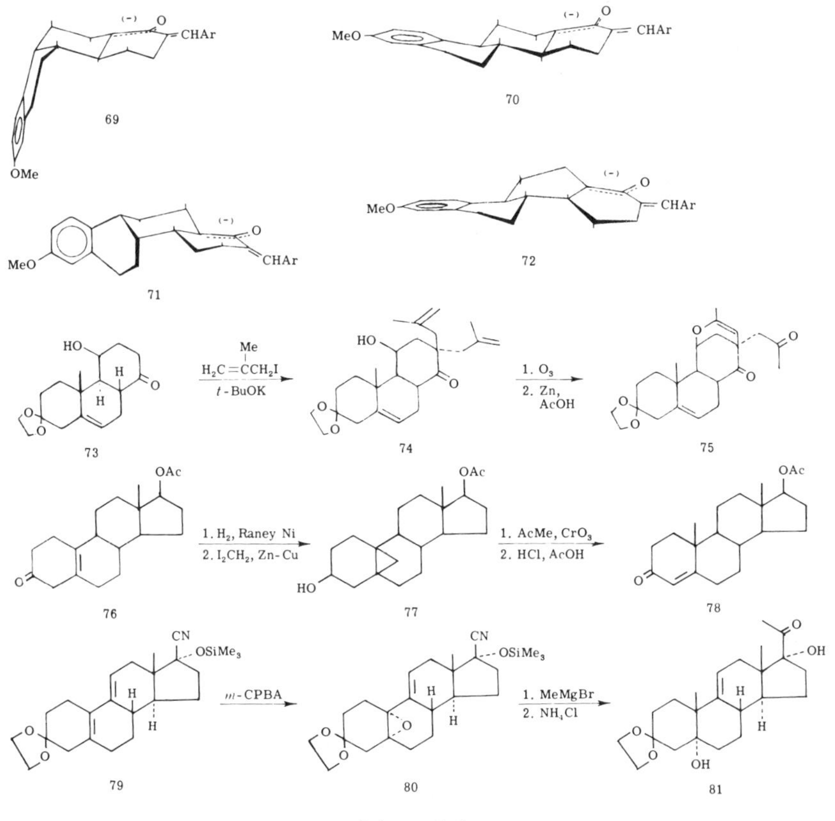

Scheme 1-7

transition states also play a role, and these can be severely altered by double bonds at different positions relative to the reaction center [73,74].

The problem of obtaining only the desired geometry in the creation of a new asymmetric center was solved in an ingenious fashion in Wettstein's synthesis of aldosterone. The tricyclic ketone **73** (Sarett's ketone) was alkylated with two moles of methallyl iodide to produce **74** in 89% yield. Upon ozonolysis of the two methylene groups, the β-2-oxo-1-propyl group spontaneously cyclized to the enol ether **75**, the carbons of the α-2-oxo-1-propyl group were ultimately transformed into ring D, and thus *complete* stereospecificity with regard to C-13 was achieved [75].

Two methods for introducing the C-10 angular methyl group in a highly stereospecific fashion have become available recently. For example, the acetate **76** of the product of Birch reduction and hydrolysis of estradiol 3-methyl ether was reduced in good yield with Raney nickel to the corresponding 3β-ol. The latter was converted to the 5β,10β-methylene derivative **77** with the Simmons-Smith reagent. Jones oxidation to the 3-ketone and acid-catalyzed rearrangement gave testosterone acetate (**78**) [76]. In another series, the diene **79** was oxidized with *m*-chloroperbenzoic acid to the epoxide **80**; small amounts of the 5β,10β- and 9α,11α-epoxides were formed also. Methylmagnesium bromide reacted with both the oxirane ring and the nitrile group, and ammonium chloride hydroylsis of the imine and trimethylsilyl ether groups gave **81** [77].*

1.3.4 Acylation

In the Robinson synthesis of epiandrosterone the tricyclic ketone **82** (Scheme 1-8) was carboxylated with carbon dioxide and triphenylmethyl sodium; esterification with diazomethane gave a 1:2 mixture of **83** and the unwanted 8-carbomethoxy isomer [20]. Another acylation method for introducing C-17 is the use of ethyl formate prior to methylation, as in the conversion of **85** to the hydroxymethylene derivative **84** [78]. The same ketone **85** was alternatively acylated by the Bachmann method of condensation with dimethyl oxalate to give the intermediate 13-oxalomethoxy derivative, which was decarbonylated to **86** by heating with powdered glass [79]. Bachmann had used this method on the tricyclic ketone **87** in the synthesis of equilenin [17]. The keto ester **88**, obtained in 90% yield, was racemic (as was **86**).

---

* For alkylation via cyclopropylamines, see: M. E. Kuehne and J. C. King, *J. Org. Chem.* **38**, 304 (1973).

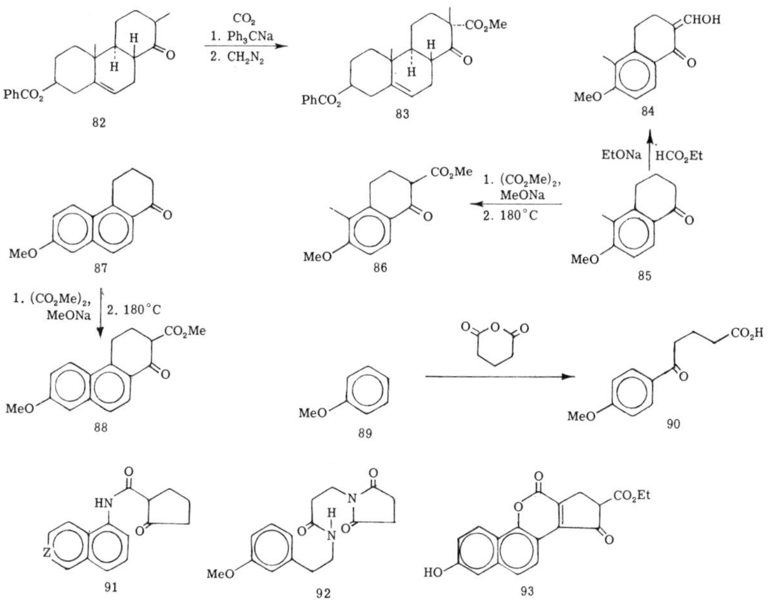

1.3 Methods and Reactions

Scheme 1-8

Acylation of anisole (**89**) with glutaric anhydride constituted the first step in Johnson's second synthesis of estrone [80]. 11-Azasteroids were synthesized beginning with the acylation of 1-naphthyl amines with 2-carboethoxycyclopentanone, which gave rise to **91** (Z = HC or MeOC) [81]. The precursor **91** (Z = N) to a 3,11-diazasteroid was prepared similarly from 5-aminoisoquinoline [82]. Acylation of 2-(*m*-methoxyphenyl)ethylamine gave the AD intermediate **92**, precursor to 8,13-diazasteroids [83]. 11-Oxasteroids, **93** and the 2-hydroxy and 4-hydroxy isomers, were prepared by acylation of the requisite naphthalenediols [84].

1.3.5 Reformatsky Reaction

The Reformatsky reaction was utilized in the Bachmann synthesis of equilenin to introduce C-15 and C-16. The keto ester **94** (Scheme 1-9), obtained by methylation of **88** (Scheme 1-8), was condensed with methyl bromoacetate and zinc; the asymmetry at C-14 was subsequently destroyed by a dehydration step [17]. The Reformatesky reaction has been used frequently in just this way to introduce C-15 and C-16. Furthermore, it has been used to add C-11 and C-12 [85], C-13 and C-17 [85a], C-20 and C-21 [86], as well as three [87] and four-carbon chains [88], both of the latter having been carried out on the 13-methyl homolog of **87** (Scheme 1-8).

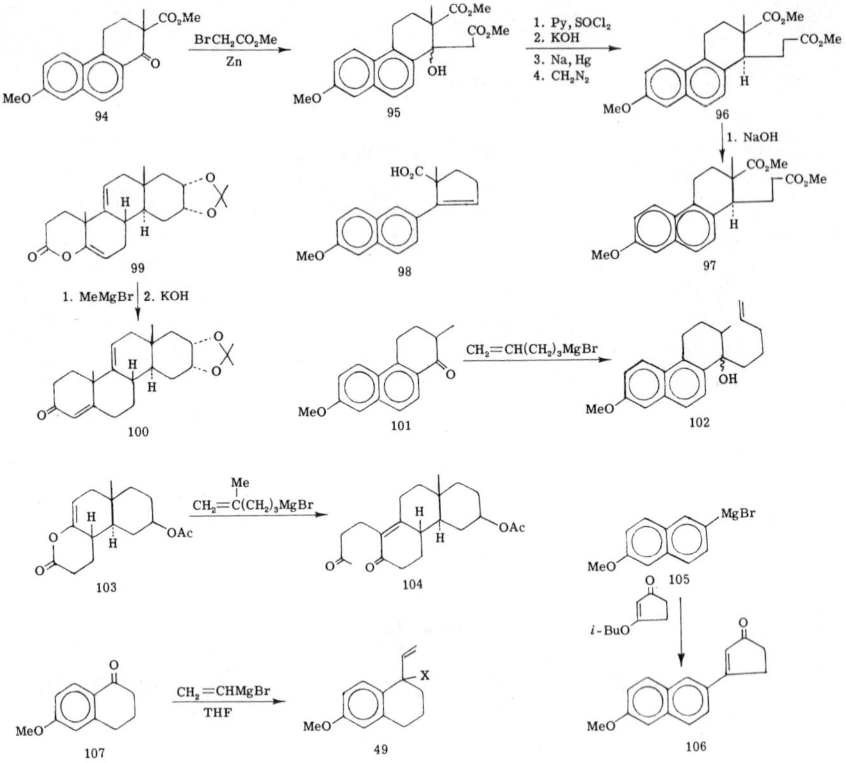

Scheme 1-9

1.3.6 Arndt-Eistert

Many steroid syntheses have included an Arndt-Eistert sequence to add on C-17 in the construction of the D ring. As in the synthesis of equilenin [17], the Reformatsky product **95** (Scheme 1-9) was dehydrated, the ester groups hydrolyzed, the 14,15-double bond reduced, and the acid groups re-esterified giving **96** and the CD-*cis* isomer (not shown), which were separated and used individually. After selective hydrolysis of only the C-16 ester, the Arndt-Eistert procedure gave the diester **97**. A similar sequence added C-11 to **98** [89].

1.3.7 Grignard Reaction

A wide variety of Grignard reagents, reacting with either ketones or lactones, have been utilized in steroid syntheses. The first step in the Fujimoto-Belleau reaction, employed in the Woodward synthesis of cortisol [19], consisted of reaction of the lactone **99** with methyl-

1.3 Methods and Reactions

magnesium bromide. The resulting 4,5-seco-3,5-dioxosteroid was subjected to intramolecular base-catalyzed condensation giving **100** in 58% yield from the lactone. The 13-methyl homolog, **101**, of **87** (Scheme 1-8) reacted with 4-penten-1-ylmagnesium bromide to produce a 90% yield of the *cis* and *trans* isomers **102** [90]. The lactone **103** (prepared by ozonolysis of the tricyclic ketone **61**) (Scheme 1-6) reacted with 4-methyl-4-penten-1-ylmagnesium bromide to give the secosteroid **104** [44]. The naphthalene Grignard reagent **105** reacted with the isobutyl enol ether of 2-methylcyclopentane-1,3-dione to give a good yield of the ABD intermediate **106** [91]. A large number of vinyl alcohols have been prepared for the Torgov synthesis by the reaction of vinylmagnesium bromide with the appropriate ketone, as in the preparation of **49** (Scheme 1-5, X = OH) from **107** [59].

1.3.8 Wittig Reaction

The Wittig reaction has not often been used in steroid synthesis, but Inhoffen found it indispensible in one synthesis of cholecalciferol (**110**, Scheme 1-10), utilizing it no fewer than three times. Allyltriphenylphosphonium bromide reacted with the ketone **108** to give a 40% yield of **109a**, while the corresponding methyltriphenylphosphonium bromide reacted with **109b** in 22% yield to produce **110** [25]. In a synthesis of retroprogesterone, ethylidenetriphenylphosphorane was used to prepare **111** from the corresponding 17-ketone [92]. A 46% yield of **113** was obtained in the Wittig reaction shown, leading to the synthesis of estrone methyl ether [93].

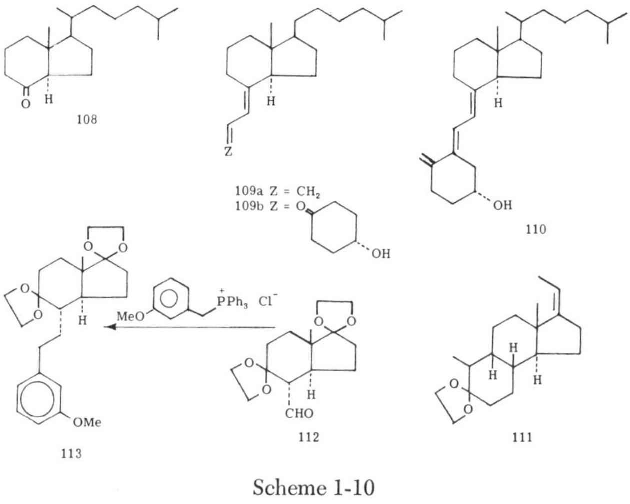

Scheme 1-10

1.3.9 Stobbe Reaction

The Stobbe reaction has been used to introduce two-, three-, and four-carbon segments. In the synthesis of bisdehydrodoisynolic acid, dimethyl succinate was condensed with the ketone **114** (Scheme 1-11); subsequent hydrogenation gave the half ester **115** in 52% yield [94]. One of the key intermediates **117** in a Johnson synthesis of estrone was synthesized by Stobbe condensation of **116** with dimethyl succinate, followed by hydrogenation and angular methylation to give **117** [95,96].

1.3.10 Ethynylation

In an alternative to the Reformatsky reaction, two-carbon chains have been introduced with acetylenic reagents. The lithium salt of methoxyacetylene added to **117** and the resulting *tert*-alcohol was converted directly to **118** by the action of base [96]. Diene intermediates such as **121** were prepared for Diels-Alder type syntheses of steroids by reaction of salts of acetylene with the appropriate ketone, followed by reduction and dehydration (**119** → **120** → **121**) [97].

1.3.11 Hydrogenation and Reduction of Double Bonds

Many syntheses produce unsaturated compounds in the ring closure reaction, so that usually at a subsequent stage of the synthesis double

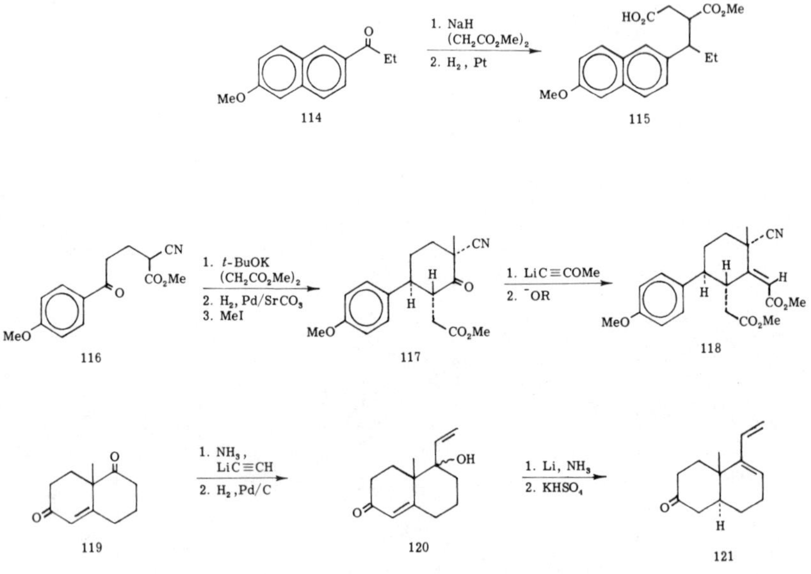

Scheme 1-11

1.3 Methods and Reactions

bonds are reduced, giving rise to new asymmetric centers. The stereochemistry of the product is determined by the position of the double bond, other substituents in the molecule, and whether catalytic hydrogenation or chemical reduction is employed.

The 14-double bond of tetracyclics is readily hydrogenated selectively, normally to a CD-*trans* product if there is a 13β-substituent, but other groups in the molecule can reverse that. Thus, the 17β-ester **122** (Scheme 1-12) was converted to the 14α-dihydro derivative **123**, but the epimeric 17α-ester **124** gave the 14β-dihydro derivative **125** [98]. This implies that the normal steric inhibition by the 13β-methyl group is overshadowed by the bulkier carbomethoxy group. The 16α-methoxyl group of **126** (X = CH$_2$, R = OMe), however, did not block α-hydrogenation [99]. The 6-oxasteroid **126** (X = O, R = H) was hydrogenated to the 14α-H derivative as usual [100], but the aza analog **126** (X = NTs, R = H) [101] and the thia analog **126** (X = S, R = H) [48], as well as **127** [102] and **128** [103] all gave mixtures of 14-epimers. The aza- and thiasteroids and **128** were hydrogenated to 14α-H derivatives exclusively

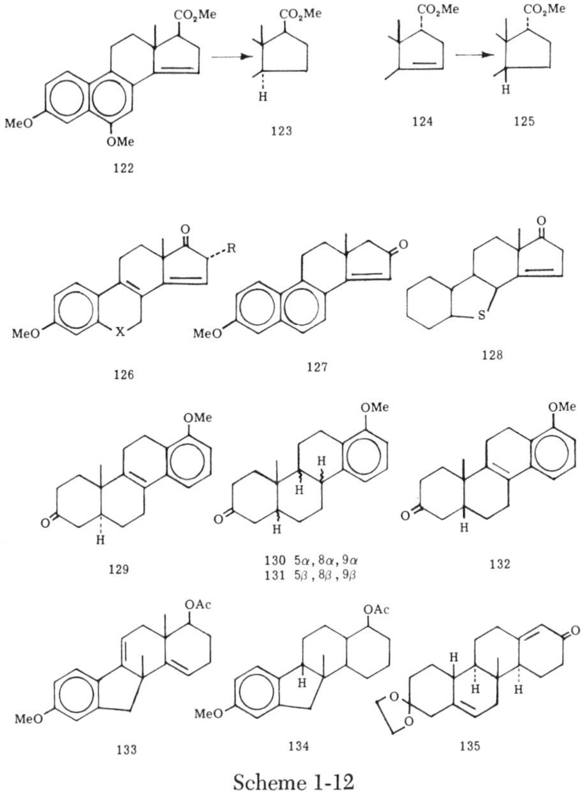

Scheme 1-12

when the 17-oxo group was first reduced to the corresponding 17β-ol, while **127** was hydrogenated exclusively *cis* in the presence of alkali.

The stereochemical outcome of catalytic hydrogenation being determined by the accessibility of the double bond to the catalyst surface [104], it was understandable that **129** would be hydrogenated to **130** (19 methyl blockage) and **132** would be hydrogenated to **131** (C-4 methylene blockage) [46, 105]. The hydrogenation of **133** with its two β-methyl groups to the 9β-derivative **134** is inexplicable, however [106].

Exhaustive hydrogenation of the two double bonds in compounds such as **126** (R=H) gave 8α, 9α, 14α-derivatives, so that other approaches to 8β-derivatives were sought. One method was to hydrogenate only the 14-double bond (as discussed above), isomerize the 8-double bond to 9(11) under acid catalysis, which gave 8β-geometry, then hydrogenate the 9(11)-double bond to a 9α-derivative [107]. More often, though, Δ^8-intermediates were reduced directly to 8β,9α-derivatives with lithium in liquid ammonia [59]. Although occasionally *cis* products were obtained in metal–ammonia reductions of double bonds conjugated to aromatic rings or carbonyl groups, the usual result was the more stable *trans* product, as in the reduction of **135** to the corresponding 13β-dihydro derivative [108].

1.3.12 Reduction of Aromatic Rings, Metal–Ammonia Reductions

Use of the Birch reduction for synthesis of 19-norsteroids is well known outside the area of total synthesis as a result initially of the pioneering work of Birch [109] and of refinements by Wilds [110]; it has been applied to both A and D rings in the total synthesis of steroids. Birch reduction of **136** (Scheme 1-13), followed by hydrolysis gave the

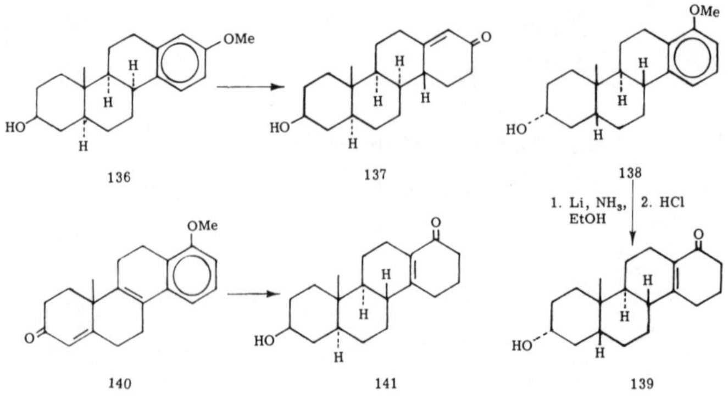

Scheme 1-13

1.4 Resolutions

unsaturated ketone **137** [111]. The *trans* relationship of the 8- and 14-hydrogens is typical of these reductions. Birch reduction and hydrolysis of **138** took place in both possible directions, and gave rise predominantly to **139**, accompanied by a lesser amount of the Δ^{16}-isomer [112]. This mixture of isomers was no problem as its hydrogenation in alkaline medium gave one product, the 13β,14α-dihydro derivative. Similarly Birch reduction of the 3β,5α-isomer of **138** gave a mixture of **141** and its Δ^{16}-isomer, which was hydrogenated in the presence of potassium hydroxide to the corresponding 13β,14α-dihydro derivative. In fact, it was possible to begin the sequence with **140** and reduce the 3-keto and 8-double bond simultaneously with the aromatic ring. In this instance two simple steps created six new asymmetric centers, a remarkable feat of stereochemical control.

Numerous other methods and procedures frequently used in total synthesis could be discussed here, such as the Dieckmann cyclization or the protection of keto groups by ethylene ketal formation, but the reader will become amply aware of them in succeeding chapters. There is also a great body of steroid chemistry, some of which (for example, photolytic functionalization of angular methyl groups) is applicable to the synthesis of one class of steroid from another. Much of this is beyond the scope of the present book, and the reader is referred to, in addition to the well known volume by the Fiesers [15], works by Kirk and Hartshorn [113], Djerassi [114], and Fried and Edwards [115]. Attention is also directed to a chapter by Lednicer on latent functionality in organic synthesis [115a].

1.4 Resolutions

Chemical syntheses of steroids of necessity produce racemic mixtures of products unless a resolution step is introduced at some stage. This means, of course, that a synthesis that produces estrone (**142**, Scheme 1-14) produces at the same time an equimolar amount of its enantiomer, 8α,9β,13α,14β-estrone (**143**). A major goal of many of the synthetic plans to be described in this book has been to achieve the highest possible degree of stereospecificity at each stage where a new asymmetric center was generated. Such steric control is possible only after an asymmetric center already exists in the molecule (unless an asymmetric catalyst is used), and, as a racemic pair is always produced in creation of the first asymmetric center, succeeding steps, even if 100% stereospecific, produce both enantiomers equally. The naturally occurring enantiomer of

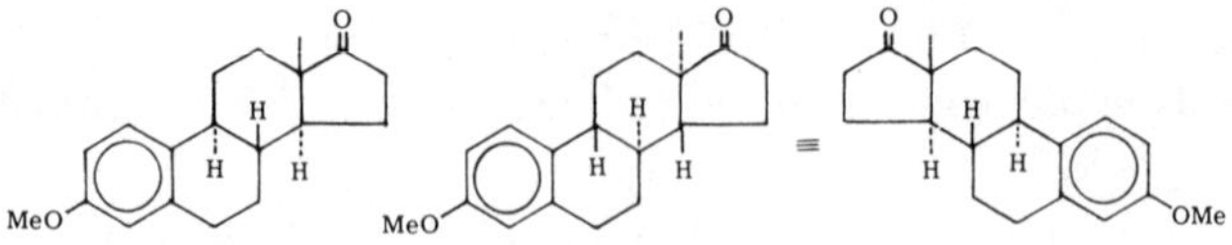

Scheme 1-14

estrone has the absolute configuration depicted by **142**, and is referred to as (+)-estrone or *d*-estrone to distinguish it from **143** and from the racemate, (±)-estrone or *dl*-estrone. Similarly, natural *d*-cholesterol has the absolute configuration shown in sturcture **2** (Scheme 1-1, in which the configuration at C-10 is R). All steroids corresponding in configuration at C-10 to cholesterol, as depicted by the tetrahedron **144**, are said to belong to the *d*-series [116]; this is true even if substitution (such as replacing C-1 with N, O, or S, or attaching an hydroxyl to C-1) changes the configurational notation of C-10 to S. All steroids with the configuration at C-10 represented by tetrahedron **145** belong to the *l*-series. If C-10 in a steroid is trigonal rather than tetrahedral, C-13 in cholesterol becomes the point of reference. Greek letters α and β refer to substituents on opposite or same sides, respectively, as the angular methyl groups (on C-10 and C-13 of cholesterol), and this convention applies equally to *d*- and *l*-steroids. The cumbersome name for **143** in the second sentence of this paragraph is appropriately replaced with *l*-estrone.

The reader should assume that unless there is a resolution step at some stage in the syntheses described in this book, the products (and appropriate intermediates) are racemic, but that this usually will not be restated. Furthermore, structures such as **142** are used to depict racemic products without their being labeled *dl*, and outside this section very little is said about resolution. This has been done not to diminish the importance of resolutions, but because the chemistry of constructing the molecule is the same whether it is carried out on an optically pure compound or on a racemic mixture, which may be resolved at completion of the synthesis (the economics, of course, might be quite different in the two cases).

Traditionally the justification for carrying out a resolution of a racemic steroid has been twofold: (1) to compare physical properties with a

1.4 Resolutions

naturally occurring counterpart (if one exists), and (2) to test each enantiomer for biological activity. It is now believed that often only steroids belonging to the same absolute configuration series as **142** are biologically active; consequently, it is quite feasible to assess the biological activity of racemic mixtures on the assumption that half of the mixture is inert [117]. Nevertheless, it often has been and at times still is, important to resolve racemates, and the methods for doing so are hereby set down.

All practical resolutions are based on one or the other of two basic methods: (1) reaction of the racemate with a chiral reagent to produce diastereomers that can be separated by physical means, and (2) reaction with a chiral reagent to take advantage of its faster reaction with one enantiomer than with the other [118]. The first resolution ever performed was achieved when Pasteur manually separated crystals whose physical appearance indicated a mirror-image relationship between the two types present. A more common method of resolution employs selectively seeding a supersaturated solution of a racemate with one of the enantiomers. We are unaware of any reports of either of these last two methods used with steroids.

The two components of a racemic mixture, having identical physical properties except for their interaction with polarized light, cannot be separated by ordinary physical means. But when the two enantiomers combine with a chiral reagent, which is itself a single enantiomer, diastereomers are created which usually have differing physical properties. That being so, the customary methods of separation such as fractional crystallization, distillation, extraction, and chromatography, can be employed to separate the two diastereomers. The separation having been accomplished, each portion is cleaved to regenerate the starting compound to complete the resolution. In steroid syntheses, fractional crystallization has been the preferred separation method for this type of resolution.

In some instances the resolution was carried out early in the synthesis, in other instances it was performed on the final product. With carboxylic acids, the most common practice has been the formation of diastereomeric salts with optically active amines. For example, in a synthesis leading to estrone, racemic **146** (Scheme 1-15) was resolved with ephedrin [119]. Racemic **147** was resolved with brucine, the d-enantiomer being recognized by its transformation to an oxidation product of calciferol [25]. Chloroamphenicol [54] and strychnine [120] also have been used for resolution of carboxylic acid intermediates. 14β-Equilenin was obtained optically pure by resolving precursor **148** by means of its

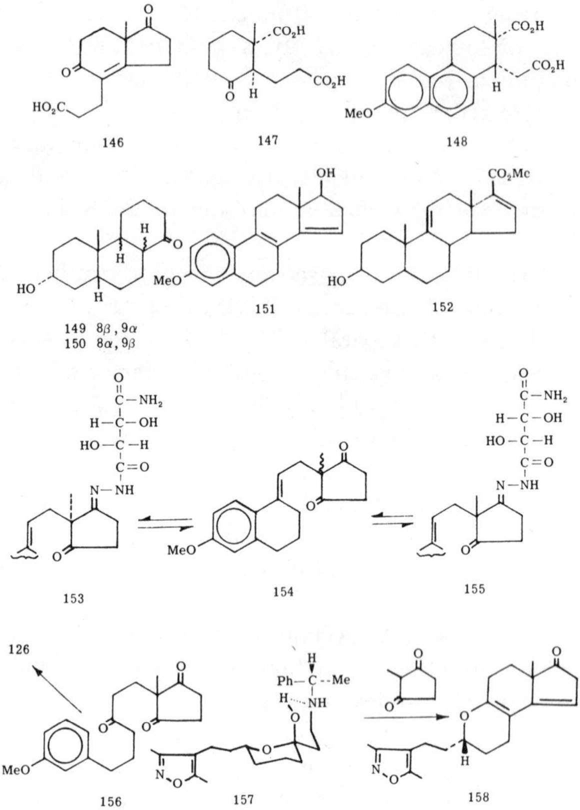

Scheme 1-15

l-menthyl ester [17]. Hydroxy intermediates and final products have been resolved via their *l*-menthoxyacetate esters or as mono esters of diacids. For example, after **149** and **150** were separated by fractional crystallization of their hemisuccinates, **149** was resolved as the *l*-menthoxyacetate and **150** as the brucine salt of the hemisuccinate [53]. Conversion to the hemisuccinate permitted resolution of **151** with dehydroabietylamine or with ephedrine [121]. The 3-methyl ether of estradiol was resolved via ester formation at C-17 with (+)-diacetyltartaric acid methyl ester [121,122]. 5-Hydroxy-10-methyl-$\Delta^{1(9)}$-2-octalone (dihydro-**119**, Scheme 1-11), and the 10-ethyl and 10-isopropyl homologs, were converted to their corresponding phthalate half esters and resolved by means of their brucine salts [123]. The 13-methyl homolog of estradiol 3-methyl ether was converted to its 17-sulfate ester and resolved with (−)-2-amino-1-butanol [124]. In the Woodward synthesis of cortisone, **152** was sepa-

1.4 Resolutions

rated from its 3α-epimer by precipitation with digitonin; a resolution (differing from those above) was accomplished simultaneously since the enantiomer of **152** did not precipitate [19]. Several steroidal ketones also have been resolved. (−)-2-Amino-1-butanol was condensed with the 13-methyl homolog of estrone methyl ether, permitting resolution of the resulting imine [124]. The intermediate **154** was resolved by hydrazone formation with the monohydrazide monoamide of L-tartaric acid. The diastereomer **155** crystallized out fortuitously, shifting the equilibrium to the right, and permitting isolation of **155** in 75% yield. Anhydrous hydrogen chloride in dioxane cyclized **155** and cleaved the hydrazone to give the corresponding tetracyclic steroid [125]. Optically active 3-methoxy-1,3,5(10),8,14-estrapentaen-17-one (**126**, X = CH_2, R = H, Scheme 1-12) was obtained by cyclization of **156** with a chiral amino acid such as (S)-phenylalanine [126]. Employing the principle of asymmetric induction (see Chapter 12, Section 12.6), Saucy and colleagues condensed 2-methyl-1,3-cyclopentanedione with the chiral Mannich base **157** (shown in its spirohemiketal form, in which it is most stable) to obtain predominately **158** and very little of the corresponding 13α-diastereomer [66]. Asymmetric induction is much more efficient than the typical resolution because in the latter, fully one-half of the product is useless. And obviously, the earlier in a synthesis a resolution is performed, the better from the point of view of conserving reagents.

Resolution by the kinetic method has not been applied to total synthesis of steroids except in the special sense of enzymic resolutions. Enzymes are chiral reagents that frequently make an absolute differentiation between enantiomers by being inert toward an "unnatural" one. This fact was first exploited with respect to steroids in the Wettstein synthesis of aldosterone. The racemic diketo lactone **159** (Scheme 1-16) was hydroxylated at C-21 by *Ophiobolus herpotrichus*. Only d-**159** reacted, and the newly formed d-21-hydroxy-**159** was separated from unreacted l-**159** [127]. Similarly, androstenedione has been resolved with *Saccharomyces cerevisae* (17-carbonyl reduction) [112], **160** with *Pseudomonas testosteroni* (oxidation of hydroxyl) [128], **161** (R=O) with *Saccharomyces cerevisiae* (17-carbonyl reduction) [129], **162** (R=O, R'=α-H,β-OH) with *Saccharomyces uvarum* (keto reduction) [130], **162** (R=-OH, -H, R'=O) with *Bacillus thuringiensis* [130], **163** and **160**-acetate with protaminase (hydrolysis of acetate) [129], D-homo-**162** (R=R'=O) with *Saachromyces carisbergensis* (14- and 17a-carbonyl reduction) [131], estradiol with placental 17β-hydroxysteroid dehydrogenase [132], **164** with *Rhizopus arrhizus* [133], **165** (R=Et or n-Pr, R'=H; R=R'=Et) with *Corynebacterium simplex* [134], and **166** with a

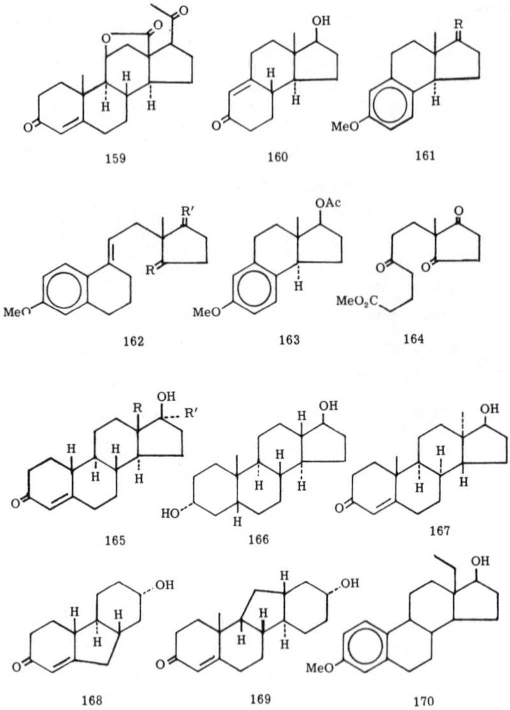

Scheme 1-16

3β,17β-hydroxysteroid dehydrogenase from *Pseudomonas testosteroni* [135].

Racemic **167** was resolved with *Arthrobacter simplex*, which contains a steroid 1,2-dehydrogenase. Similar resolution of **168** (with A-ring aromatization) and **169** helped to define the stereo requirement of the enzyme for an R configuration at C-10. The reaction can be used diagnostically, therefore, to establish absolute configuration of a synthetic product [136].

Not all biological resolutions are successful, however, as in some cases both enantiomers are transformed and a racemic product results; examples are the action of *Flavobacterium dehydrogenans* on **170** [134], of *Cunninghamella blakeseeana* on 8-aza-D-homoestradiol 3-methyl ether [137], and of the 3α-hydroxysteroid dehydrogenase from *Pseudomonas testosteroni* on **166** [135]. Following the pioneering work on C-11-hydroxylation by Peterson and colleagues [138], a large number of microbiological transformations of steroids have been performed [139], some of which may yet be adapted to resolutions.

References

1. I. V. Torgov, *Pure Appl. Chem.* **6**, 525 (1963); *Recent Develop. Chem. Natur. Carbon Compounds* **1**, 235 (1965).
2. L. Velluz, J. Mathieu, and G. Nominé, *Angew. Chem.* **72**, 725 (1960); *Tetrahedron* **22**, Suppl. 8, 495 (1966).
3. W. S. Johnson, J. A. Marshall, J. F. W. Keana, R. W. Franck, D. G. Martin, and V. J. Baeur, *Tetrahedron* **22**, Suppl 8, 541 (1966).
4. J. Mathieu, *Proc. Int. Symp. Drug Res., 1967* p. 134 (1967).
5. H. Smith, *Proc. Int. Symp. Drug Res., 1967* p. 221 (1967).
6. A. A. Akhrem and Yu. A. Titov, "Total Steroid Synthesis." Plenum, New York, 1970.
7. P. Morand and J. Lyall, *Chem. Rev.* **68**, 85 (1968).
8. V. N. Gogte, *J. Sci. Ind. Res.* **27**, 353 (1968).
9. H. O. Huisman, *Bull. Soc. Chim. Fr.* [1] p. 13 (1968).
10. R. Pappo, *Intra-Sci. Chem. Rep., Santa Monica, Cal.* **3**, 123 (1969).
11. G. Stork, *Proc. Int. Congr. Horm. Steroids, 3rd, 1970* p. 101 (1971).
12. H. O. Huisman, *Angew. Chem., Int. Ed. Engl.* **10**, 450 (1971).
13. I. Ninomiya, *J. Syn. Org. Chem. (Jap.)* **30**, 318 (1972).
14. G. Saucy and N. Cohen, in "Steroids," (W. Johns, ed.), MTP International Review of Science: Organic Chemistry, Vol. 8, p. 1. Butterworths, London, 1973.
15. L. F. Fieser and M. Fieser, "Steroids," Chapters 2 and 3. Van Nostrand-Reinhold, Princeton, New Jersey, 1959.
16. A. Cohen, J. W. Cook, and C. L. Hewett, *J. Chem. Soc., London*, p. 445 (1935).
17. W. E. Bachmann, W. Cole, and A. L. Wilds, *J. Amer. Chem. Soc.* **61**, 974 (1939); **62**, 824 (1940).
18. G. Anner and K. Miescher, *Experientia* **4**, 25 (1948); *Helv. Chim. Acta* **31**, 2173 (1948); **32**, 1957 (1949).
19. R. B. Woodward, F. Sondheimer, D. Taub, K. Heusler, and W. M. McLamore, *J. Amer. Chem. Soc.* **74**, 4223 (1952).
20. H. M. E. Cardwell, J. W. Cornforth, S. R. Duff, H. Holtermann, and R. Robinson, *Chem. Ind., London*, p. 389 (1951); *J. Chem. Soc., London*, p. 361 (1953).
21. A. L. Wilds, J. W. Ralls, D. A. Tyner, R. Daniels, S. Kraychy, and M. Harnik, *J. Amer. Chem. Soc.* **75**, 4878 (1953).
22. R. B. Woodward, A. A. Patchett, D. H. R. Barton, D. A. J. Ives and R. B. Kelly, *J. Amer. Chem. Soc.* **76**, 2852 (1954); *J. Chem. Soc., London*, p. 1131 (1957).
23. W. S. Johnson, R. Pappo, and J. E. Pike, *J. Amer. Chem. Soc.* **77**, 817 (1955); W. S. Johnson, B. Bannister, R. Pappo, and J. E. Pike, *ibid.*, **78**, 6354 (1956).
24. J. Schmidlin, G. Anner, J. R. Billeter, and A. Wettstein, *Experientia* **11**, 365 (1955); E. Vischer, J. Schmidlin, and A. Wettstein, *ibid.* **12**, 50 (1956).
25. H. H. Inhoffen, G. Quinkert, and S. Schütz, *Chem. Ber.* **90**, 1283 (1957); **91**, 2626 (1958); H. H. Inhoffen, K. Irmscher, H. Hirscheld, U. Stache, and A. Kreutzer, *ibid.* p. 2309.
26. E. J. Corey and W. R. Hertler, *J. Amer. Chem. Soc.* **80**, 2003 (1958).

27. N. Danieli, Y. Mazur, and F. Sondheimer, *Chem. Ind. (London)* pp. 1724 and 1725 (1958); Y. Mazur and F. Sondheimer, *J. Amer. Chem. Soc.* **81**, 3161 (1959); Y. Mazur, N. Danieli, and F. Sondheimer, *ibid.* **82**, 5889 (1960).
28. N. Danieli, Y. Mazur, and F. Sondheimer, *J. Amer. Chem. Soc.* **84**, 875 (1962).
29. D. H. R. Barton, *Experientia* **6**, 316 (1950); *J. Chem. Soc., London* p. 1027 (1953); D. H. R. Barton and R. C. Cookson, *Quart. Rev., Chem. Soc.* **10**, 44 (1956).
30. F. C. Uhle, *J. Amer. Chem. Soc.* **76**, 4245 (1954); *J. Org. Chem.* **27**, 656 (1962).
31. F. Sondheimer, R. Yashin, G. Rosenkranz, and C. Djerassi, *J. Amer. Chem. Soc.* **74**, 2696 (1952); F. Sondheimer, O. Mancera, J. Rosenkranz, and C. Djerassi, *ibid.* **75**, 1282 (1953); G. Rosenkranz, O. Mancera, and F. Sondheimer, *ibid.* **76**, 2227 (1954).
32. F. C. Uhle and J. A. Moore, *J. Amer. Chem. Soc.* **76**, 6412 (1954).
33. D. H. R. Barton and R. L. Morgan, *Proc. Chem. Soc., London* p. 206 (1961).
34. W. S. Johnson, H. A. P. deJongh, C. E. Coverdale, J. W. Scott, and U. Burckhardt, *J. Amer. Chem. Soc.* **89**, 4523 (1967); W. S. Johnson, J. M. Cox, D. W. Graham, and H. W. Whitlock, Jr., *ibid.* p. 4524.
35. F. Sondheimer, W. McCrae, and W. G. Salmond, *J. Amer. Chem. Soc.* **91**, 1228 (1969).
36. W. S. Johnson, K. Wiedhaup, S. F. Brady, and G. L. Olson, *J. Amer. Chem. Soc.* **90**, 5277 (1968).
37. W. S. Johnson, M. B. Gravestock, and B. E. McCarry, *J. Amer. Chem. Soc.* **93**, 4332 (1971).
38. S. N. Ananchenko and I. V. Torgov, *Dokl. Akad. Nauk SSSR* **127**, 553 (1959).
39. P. A. Robins and J. Walker, *J. Chem. Soc., London* p. 177 (1957).
40. A. J. Birch and H. Smith, *J. Chem. Soc., London* p. 1882 (1951); A. J. Birch, H. Smith, and R. E. Thornton, *ibid.* p. 1338 (1957).
41. W. S. Johnson, D. K. Banerjee, W. P. Schneider, and C. D. Gutsche, *J. Amer. Chem. Soc.* **72**, 1426 (1950).
42. T. Hiraoka and I. Iwai, *Chem. Pharm. Bull.* **14**, 262 (1966).
43. W. Nagata, T. Terasawa, and T. Aoki, *Tetrahedron Lett.* pp. 865 and 869 (1963).
44. G. Stork, H. J. E. Lowenthal, and P. C. Mukharji, *J. Amer. Chem. Soc.* **78**, 501 (1956).
45. S. I. Zav'yalov, G. V. Kondrat'eva, and L. F. Kudryavtseva, *Med. Prom. SSSR* **15**, (1961); *Izv. Akad. Nauk SSSR, Otd. Khim. Nauk* p. 529 (1961).
46. W. S. Johnson, E. R. Rogier, J. Szmuszkovicz, H. I. Hadler, J. Ackerman, B. K. Bhattacharyya, B. M. Bloom, L. Stalmann, R. A. Clement, B. Bannister, and H. Wynberg, *J. Amer. Chem. Soc.* **78**, 6289 (1956).
47. H. Smith, G. H. Douglas, and C. R. Walk, *Experientia* **20**, 418 (1964).
48. J. G. Westra, W. N. Speckamp, U. K. Pandit, and H. O. Huisman, *Tetrahedron Lett.* p. 2781 (1966).
49. W. N. Speckamp, U. K. Pandit, and H. O. Huisman, *Rec. Trav. Chim. Pays-Bas* **82**, 39 and 898 (1963).
50. S. Barcza, U. S. Patent 3,637,782 (1972) *Chem. Abstr.* **76**, 154021p (1972).
51. E. C. du Feu, F. J. McQuillin, and R. Robinson, *J. Chem. Soc., London* p. 53 (1937); R. Robinson and F. Weygand, *ibid.* p. 386 (1941).

References

52. G. I. Poos, G. E. Arth, R. E. Beyler, and L. H. Sarett, *J. Amer. Chem. Soc.* **75**, 422 (1953).
53. J. W. Cornforth and R. Robinson, *J. Chem. Soc., London* p. 676 (1946); p. 1855 (1949).
54. L. Velluz, G. Nominé, and J. Mathieu, *Angew. Chem.* **72**, 725 (1960); L. J. Chinn and H. L. Dryden, *J. Org. Chem.* **26**, 3904 (1961).
55. A. R. Daniewski and M. Kokor, *Bull. Acad. Pol. Sci.* **20**, 395 (1972).
56. L. B. Barkley, W. S. Knowles, H. Raffelson, and Q. E. Thompson, *J. Amer. Chem. Soc.* **78**, 4111 (1956).
57. N. N. Gaidamovich and I. V. Torgov, *Izv. Akad. Nauk SSSR, Otd. Khim. Nauk* p. 1803 (1961).
58. S. Danishefsky and A. Nagal, *Chem. Commun.* p. 373 (1972); A. Nagal, Ph.D. Thesis, University of Pittsburgh, Pittsburgh, Pennsylvania, 1971.
59. S. N. Ananchenko and I. V. Torgov, *Tetrahedron Lett.* p. 1553 (1963).
60. W. E. Bachmann, S. Kushner, and A. C. Stevenson, *J Amer. Chem. Soc.* **64**, 974 (1942); A. L. Wilds and T. L. Johnson, *ibid.* **70**, 1166 (1948).
61. P. Rosen, Ph.D. Dissertation, Columbia University, New York, 1962; G. Stork, *Pure Appl. Chem.* **9**, 131 (1964).
62. Z. G. Hajos, R. A. Micheli, D. R. Parrish, and E. P. Oliveto, *J. Org. Chem.* **32**, 3008 (1967).
63. J. W. ApSimon, P. Baker, J. Ruccini, J. W. Hooper, and S. Macaulay, *Can. J. Chem.* **50**, 1944 (1972).
64. S. Stournas, Ph.D. Dissertation, Columbia University, New York, 1970.
65. G. Stork and J. E. McMurry, *J. Amer. Chem. Soc.* **89**, 5461, 5463, and 5464 (1967).
66. J. W. Scott and G. Saucy, *J. Org. Chem.* **37**, 1652 (1972); J. W. Scott, R. Borer, and G. Saucy, *ibid.* pp. 1659 and 1664.
67. P. T. Lansbury, *Accounts Chem. Res.* **5**, 311 (1972).
68. G. Stork and P. L. Stotter, *J. Amer. Chem. Soc.* **91**, 7780 (1969).
69. H. C. Brown and E. Negishi, *Chem. Commun.* p. 594 (1968).
70. B. M. Trost and M. J. Bogdanowicz, *J. Amer. Chem. Soc.* **94**, 4777 (1972).
71. W. S. Johnson, D. K. Banerjee, W. P. Schneider, C. D. Gutsche, W. E. Shelberg, and L. J. Chinn, *J. Amer. Chem. Soc.* **74**, 2832 (1952).
72. W. S. Johnson, I. A. David, H. C. Dehm, R. J. Highet, E. W. Warnhoff, W. D. Wood, and E. T. Jones, *J. Amer. Chem. Soc.* **80**, 661 (1958).
73. J. E. Cole, W. S. Johnson, P. A. Robins, and J. Walker, *J. Chem. Soc., London* p. 244 (1962).
74. W. S. Johnson, D. S. Allen, R. R. Hindersinn, G. N. Sausen, and R. Pappo, *J. Amer. Chem. Soc.* **84**, 2181 (1952).
75. A. Wettstein, P. Desaulles, K. Heusler, R. Neher, J. Schmidlin, H. Ueberwasser, and P. Wieland, *Angew. Chem.* **69**, 689 (1957); K. Heusler, P. Wieland, H. Uberwasser, and A. Wettstein, *Chimia. (Buenos Aires)* **12**, 121 (1958).
76. H. D. Berndt and R. Wiechert, *Angew, Chem., Int. Ed. Engl.* **8**, 376 (1969).
77. J. C. Gasc and L. Nédélec, *Tetrahedron Lett.* p. 2005 (1971).
78. D. K. Banerjee, S. Chatterjee, C. N. Pillai, and M. V. Bhatt, *J. Amer. Chem. Soc.* **78**, 3769 (1956); D. K. Banerjee and S. K. Balasubramanian, *J. Org. Chem.* **23**, 105 (1958); D. K. Banerjee, V. Paul, S. K. Balasubramanian, and P. S. Murthy, *Tetrahedron* **20**, 2487 (1964).
79. H. Martin and R. Robinson, *J. Chem. Soc., London* p. 491 (1943).

80. W. S. Johnson, R. G. Christiansen and R. E. Ireland, *J. Amer. Chem. Soc.* **79**, 1995 (1957); W. S. Johnson and R. G. Christiansen, *ibid.* **73**, 5511 (1951).
81. G. R. Clemo and L. K. Mishra, *J. Chem. Soc., London* p. 192 (1953).
82. F. D. Popp, W. R. Schleigh, P. M. Froelich, R. J. Dubois, and A. C. Casey, *J. Org. Chem.* **33**, 833 (1968).
83. E. C. Taylor and K. Lenard, *Chem. Commun,* 97 (1967); G. Reduilh and C. Viel, *Bull. Soc. Chim. Fr.* p. 3115 (1969).
84. N. P. Buu-Hoi and D. Lavit-Lamy, *Bull. Soc. Chim. Fr.* p. 773 (1962).
85. G. Haberland, *Ber. Deut. Chem. Ges.* B **69**, 1380 (1936); J. Hoch, *Bull. Soc. Chim. Fr.* [5] **5** p. 264 (1938).
85a. W. E. Bachmann and R. D. Morin, *J. Amer. Chem. Soc.* **66**, 553 (1944).
86. R. Robinson and S. N. Slater, *J. Chem. Soc., London* p. 376 (1941).
87. G. Haberland, *Angew. Chem.* **51**, 499 (1938); *Ber. Deut. Chem. Ges.* B **76**, 621 (1943); G. Haberland and E. Heinrich, *ibid.* **72**, 1222 (1939); W. E. Bachmann and R. E. Holmen, *J. Amer. Chem. Soc.* **73**, 3660 (1951).
88. Chang Chinn, *Acta Chim. Sinica* **21**, 190 (1955); *Scientia (Peking)* **4**, 547 (1955).
89. G. Eglington, J. C. Nevenzel, A. I. Scott, and M. S. Newman, *J. Amer. Chem. Soc.* **78**, 2331 (1956).
90. V. C. E. Burnop, G. H. Elliott, and R. P. Linstead, *J. Chem. Soc., London* p. 727 (1940).
91. H. Lapin, *Chimia* **18**, 141 (1956).
92. A. M. Krubiner, G. Saucy, and E. P. Oliveto, *J. Org. Chem.* **33**, 3548 (1968).
93. G. S. Grinenko, E. V. Popova, and V. I. Maksimov, *Zh. Org. Khim.* **5**, 1329 (1969); **7**, 935 (1971).
94. D. L. Turner, B. K. Bhattacharyya, R. P. Graber, and W. S. Johnson, *J. Amer. Chem. Soc.* **72**, 5654 (1950).
95. W. S. Johnson and R. G. Christiansen, *J. Amer. Chem. Soc.* **73**, 5511 (1951); W. S. Johnson, R. G. Christiansen, and R. E. Ireland, *ibid.* **79**, 1995 (1957).
96. R. E. Tarney, Ph.D. Thesis, University of Wisconsin, Madison, 1958; *Diss. Abstr.* **21**, 2901 (1961).
97. I. N. Nazarov and I. A. Gurvich, *Zh. Obshch. Khim.* **25**, 956 (1955); I. N. Nazarov, S. I. Zav'yalov, M. S. Burmistrova, I. A. Gurvich, and L. I. Shmonina, *ibid.* **26**, 441 (1956).
98. M. Harnik and E. V. Jensen, *Isr. J. Chem.* **3**, 173 (1965).
99. K. Hiraga, T. Asako, and T. Miki, *Chem. Commun.* p. 1013 (1969).
100. H. Smith, British Patent 1,069,844 (1967); *Chem. Abstr,* **68**, 114835k (1968).
101. W. N. Speckamp, H. de Koning, U. K. Pandit, and H. O. Huisman, *Tetrahedron* **21**, 2517 (1965).
102. A. L. Wilds, R. H. Zeitschel, R. E. Sutton, and J. A. Johnson, Jr., *J. Org. Chem.* **19**, 255 (1954).
103. R. B. Mitra and B. D. Tilak, *J. Sci. Ind. Res., Sect.* B **15**, 573 (1956).
104. R. P. Linstead, W. E. Doering, S. B. Davis, P. Levien, and R. R. Whetstone *J. Amer. Chem. Soc.* **64**, 1985 (1942). H. I. Hadler, *Experientia* **11**, 175 (1955).
105. W. S. Johnson, J. Ackerman, J. F. Eastham, and H. A. DeWalt, Jr., *J. Amer. Chem. Soc.* **78**, 6302 (1956).
106. D. J. France, J. J. Hand, and M. Los, *Tetrahedron* **25**, 4011 (1969).
107. G. H. Douglas, J. M. H. Graves, D. Hartley, G. A. Hughes, B. J. McLoughlin, J. Siddall, and H. Smith, *J. Chem. Soc., London* p. 5072 (1963).

References

108. G. Stork, H. N. Khastgir, and A. J. Solo, *J. Am. Chem. Soc.* 80, 6457 (1958).
109. A. J. Birch and S. M. Mukherji, *J. Chem. Soc., London* p. 2531 (1949).
110. A. L. Wilds and N. A. Nelson, *J. Amer. Chem. Soc.* 75, 5360 and 5366 (1953).
111. W. Nagata, S. Hirai, T. Terasawa, and K. Takeda, *Chem. Pharm. Bull.* 9, 769 (1961).
112. W. S. Johnson, W. A. Vredenburg, and J. E. Pike, *J. Amer. Chem. Soc.* 82, 3409 (1960).
113. D. N. Kirk and M. P. Hartshorn, "Steroid Reaction Mechanisms." Amer. Elsevier, New York, 1968.
114. C. Djerassi, "Steroid Reactions." Holden-Day, San Francisco, California, 1963.
115. J. Fried and J. A. Edwards, "Organic Reactions in Steroid Chemistry," Vols. I and II. Van Nostrand-Reinhold, Princeton, New Jersey, 1972.
115a. D. Lednicer, *Advan. Org. Chem.* 8, 179 (1972).
116. A. Lardon, O. Schindler, and T. Reichstein, *Helv. Chim. Acta* 40, 676 (1957).
117. H. Smith, G. A. Hughes, G. H. Douglas, G. R. Wendt, G. C. Buzby, Jr., R. A. Edgren, J. Fisher, T. Foell, B. Gadsby, D. Hartley, D. Herbst, A. B. A. Jansen, K. Ledig, B. J. McLoughlin, J. McMenamin, T. W. Pattison, P. C. Phillips, R. Rees, J. Siddall, J. Suida, L. L. Smith, J. Tokolics, and D. H. P. Watson, *J. Chem. Soc., London* p. 4472 (1964).
118. P. H. Boyle, *Quart. Rev., Chem. Soc.* 25, 323 (1971).
119. G. Nomine, G. Amiard, and V. Torelli, *Bull. Soc. Chim. Fr.* p. 3664 (1968).
120. J. Schmidlin, G. Anner, J. R. Billeter, K. Heusler, H. Ueberwasser, P. Wieland, and A. Wettstein, *Helv. Chim. Acta* 40, 2291 (1957); L. H. Sarett, G. E. Arth, R. M. Lukes, R. E. Beyler, G. I. Poos, W. F. Johns, and J. M. Constantin, *J. Amer. Chem. Soc.* 74, 4974 (1952).
121. G. C. Buzby, Jr., D. Hartley, G. A. Hughes, H. Smith, B. W. Gadsby, and A. B. A. Jansen, *J. Med. Chem.* 10, 199 (1967).
122. M. Hubner, K. Ponsold, H.-J. Sieman, and S. Schwarz, *Z Chem.* 8, 380 (1968).
123. G. R. Newkome, L. C. Roach, R. C. Montelaro, and R. K. Hill, *J. Org. Chem.* 37, 2098 (1972).
124. E. W. Cantrall, C. Krieger, and R. B. Brownfield, German Patent 1,942,453 (1970); *Chem. Abstr.* 72, 111702m (1970).
125. R. Bucourt, L. Nédélec, J. C. Gasc, and J. Weill-Raynal, *Bull. Soc. Chim. Fr.* [2] p. 561 (1967).
126. U. Eder, G. Sauer, and R. Wiechert, *Angew. Chem.* 83, 492 (1971); Z. G. Hajos and D. R. Parrish, German Patent 2,102,632 (1971); *Chem. Abstr.* 75, 129414r (1971).
127. E. Vischer, J. Schmidlin, and A. Wettstein, *Experientia* 12, 50 (1956).
128. C. J. Sih and K. C. Wang, *J. Amer. Chem. Soc.* 85, 2135 (1963).
129. Y. Kurosawa, H. Shimojima, and Y. Osawa, *Steroids, Suppl.* 1, 185 (1965).
130. H. Gibian, K. Kieslich, H. J. Koch, H. Kosmol, C. Rufer, E. Schroder, and R. Vossing, *Tetrahedron Lett.* p. 2321 (1966).
131. L. M. Kogan, V. E. Gulaya, and I. V. Torgov, *Tetrahedron Lett.* p. 4673 (1967).
132. G. M. Segal, A. N. Cherkasov, and I. V. Torgov, *Khim. Prir. Soedin.* 3, 304 (1967).
133. P. Bellet, G. Nominé, and J. Mathieu, *C. R. Acad. Sci., Ser. C* 263, 88 (1966).

134. G. Greenspan, L. L. Smith, R. Rees, T. Foell, and H. E. Alburn, *J. Org. Chem.* **31**, 2512 (1966).
135. K. V. Yorka, W. L. Truett, and W. S. Johnson, *J. Org. Chem.* **27**, 4580 (1962).
136. J. Fried, M. J. Green, and G. V. Nair, *J. Amer. Chem. Soc.* **92**, 4136 (1970).
137. P. J. Curtis, *Biochem. J.* **97**, 148 (1965).
138. D. H. Peterson, H. C. Murray, S. H. Eppstein, L M. Reineke, A. Weintraub, P. D. Meister, and H. M. Leigh, *J. Amer. Chem. Soc.* **74**, 5933 (1952); S. H. Eppstein, P. D. Meister, D. H. Peterson, H. C. Murray, H. M. Leigh, D. A. Lyttle, L. M. Reineke, and A. Weintraub, *ibid* **75**, 408 (1953).
139. W. Charney and H. L. Herzog, "Microbial Transformations of Steroids." Academic Press, New York, 1967.

2
Biogenetic-like Steroid Synthesis

The emergence of the correct scheme for the biological conversion of squalene into lanosterol and thence into cholesterol [1–4] was the starting point for intensive investigations by many workers into the more intimate details of this and similar cyclizations. Particularly intriguing to synthetic chemists was the stereospecificity of this polyolefin cyclization in which a compound (squalene) containing no asymmetric centers was converted into another (lanosterol) possessing seven centers of asymmetry. Theoretically this latter material may exist in 128 different isomeric forms; however, only one is actually produced. A stereoelectronic theory which correctly explained these results was presented independently in 1955 by two groups and has come to be known as the Stork-Eschenmoser hypothesis [5,6]. In its simplest terms the theory predicts that cationic cyclization of *trans* olefin **1** will lead to the *trans* fused product **2** via a concerted mechanism involving the entering electrophile Y^+ and the nucleophile X^-. The product obtained should therefore be a result of an antiparallel addition mechanism and should possess substituents X and Y in *cis* diequatorial relationship. Extension of the olefin chain such that X^- is another π-bond will lead to an analogous *trans-anti-trans* tricycle. Also illustrated in Scheme 2-1 is the related cyclization of a *cis* olefin **3** to give the *cis* product **4**. Interestingly, when this theory was first tested by Eschenmoser, negative results were

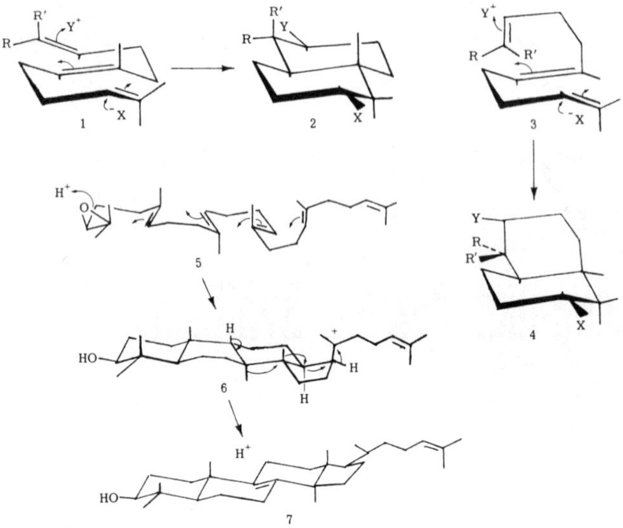

Scheme 2-1

obtained [7]. Thus acid-catalyzed cyclization of the methyl ester of *trans*-demethylfarnesic acid (1, R=CO$_2$Me, R'=H) gave the expected *trans*-decalin derivative; but similar treatment of the *cis* compound 3 (R=CO$_2$Me, R'=H) did not provide the predicted *cis*-fused decalin. This apparent anomaly was resolved when it was discovered that, under the conditions employed for these cyclizations, the reaction was, in fact, nonconcerted and an intermediate cyclohexane derivative could be isolated (and further converted into the *trans*-decalin) [6]. Subsequent investigations by numerous workers have solidly established the validity of the Stork-Eschenmoser hypothesis.

Although this explanation clarifies much of the biosynthesis of natural terpenoids and sterols, it does not adequately account for the formation of lanosterol (7) from squalene oxide (5). It is at this point that the role of the enzyme is presumed important in biological cyclizations. Thus, the conformation illustrated in 5 (potential B ring in boat form) is suggested to result from the template-like action of the lanosterol cyclase enzyme; a concerted cyclization would then provide an intermediate such as 6 which possesses the stereochemistry necessary to yield 7 upon appropriate methyl-hydrogen shifts. Alternatively, if the enzyme holds the polyolefin in the all-chair conformation, there results various terpenoids of the euphol class. Recent work by van Tamelen has elegantly demonstrated the correctness of these theories [8–10].

2.1 Cyclization of Terminal Epoxides

An extensive literature has developed concerning biogenetic-type synthesis in general, but for the present purposes only that directly relating to steroid total synthesis is discussed. However, the interested reader is directed to reviews by three of the leading investigators in the field [11–14].

Although the feasibility of acid-catalyzed cyclizations of various polyolefinic substrates had been established, yields were very poor and a multitude of unidentified by-products resulted. These problems arose, apparently, because the substrates such as squalene or the trienes 1 and 3 contained several very similar unsaturated centers (i.e., trisubstituted double bonds) susceptible to nonpreferential attack by the proton. Furthermore, the acid conditions used were of such strength as to foster this indiscriminate protonation. It was these difficulties to which both Johnson and van Tamelen addressed themselves in the early 1960's, and which they continue to investigate. Whereas Johnson's major emphasis has been the development of such cyclizations with the aim of applying them in a preparative fashion to the total synthesis of steroids, van Tamelen has concentrated more on the biochemical aspects with the ultimate goal of achieving in the laboratory by purely chemical means what has been accomplished by nature biosynthetically.

The difficulty alluded to above, i.e., the close steric and chemical similarity of the double bonds in squalene and most other acyclic terpenes initially presented a major synthetic problem. Since it was known that squalene (8, Scheme 2-2) was enzymically converted to squalene 2,3-oxide prior to cyclization, it was necessary to duplicate this feat chemically. An in depth study of various procedures for effecting this selective epoxidation revealed that N-bromosuccinimide in aqueous glyme produced good yields of the terminal bromohydrin which was readily transformed into squalene oxide (9) by treatment with base [15,16]. This high selectivity for terminal attack (>95%) was partially explained on conformational grounds. Presumably, in a highly polar solvent squalene exists in a tightly coiled, more compact conformation such that the internal sites of unsaturation are sterically less available for attack. Indeed, this high degree of selectivity was lost when nonpolar solvents such as petroleum ether were employed.

Having established methods for preparing squalene oxide from squalene (recent synthetic methods have made squalene more readily available [17–19]), and its further enzymic conversion into lanosterol and cholesterol [20,21], it was of interest to attempt the analogous non-

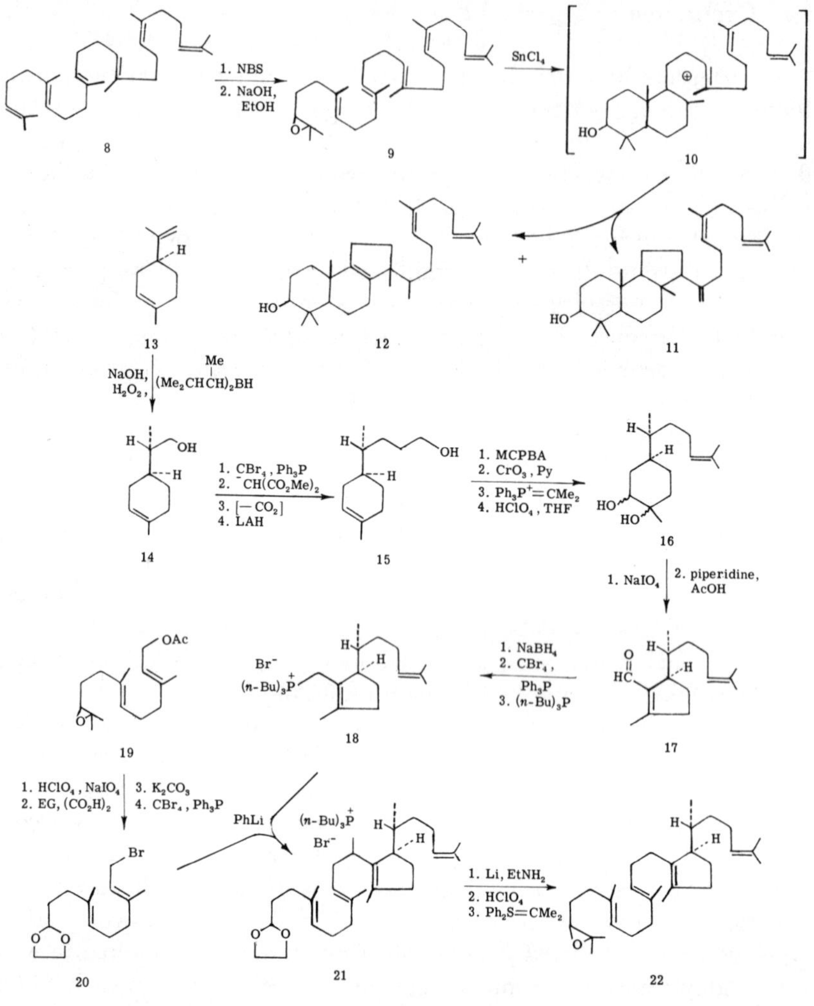

Scheme 2-2

enzymic cyclization. Treatment of **9** with stannic chloride in benzene at 10° for five minutes provided a mixture of compounds **11** and **12** in unspecified yield along with lesser amounts of other products [22]. The five-membered C ring was to be expected on purely chemical grounds (thus, cyclization of **10** to give a six-membered C ring would have to proceed through a secondary carbonium ion at C-13, whereas the isolated **11** and **12** presumably arise from a more stable tertiary carbonium ion); its formation *in vitro* pointed out the importance of

2.1 Cyclization of Terminal Epoxides

enzymic control in the formation of natural products with six-membered C rings. However, it has been suggested [23] that the enzymic conversion actually proceeds through an intermediate possessing a five-membered C ring, although substantiating evidence is still lacking.

To overcome this propensity for C-nor ring formation, van Tamelen synthesized various sterol precursors having a preformed ring D and subjected them to cyclization procedures [8–10, 24–26]. Thus (S)-(−)-limonene (13) was converted to the alcohol 14 by hydroboration with disiamylborane followed by separation of the mixture of diastereoisomers (fractional crystallization of the corresponding 3,5-dinitrobenzoates) [8,27]. This material was transformed into the corresponding bromide and reacted with the anion of dimethyl malonate. Decarboxylation and hydride reduction provided the homologous alcohol 15. Epoxidation with m-chloroperbenzoic acid gave a mixture of α- and β-epoxides which was oxidized with Collins reagent to the aldehyde. A Wittig reaction with triphenylphosphonium isopropylide gave a mixture of epoxides with the sought for side chain, and this mixture was hydrated with perchloric acid in tetrahydrofuran to a mixture of diastereomeric *trans* glycols 16. Cleavage with sodium metaperiodate provided the expected keto aldehyde which, on heating with piperidine–acetic acid, underwent ring closure to 17. The phosphonium salt 18 was prepared therefrom by a sequence of hydride reduction, bromination, and displacement. Coupling was then performed [28] with the acetal 20 which was prepared as shown from the terminal epoxide of farnesyl acetate (19) [29]. The initial coupling product 21 was reduced with lithium in ethylamine to remove the phosphorus substituent, and subsequent acid hydrolysis regenerated the aldehyde moiety. Utilization of Corey's epoxidation procedure [30] then produced the desired polyene 22 as a mixture (ca. 1:1) of α- and β-epoxides. This mixture was produced by the indicated sequence in an overall yield of less than 6% based on optically active alcohol 14. In a related sequence the 24,25-dihydro analog [23; $R = CH_2CH_2CH_2CH(CH_3)_2$] was prepared [8]. Attention is directed to the similarity of 22 and 9.

Structures 23 and 24 in Scheme 2-3 depict two possible conformations of the same compound which are invoked to explain the results obtained on cyclization. Thus it has been hypothesized [5,6] that in biogenesis when squalene oxide is folded in the all-chair conformation (presumably this folding is the primary role of the enzyme), subsequent cyclization gives rise to various sterols belonging to the euphol class; alternatively, when the potential ABC rings are held in the chair-boat-chair conformation (as in 23), the products are those of the lanosterol series (euphol differs from lanosterol only in having the opposite configurations at C-13,

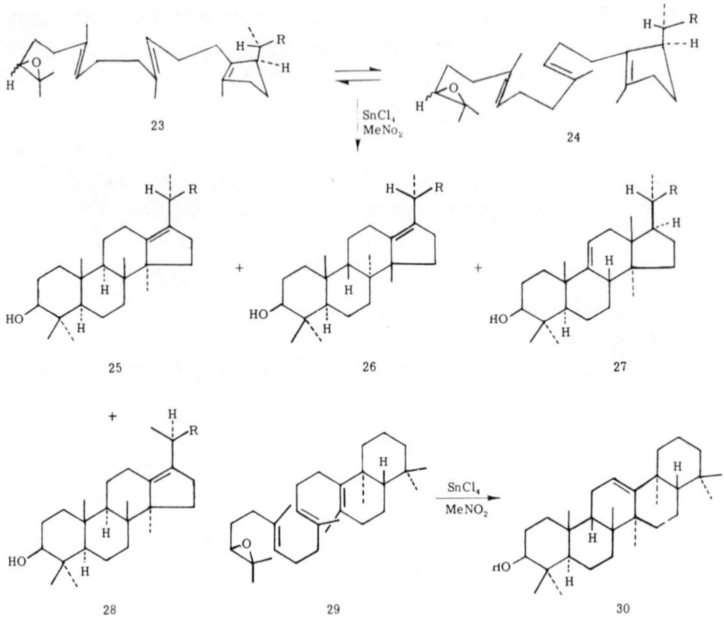

Scheme 2-3

C-14, and C-17). Thus, cyclization of **23** would be expected to provide 9β-products (e.g., **26**) whereas **24** should lead to the 9α-configuration (e.g., **25**).

On treating the mixture of epoxides [**23** and **24** with approximately equal amounts of the C-3α hydrogen configuration and the C-3β hydrogen configuration; $R = CH_2CH_2CH_2CH(CH_3)_2$] with 2.5 equivalents of stannic chloride in nitromethane for 1½ hr at 0°, a complex mixture of products was obtained from which could be recovered a tetracyclic fraction (30%) [8]. The products purified from this fraction were isoeuphenol (**25**, 3.5%), 24,25-dihydro-$\Delta^{13(17)}$-protolanosterol (**26**, 2%), 24,25-dihydroparkeol (**27**, 3.5%), and (−)-isotirucallenol (**28**, 43%). The first three products were considered to have arisen from the C-3α hydrogen epimer, while **28** presumably derived from the C-3β hydrogen epimer. Furthermore, it was shown that **26** was readily converted into **27** by treatment with boron trifluoride etherate in nitromethane. These transformations very closely parallel the biosynthetic transformations of squalene oxide and suggest that enzymic control of the reaction may be much less important than formerly suspected, and may primarily function to optimize a process which has now been shown to have as its basis purely chemical principles.

Further effort in this area has led to the total synthesis of several penta-

2.2 Total Synthesis from Nonepoxide Precursors

cyclic triterpenoids including dl-Δ^{12}-dehydrotetrahymanol (**30**) from **29** as indicated in Scheme 2-3 [9,10]. This elegant work has done much to clarify the biochemical pathways employed by nature for sterol and terpenoid synthesis and, in addition, has provided relatively efficient means for producing exceptionally complex natural products. The applicability to actual steroid synthesis is, to date, limited; but the potential is tremendous and, when coupled with Johnson's work described below, provides a basis for what is certain to become a widely explored route to total synthesis.

2.2 Total Synthesis from Nonepoxide Precursors

Concurrent with the biogenetic-like cyclizations, W. S. Johnson and his group were investigating similar cyclizations with the more immediate goal of elucidating new pathways to synthetic steroids [12]. Rather than attempting to closely duplicate the natural cyclization of squalene oxide, they attempted to apply the thermodynamic, stereochemical and practical implications learned from natural, enzymic processes to purely organic reactions with starting materials only roughly approximating those of natural sources. Various functional groups were investigated for their potential in initiating the cationic cyclizations including sulfonate esters [31–33], acetals, and allylic alcohols. Thus, solvolysis of **31** (R = $SO_2C_6H_4NO_2$-p) followed by removal of the acetate group provided **32** (2.8%) as the only tricyclic material (Scheme 2-4) [12,33]. Though this method provided a product with the desired stereochemistry (it was a mixture epimeric at C-13), the yield was obviously too low to be of synthetic value and, furthermore, the product possessed no functionality at C-3, a distinct disadvantage for potential steroid synthesis. In an effort

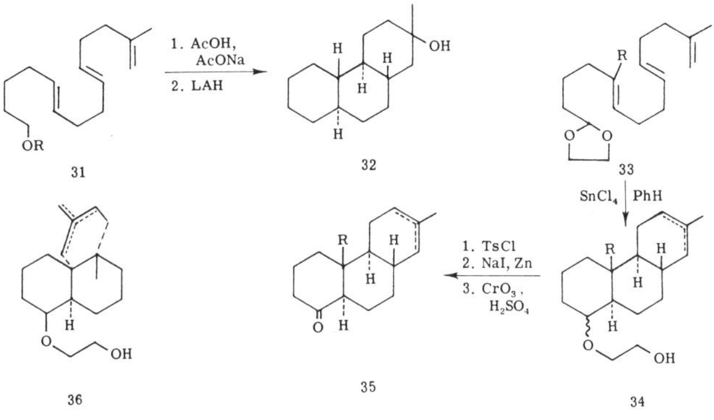

Scheme 2-4

to circumvent these limitations, the cyclization of several polyenic acetals was studied with gratifying results. After establishing the feasibility of the procedure in the preparation of *cis* and *trans* bicyclic system [34], attention was turned to tricyclic systems. Thus, the triene 33 (R=H[33], R=Me [35]) was prepared and subjected to the action of stannic chloride in benzene. In the first case (R=H), the only tricyclic product formed was 34 (R=H, 50%; this was a mixture of the isomers indicated, i.e., Δ^{12} or Δ^{13} unsaturation and an α- or β-substituent at C-4). Significantly, this reaction was appreciably slower than in the case where R=Me; the rate decrease was attributed to a lowered nucleophilicity of the disubstituted double bond (R=H) as compared to the trisubstituted case (R=Me). Conversion to 35 (R=H) was performed by the sequence; tosylation, elimination, and oxidation.

Like treatment of 33 (R=Me) similarly provided 34 (R=Me, 62%; again a mixture of isomers including those with an exocyclic double bond), which was also converted to 35. In fulfillment of the Stork-Eschenmoser hypothesis, the only isolable tricyclic products in both of the above cases possessed the natural *trans-anti-trans* configuration. In contrast, treatment of 33 (R=Me) with stannic chloride in nitromethane gave a mixture of 34 and the interesting 36, the proportions of which could be changed as the amount of stannic chloride varied. Presumably, 36 was a result of 1,2-hydride shift followed by a 1,2-methyl shift.

A milestone was reached in 1968 when Johnson and his group announced the conversion of an acyclic tetraenic acetal into a tetracyclic material in 30% yield [36]. This remarkable cyclization [the specific transformation is illustrated in Scheme 1-2 in Chapter 1 (7 → 8) and will not be detailed here since most of the intermediate synthetic procedures involved have been more recently employed in alternate steroid syntheses and are treated below] was stereospecific, providing the natural *trans-anti-trans-anti-trans* configuration. Thus a compound with no centers of asymmetry was converted in one, nonenzymic step into a D-homosteroid possessing numerous such centers (only at C-4 were both epimers formed; the substituent was later removed thereby rendering the cyclization completely stereospecific). Although this work with acetals would appear promising, it has more recently been supplanted by allylic alcohol cyclizations.

The use of the allylic alcohol functionality for initiating the cationic cyclization was developed in early work leading to bicyclic and tricyclic materials [12,37,38]. Advantages of this method include much milder conditions necessary to induce ionization with resulting increases in yields. As outlined in Scheme 2-5, *dl*-16-dehydroprogesterone (55) has been obtained by the use of the tetraenic alcohol 51 which was, in turn,

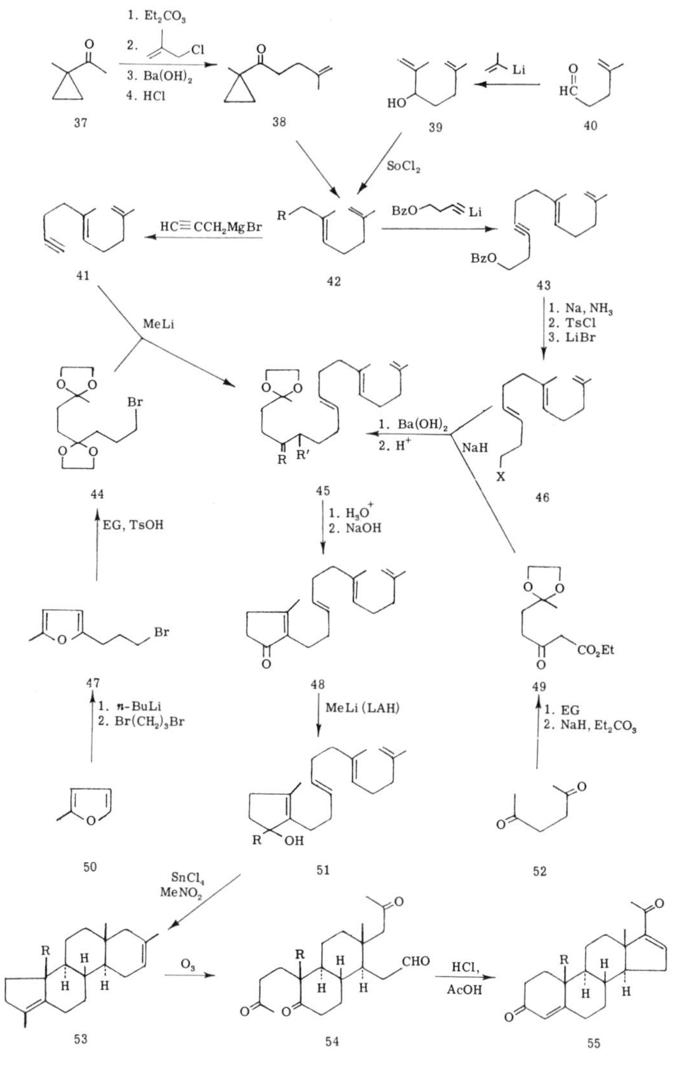

2.2 Total Synthesis from Nonepoxide Precursors

Scheme 2-5

prepared by multiple routes [39–41]. In the initial preparation [39], the cyclopropyl ketone **37** was converted in 88% yield into **38** by condensing with diethyl carbonate (to give a keto ester), alkylating with methylallyl chloride, and then hydrolyzing and decarboxylating by successive treatment with barium hydroxide and hydrochloric acid [42]. This product was readily reduced to the corresponding alcohol which, by a modification of the Julia olefin synthesis [43], was transformed into the dienic bromide **42** (R = CH$_2$Br). The latter rearrangement proceeded with a high degree of stereoselectivity, since only trace amounts of cis-**42** could be detected. The overall yield for this bromination–isomerization sequence was 85%. Further reaction with potassium acetate in dimethylformamide and subsequent basic hydrolysis afforded **42** (R=CH$_2$OH; 67%).

A later preparation [41] of **42** (R=Cl) was much more direct and entailed the reaction of aldehyde **40** with isopropenyllithium to give **39**. Exclusive formation of the *trans* double bond of **42** (R=Cl) resulted from treatment of **39** with thionyl chloride. Although the yields for these two steps were not specifically reported, this route was felt to be much more attractive than the preceeding one [41].

For the preparation of the key intermediate **48** two different pathways from **42** were followed. In the first the tosylate **42** (R=CH$_2$OTs) was treated with the anion of 4-benzyloxy-1-butyne to give **43**. Sodium–ammonia treatment concurrently reduced the acetylenic bond and removed the protecting benzyl group to give **46** (X=OH) which was first tosylated and then treated with lithium bromide to form **46** (X=Br). This halide was used to alkylate the sodium salt of **49** (the latter was obtained from 2,5-hexanedione via monoketalization followed by condensation with diethyl carbonate), and the product obtained in unspecified yield was the triene **45** (R=O, R'=CO$_2$Et). The carbethoxy group was removed by reaction with aqueous barium hydroxide and acidification to give **45** (R=O, R'=H). And finally, hydrolysis of the ketal group produced the acyclic diketone which was cyclized to **48** by treatment with 2% sodium hydroxide in aqueous ethanol. Unfortunately, the yields in most of these intermediate steps were not reported.

A second route to **48** is also recorded in Scheme 2-5 [41]. A Grignard reaction between propargylmagnesium bromide [44] and the allylic chloride **42** (R=Cl) gave **41** (ca. 50% from **39**); this material was converted to the lithium salt (methyllithium) and condensed with **44**. To prepare this second fragment, 2-methylfuran (**50**) was alkylated with 1,3-dibromopropane (59%) and the product treated with ethylene glycol and *p*-toluenesulfonic acid. This procedure effected simultaneous hydrolysis and bisketalization and afforded **44** in 76% yield. Alkylation of **41** then provided an intermediate acetylene (50%) which was reduced with sodium–ammonia to **45** (R=OCH$_2$CH$_2$O, R'=H); **48** was then obtained by ketal hydrolysis and cyclization as above.

Reaction of **48** with methyllithium gave the necessary allylic alcohol **51** (R=Me), which was unstable and utilized in the crude form [39,41]. Alternatively, reduction of **48** with either lithium aluminum hydride or sodium borohydride provided **51** (R=H), also air sensitive [40,41]. Stereospecific cyclization to **53** (R=Me or R=H) was achieved initially

2.2 Total Synthesis from Nonepoxide Precursors

with excess trifluroacetic acid in methylene chloride [39,40]; however, later work [41] showed that these yields (ca. 30%) could be dramatically improved (ca. 70%) by using four equivalents of stannic chloride in nitromethane at −23°. Interestingly, cyclization of **51** (R=H) proceeded to give the indicated 19-nor compound **53** (R=H) rather than the isomeric material with an angular methyl group. This behavior was, however, to be expected from similar results in the bicyclic series [45]. The remaining steps in the synthesis proceeded readily, and **54** was prepared from **53** either by ozonolysis [40] or by a more complex sequence involving osmium tetroxide in pyridine, hydrogen sulfide–dimethyl sulfoxide, and lead tetraacetate [39]. The final cyclization to **55** could be accomplished under either acidic [40] or basic [39] conditions. Although the yields for several of the reactions outlined in Scheme 2-5 have not been reported, and presumably are subject to improvement, it has been stated that approximately ten grams of *dl*-16-dehydroprogesterone (**55**, R=Me) is obtainable starting from one hundred grams of methallyl alcohol (the aldehyde **40** was prepared from the vinyl ether of methallyl alcohol via a Claisen rearrangement). Considering the number of steps involved, this is a remarkable achievement.

Having established the utility of the polyolefinic cyclization in steroid synthesis, it was of interest, from both a theoretical and practical viewpoint to determine if an acetylenic bond could be made to participate in the cyclization. Such has, indeed, been found to be the case [46,47], and Scheme 2-6 illustrates the total synthesis of *dl*-progesterone by this method [48]. A Grignard reaction between 1-bromo-3-pentyne and methacrolein gave the allylic alcohol **58** which was immediately transformed into **59** by an orthoacetate Claisen rearrangement [49] (to give **59** with the CHO group replaced with CO_2Et) followed by hydride reduction and Collins oxidation (42%).

The Wittig reagent **62** was obtained from **50** by the aforementioned sequence of alkylation (1,4-dibromobutane, to give **65**) and hydrolysis–bisketalization to give **62** (as the bromide rather than the phosphorane). The ylid **62** was then prepared by treating the corresponding iodide with triphenylphosphine followed by phenyllithium. Combination of **59** and **62** under special conditions [50] gave **61** contaminated with 3% of the 8,9-*cis*-dienyne. Acid hydrolysis, basic cyclization, and attack with methyllithium in a manner similar to that described for Scheme 2-5 gave **60**, again quite unstable. The yield of **60** (as the ketone prior to reaction with methyllithium) was 40% from the aldehyde **59**. Cyclization with trifluoroacetic acid in the presence of excess ethylene carbonate (to trap the intermediate vinyl cation) and subsequent hydrolysis with potassium carbonate gave a 71% yield of **63** (actually a 5:1 mixture of **63**

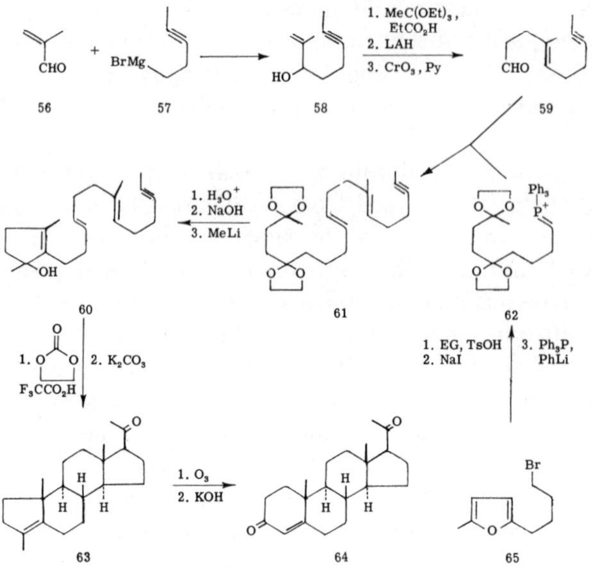

Scheme 2-6

and the corresponding 17α-epimer). As before, ozonization and basic cyclization led to the desired product **64** (45%). This material contained 15% of the 17α-isomer which was removed by repeated recrystallizations. Thus pure dl-progesterone was isolated in about 5% yield from **56** and **57**.

It is obvious that polycyclizations as a route to totally synthetic steroids are still in their infancy; however, the merit of this method has been amply demonstrated. Furthermore, the ability of this method to stereoselectively generate multiple asymmetric centers from a precursor possessing no such centers is exceedingly attractive. Based on some initial experiments in the bicyclic series [51], future work may be expected to delineate methods for utilizing asymmetric induction in polyolefinic cyclizations as a route to optically active steroids.

References

1. R. B. Woodward and K. Bloch, *J. Amer. Chem. Soc.* **75**, 2023 (1953).
2. W. G. Dauben, S. Abraham, S. Hotta, I. L. Chaikoff, H. L. Bradlow, and A. H. Soloway, *J. Amer. Chem. Soc.* **75**, 3038 (1953).
3. W. Voser, M. V. Mijović, H. Heusser, O. Jeger, and L. Ruzicka, *Helv. Chim. Acta* **35**, 2414 (1952).
4. L. Ruzicka, *Experientia* **9**, 357 (1953).

References

5. A. Eschenmoser, L. Ruzicka, O. Jeger, and D. Arigoni, *Helv. Chim. Acta* **38**, 1268 (1955).
6. G. Stork and A. W. Burgstahler, *J. Amer. Chem. Soc.* **77**, 5086 (1955).
7. A. Eschenmoser, D. Felix, M. Gut, J. Meier, and P. Stadler, *Ciba Found. Symp., 1958 Biosyn. Terpenes Sterols*, p. 217, Little, Brown and Co., 1959.
8. E. E. van Tamelen and R. J. Anderson, *J. Amer. Chem. Soc.* **94**, 8225 (1972).
9. E. E. van Tamelen, R. A. Holton, R. E. Hopla, and W. E. Konz, *J. Amer. Chem. Soc.* **94**, 8228 (1972).
10. E. E. van Tamelen, M. P. Seiler, and W. Wierenga, *J. Amer. Chem. Soc.* **94**, 8229 (1972).
11. W. S. Johnson, *Trans. N. Y. Acad. Sci.* [2] **29**, 1001 (1967).
12. W. S. Johnson, *Accounts Chem. Res.* **1**, 1 (1968).
13. E. E. van Tamelen, *Accounts Chem. Res.* **1**, 111 (1968).
14. D. Goldsmith, *Fortschr. Chem. Org. Naturst.* **29**, 363 (1971).
15. E. E. van Tamelen and T. J. Curphey, *Tetrahedron Lett.* p. 121 (1962).
16. E. E. van Tamelen and K. B. Sharpless, *Tetrahedron Lett.* p. 2655 (1967).
17. L. Werthemann and W. S. Johnson, *Proc. Nat. Acad. Sci. U.S.* **67**, 1465 (1970).
18. L. Werthemann and W. S. Johnson, *Proc. Nat. Acad. Sci. U.S.* **67**, 1810 (1970).
19. J. F. Biellmann and J. B. Ducep, *Tetrahedron* **27**, 5861 (1971).
20. E. J. Corey, W. E. Russey, and P. R. Ortez de Montellano, *J. Amer. Chem. Soc.* **88**, 4750 (1966).
21. E. E. van Tamelen, J. D. Willett, R. B. Clayton, and K. E. Lord, *J. Amer. Chem. Soc.* **88**, 4752 (1966).
22. E. E. van Tamelen, J. Willet, M. Schwartz, and R. Nadeau, *J. Amer. Chem. Soc.* **88**, 5937 (1966).
23. E. E. van Tamelen, J. D. Willett, and R. B. Clayton, *J. Amer. Chem. Soc.* **89**, 3371 (1967).
24. E. E. van Tamelen, G. M. Milne, M. I. Suffness, M. C. R. Chauvin, R. J. Anderson, and R. S. Achini, *J. Amer. Chem. Soc.* **92**, 7202 (1970).
25. E. E. van Tamelen and J. W. Murphy, *J. Amer. Chem. Soc.* **92**, 7205 (1970).
26. E. E. van Tamelen and J. H. Freed, *J. Amer. Chem. Soc.* **92**, 7206 (1970).
27. B. A. Pawson, H.-C. Cheung, S. Gurbaxani, and G. Saucy, *J. Amer. Chem. Soc.* **92**, 336 (1970).
28. E. H. Axelrod, G. M. Milne, and E. E. van Tamelen, *J. Amer. Chem. Soc.* **92**, 2139 (1970).
29. K. B. Sharpless, R. P. Hanzlik, and E. E. van Tamelen, *J. Amer. Chem. Soc.* **90**, 209 (1968).
30. E. J. Corey, M. Jautelat, and W. Oppolzer, *Tetrahedron Lett.* p. 2325 (1967).
31. W. S. Johnson, D. M. Bailey, R. Owyang, R. A. Bell, B. Jaques, and J. K. Crandall, *J. Amer. Chem. Soc.* **86**, 1959 (1964).
32. W. S. Johnson and J. K. Crandall, *J. Org. Chem.* **30**, 1785 (1965).
33. W. S. Johnson and R. B. Kinnel, *J. Amer. Chem. Soc.* **88**, 3861 (1966).
34. W. S. Johnson, A. van der Gen, and J. J. Swoboda, *J Amer. Chem. Soc.* **89**, 170 (1967), A. van der Gen, K. Wiedhaup, J. J. Swoboda, H. C. Dunathan, and W. S. Johnson, *ibid.* **95**, 2656 (1973).
35. G. D. Abrams, W. R. Bartlett, V. A. Fung, and W. S. Johnson, *Bioorg. Chem.* **1**, 243 (1971).
36. W. S. Johnson, K. Wiedhaup, S. F. Brady, and G. L. Olson, *J. Amer. Chem. Soc.* **90**, 5277 (1968).

37. W. S. Johnson, W. H. Lunn, and K. Fitzi, *J. Amer. Chem. Soc.* **86**, 1972 (1964).
38. W. S. Johnson, N. P. Jensen, and J. Hooz, *J. Amer. Chem. Soc.* **88**, 3859 (1966).
39. W. S. Johnson, M. F. Semmelhack, M. U. S. Sultanbawa, and L. A. Dolak, *J. Amer. Chem. Soc.* **90**, 2994 (1968).
40. S. J. Daum, R. L. Clarke, S. Archer, and W. S. Johnson, *Proc. Nat. Acad. Sci. U. S.* **62**, 333 (1969).
41. W. S. Johnson, T. Li, C. A. Harbert, W. R. Bartlett, T. R. Herrin, B. Staskun, and D. H. Rich, *J. Amer. Chem. Soc.* **92**, 4461 (1970).
42. S. F. Brady, M. A. Ilton, and W. S. Johnson, *J. Amer. Chem. Soc.* **90**, 2882 (1968).
43. M. Julia, S. Julia, and S.-Y. Tchen, *Bull. Soc. Chim. Fr.* [6] 1849 (1961).
44. R. E. Ireland, M. I. Dawson, and C. A. Lipinski, *Tetrahedron Lett.* p. 2247 (1970).
45. W. S. Johnson, P. J. Neustaedter, and K. K. Schmiegel, *J. Amer. Chem. Soc.* **87**, 5148 (1965).
46. W. S. Johnson, M. B. Gravestock, R. J. Parry, R. F. Myers, T. A. Bryson, and D. H. Miles, *J. Amer. Chem. Soc.* **93**, 4330 (1971).
47. W. S. Johnson, M. B. Gravestock, R. J. Parry, and D. A. Okorie, *J. Amer. Chem. Soc.* **94**, 8604 (1972).
48. W. S. Johnson, M. B. Gravestock, and B. E. McCarry, *J. Amer. Chem. Soc.* **93**, 4332 (1971).
49. W. S. Johnson, L. Werthemann, W. R. Bartlett, T. J. Brocksom, T. Li, D. J. Faulkner, and M. R. Petersen, *J. Amer. Chem. Soc.* **92**, 741 (1970).
50. M. Schlosser and K. F. Christmann, *Angew. Chem., Int. Ed. Engl.* **5**, 126 (1966).
51. W. S. Johnson, C. A. Harbert, and R. D. Stipanovic, *J. Amer. Chem. Soc.* **90**, 5279 (1968).

3
AB → ABC → ABCD

3.1 Equilenin

3.1.1 Synthesis According to Bachmann

The classical synthesis of equilenin by Bachmann, Cole, and Wilds [1,2] in 1939 represents a milestone in the total synthesis of steroidal hormones. This early approach took advantage of the simplified stereochemistry of equilenin, which has only two asymmetric carbon atoms. The starting material was the so-called Butenandt's ketone **4**, which had already been synthesized [3–5] via different routes, each involving the cyclization of 4-(6-methoxy-1-naphthalene)butyric acid (**8**). Fusion of 1-naphthylamine-6-sulfonic acid (Cleve's acid) (**1**) with potassium hydroxide yielded an aminonaphthol, which was isolated as the N-acetate **2**. Methylation, hydrolysis, and Sandmeyer reaction led to 1-iodo-6-methoxynaphthalene (**6**). Reaction of the Grignard reagent from **6** with succinic acid half-aldehyde, followed by dehydration of the intermediate alcohol afforded **9**. Esterifcation, hydrogenation, and hydrolysis gave the acid **8**, which in the form of its acid chloride was cyclized in the presence of stannic chloride to the tricyclic ketone **4**. (Scheme 3-1).

Several modifications of the above method for the preparation of **4** have been described in the literature [2,4–11]. In one variation [2,7],

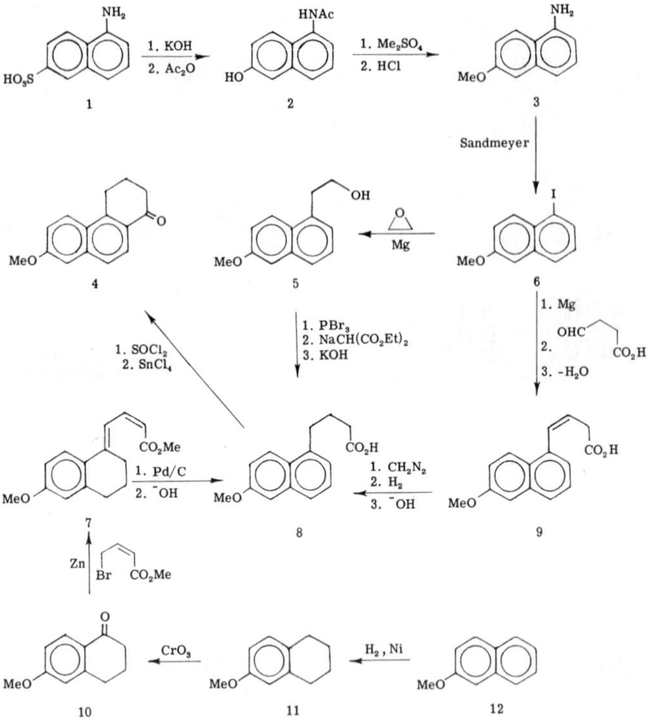

Scheme 3-1

the side chain was introduced by condensation of the Grignard reagent from **6** with ethylene oxide, followed by the malonic ester synthesis on the intermediate **5**. Later Stork [10,11] developed a more convenient synthesis of the tricyclic ketone **4**, starting with β-methoxynaphthalene (nerolin) (**12**). Successive hydrogenation and chromic acid oxidation afforded 6-methoxy-1-tetralone (**10**). A Reformatsky reaction with methyl γ-bromocrotonate then gave the unsaturated ester **7**. Isomerization by heating with palladized charcoal, followed by saponification led directly to the acid **8**.

For the construction of ring D (Scheme 3-2), Bachmann first condensed the cyclic ketone **4** with methyl oxalate in presence of sodium methoxide. The resulting keto ester **13** was then decarbonylated to **14** by heating in presence of soft glass. Angular methylation followed by Reformatsky reaction with methyl bromoacetate led to the hydroxy ester **16**. Successive dehydration, alkaline hydrolysis and reduction afforded a mixture of the two epimeric diacids **18** and **19**, in nearly equal

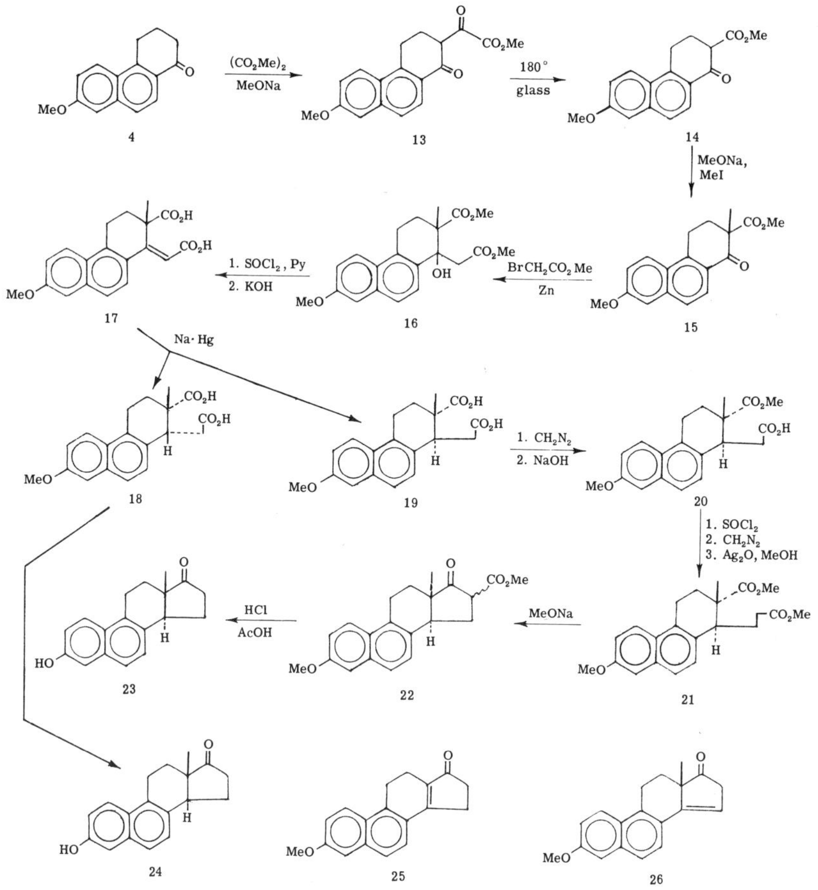

3.1 Equilenin

Scheme 3-2

amounts. After separation, the *trans* diacid **19** was esterified and selectively hydrolyzed to the half-ester **20**. This was converted by Arndt-Eistert reaction to the diester **21**. Dieckmann cyclization of the latter compound yielded the β-keto ester **22**, which on heating with a mixture of acetic acid and hydrochloric acid led to *dl*-equilenin (**23**). The overall yield of *dl*-equilenin from the tricyclic ketone **4** was about 25%. The racemic equilenin was resolved by fractional crystallization of the *l*-menthoxyacetate. A similar sequence of reactions with the *dl*-dicarboxylic acid **18** led to the synthesis of *dl*-isoequilenin (**24**).

In 1951, Bachmann and Holmen [12] described another route to the synthesis of compounds related to equilenin. Following and extending the methods described earlier by Bardhan [13,14], they synthesized

3′-oxo-7-methoxy-3,4-dihydro-1,2-cyclopentenophenanthrene (**25**). They also described the synthesis of 14,15-dehydroequilenin methyl ether (**26**). As shown by Johnson (see Scheme 3-3), the ketone **26** could be easily converted into the *dl*-equilenin by hydrogenation and subsequent demethylation.

3.1.2 Johnson's Synthesis of Equilenin

An ingenious method for the construction of ring D was developed by Johnson [15–17], who described the second synthesis of equilenin in 1945. In later years, this novel cyclization reaction has found more general applications and is often referred to as "Johnson's isoxazole method." The tricyclic ketone **4** was condensed with ethyl formate to

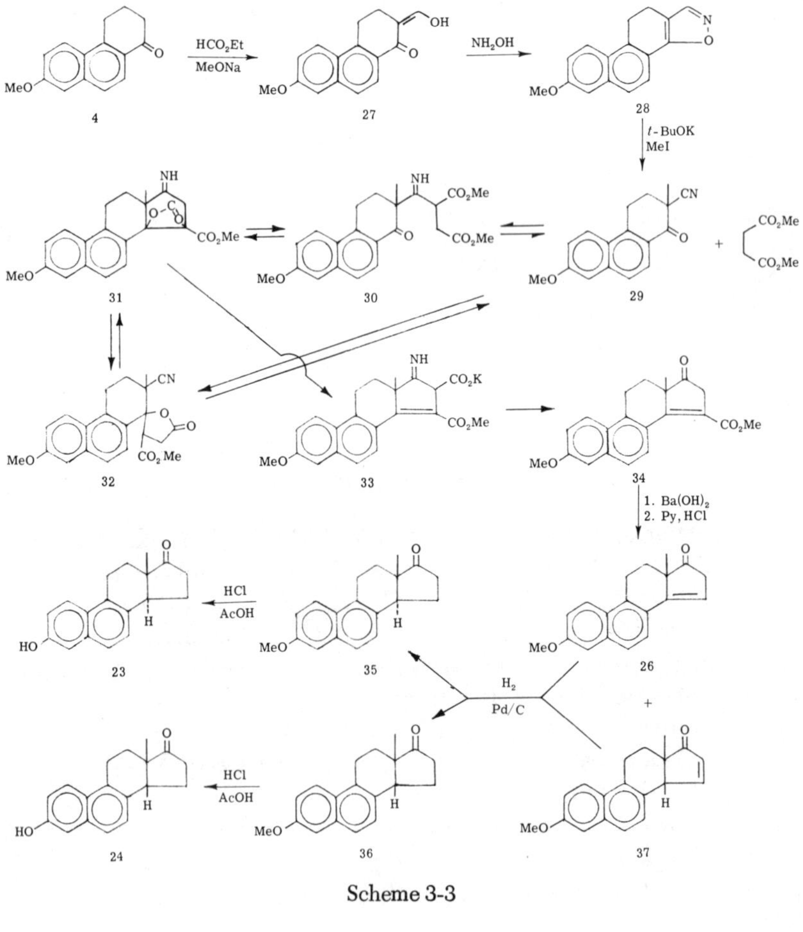

Scheme 3-3

give the formyl derivative **27** (Scheme 3-3). This, on treatment with hydroxylamine hydrochloride afforded the isoxazole **28**. Methylation of the isoxazole with methyl iodide in presence of potassium *t*-butoxide was accompanied by opening of the hetero ring and yielded the cyanoketone **29**. Stobbe condensation of the latter with dimethyl succinate in presence of potassium *t*-butoxide led directly to the cyclized keto ester **34**.

Several mechanisms could be postulated for this interesting reaction, of which two deserve special consideration. One of them could be visualized as a Thorpe-type addition of the active methylene to the cyanide group. The resulting ketimine **30** could undergo an intramolecular Stobbe condensation to give the intermediate cyclic paraconic ester **31**. An equally satisfactory mechanism might be represented by the alternate sequence, in which the first step is a Stobbe condensation to give the intermediate **32**. This compound could then undergo an intramolecular Thorpe reaction to yield **31**. The cyclic paraconic ester could now, on treatment with potassium *t*-butoxide, give the compound **33**. The latter, being the imino derivative of a keto acid, would undergo hydrolysis and decarboxylatoin, leading to the keto ester **34**.

Saponification of the keto ester **34**, followed by decarboxylation, resulted in a mixture of the stereoisomers **26** and **37**. Catalytic hydrogenation led to a 2:1 mixture of the methyl ethers of equilenin (**35**) and isoequilenin (**36**), which were separated by fractional crystallization. Demethylation of *dl*-equilenin methyl ether with hydrochloric acid and acetic acid afforded *dl*-equilenin (**23**). Resolution was easily effected via the *l*-menthoxy ester according to the method of Bachmann, Cole, and Wilds [2]. Similarly, demethylation of *dl*-isoequilenin methyl ether (**36**) led to *dl*-isoequilenin (**24**). The overall yield of *dl*-equilenin (**23**) from the tricyclic ketone 4 in eight steps was about 30%.

3.1.3 Synthesis According to Banerjee

Banerjee and co-workers [18] made a significant modification of the Johnson synthesis and achieved the first stereospecific synthesis of equilenin (Scheme 3-4). They found that by initial reduction of the unsaturated keto ester **34** with sodium borohydride, the 17β-hydroxy ester **38** was obtained in 92% yield. Thus the possibility of migration of the Δ^{14}-bond in latter stages of the synthesis (see Scheme 3-3) was eliminated. Saponification of the ester **38**, followed by decarboxylation, afforded **39**. Subsequent hydrogenation and oxidation led to *dl*-equilenin methyl ether (**35**). In later studies [19] it was found that, for large scale preparation of *dl*-equilenin methyl ether (**35**), it was more efficient to

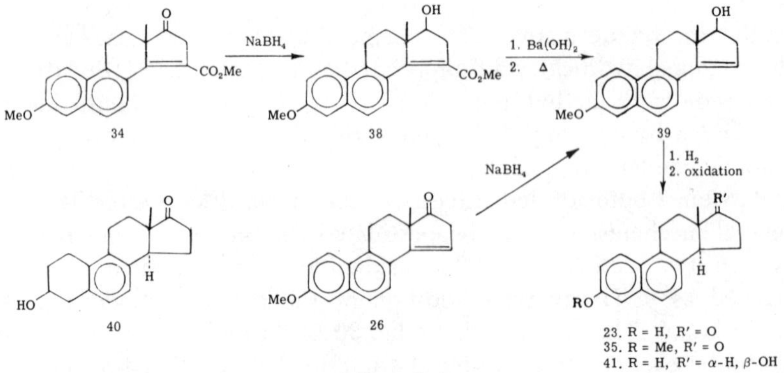

Scheme 3-4

prepare first the unsaturated ketone **26** by the method of Johnson. It was then reduced with sodium borohydride to give the unsaturated alcohol **39**. The Indian group [20] also synthesized the urinary steroid **40** and its 3α-epimer, starting from equilenin (**23**) and dihydroequilenin (**41**).

3.1.4 Johnson's Second Synthesis

Johnson and Stromberg [21] also developed a second route for the synthesis of *dl*-equilenin, starting with the ketone **46**, a methyl analog of **4** (Scheme 3-5). The latter compound had been previously obtained by Wilds [6] from the bromide **44** via the acid **45**. A number of other methods for the synthesis of **46** have been described in the literature

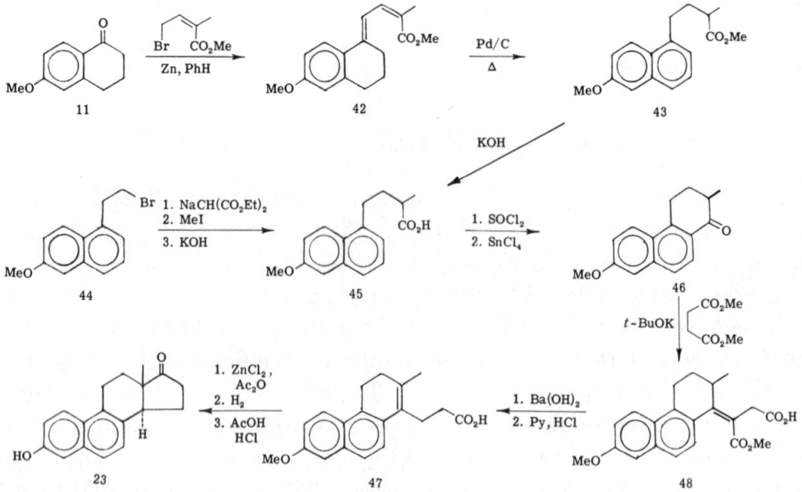

Scheme 3-5

3.1 Equilenin

[2,11,22–24]. Condensation of this ketone with succinic ester resulted in the formation of the unsaturated ester **48**. Hydrolysis and decarboxylation of **48** was accompanied by a shift of the double bond and afforded the acid **47**. Darzens cyclization, hydrogenation, and demethylation then led to *dl*-equilenin (**23**). The unsaturated acid **47** was also obtained by Haberland [24–28] as well as by Bachmann [12]. However, the products obtained by Johnson [21] had different melting points from those described by Haberland.

3.1.5 Synthesis via Homoequilenin Derivatives

The synthesis of equilenin has also been achieved via the *trans*-D-homoequilenin derivatives [29,30] (Scheme 3-6). Reformatsky reaction of the tricyclic ketone **46** with γ-bromocrotonic ester afforded the hydroxy ester **49**. Subsequent hydrogenation and saponification of **49** yielded the acid **50**. Cyclization of this acid with phosphorus pentoxide was followed by hydrogenation which proceeded stereospecifically [31] to give the methyl ether of D-homoequilenin **51**. Oxidative cleavage of the latter compound led to a dicarboxylic acid, which was esterified to

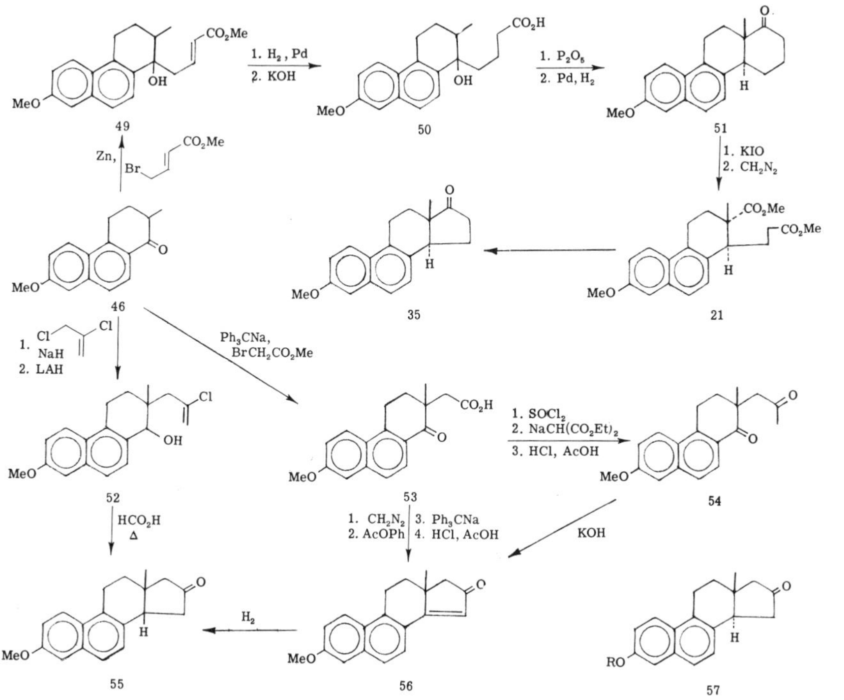

Scheme 3-6

the known compound 21. Subsequent Dieckmann cyclization, hydrolysis, and decarboxylation led to the methyl ether of equilenin (35).

3.1.6 16-Ketosteroids

A versatile, three-step method for attaching a ring D of the steroid nucleus to a tricyclic ketone was developed by Wilds and Beck [32] (Scheme 3-6). The method was later [6] extended to the synthesis of 3-hydroxy-16-equilenone (57, R=H), a structural isomer of equilenin. The enolate of the ketone 46 prepared by treatment either with triphenylmethylsodium or with sodium amide, was alkylated with methyl bromoacetate. Hydrolysis led to the acid 53, which was converted into the methyl ketone 54, via its acid chloride. Cyclization with alkali afforded the tetracyclic ketone 56. In a later study, Wilds and Johnson [33] developed an alternate method for the construction of ring D. The enolate of the ester of 53, on condensation with phenyl acetate and subsequent cyclization, led to the unsaturated ketone 56. Catalytic hydrogenation of the Δ^{14}-bond of the ketone 56 in presence of alkali yielded exclusively the cis-isomer 55. However, hydrogenation under neutral conditions led to a mixture of stereoisomers with the trans isomer 57 (R=Me) apparently predominating [34].

The above method for the construction of D ring by Wilds and co-workers had some obvious advantages. In addition to brevity, it led to Δ^{14}-16-ketosteroids which permitted the introduction of a variety of 17-substituents [35,36]. The method was also extended to the total synthesis of deoxycorticosterone [37], progesterone [38], and other nonaromatic steroids via methyl 3-ketoetiocholanate [39,40].

A facile, three-step synthesis of cis-16-equilenone methyl ether, starting with 46, was developed by Lansbury and co-workers [41]. Treatment of the ketone 46 with 2,3-dichloropropene in presence of sodium hydride and N,N-dimethylformamide was followed by reduction with lithium aluminum hydride to give 52. Cyclization by heating with formic acid led stereoselectively to cis-16-equilenone methyl ether (55). In another synthetic modification, Juday and Bukwa [42] were able to isolate the trans isomer 57 (R=Me). They also synthesized various other 16-ketosteroids.

3.1.7 Miscellaneous Syntheses

A simple approach to the synthesis of steroidal skeleton was described by Wilds, Harnik, and co-workers [43]. Condensation of naphthalene (58) with β-methyltricarballylic acid anhydride (59) afforded 60, which

3.1 Equilenin

was subsequently converted into tetracyclic ester **61** and its 17α-isomer. Later this method was extensively modified by Harnik [44–47], who achieved the synthesis of 3,6-dimethoxy-17β-acetyl-1,3,5(10),6,8-estrapentaene (**74**) and also the corresponding 17β-acetamido derivative **75** (Scheme 3-7).

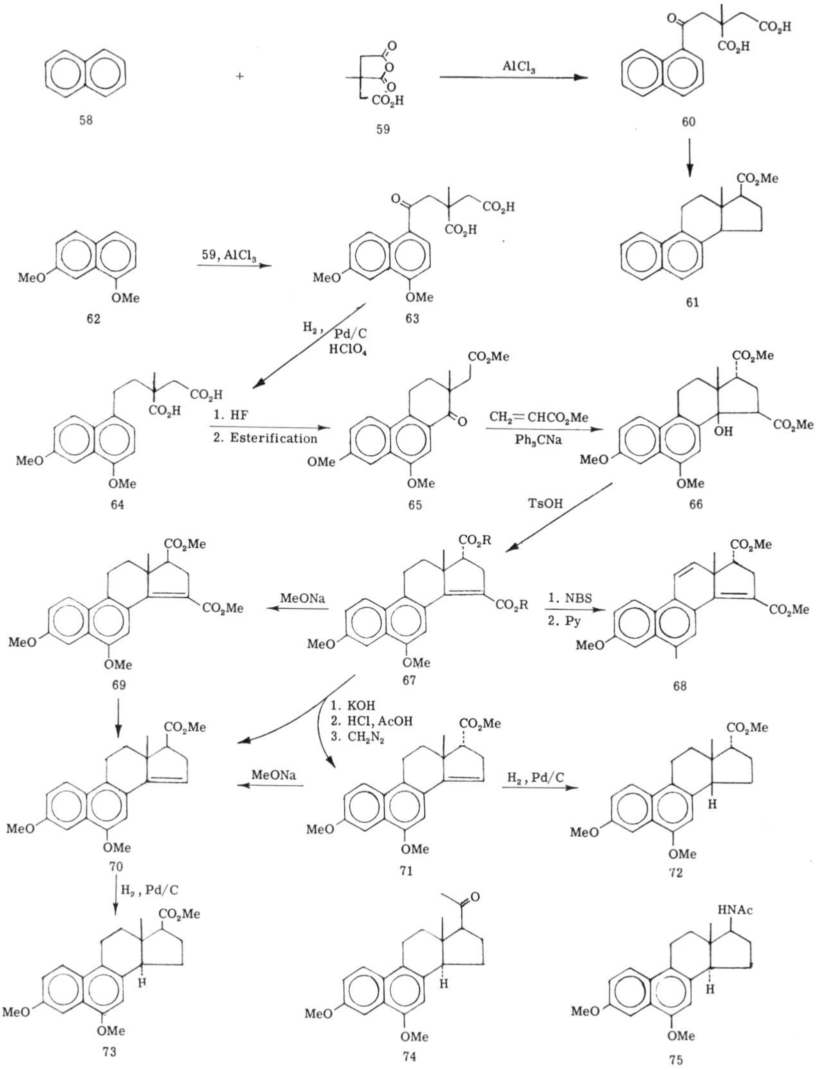

Scheme 3-7

Condensation of 2,8-dimethoxynaphthalene (62) with β-methyltricarballylic acid anhydride (59) in presence of aluminium chloride afforded the dicarboxylic acid 63. Successive catalytic reduction, cyclization, and esterification led to the keto ester 65 in 56% overall yield from 63.

For elaboration of the ring D, the ester 65 was treated with triphenylmethylsodium and methyl acrylate, whereupon Michael addition and aldol condensation occurred concomitantly to give the hydroxy ester 66. Dehydration of this hydroxy ester was followed by saponification to yield the diacid 67 (R=H). Benzylic bromination of the corresponding diester 67 (R=Me), followed by dehydrobromination led to the heptaene 68. On the other hand, decarboxylation of 67 (R=H) followed by esterification yielded a mixture of stereoisomeric monoesters which were identified as the 17β- and 17α-carbomethoxy derivatives 70 and 71, respectively. An alternative and better route to the 17β-isomer 70 involved isomerization of the diester 67 (R=Me) to the 15,17β-dicarbomethoxy derivative 69, followed by hydrolysis, decarboxylation, and esterification. Catalytic hydrogenation of 17β-isomer 70 yielded the "natural" 14α,17β-carbomethoxy isomer 73, while the 17α-isomer afforded the cis-14β,17β-isomer 72. The ester 73 was subsequently converted into 3,6-dimethoxy-17β-acetyl-1,3,5(10),6,8-estrapentaene (74) and also into 17β-acetamido-3,6-dimethoxy-1,3,5(10),6,8-estrapentaene (75) [47].

3.2 Estrone

The exceptional difficulties encountered during the early attempts for the total synthesis of steroids are well exemplified by the synthesis of the sex hormone estrone, which has four centers of asymmetry. The key intermediate, the Robinson's ketone 90 was synthesized as early as 1936 [48,49]. Bachmann and co-workers [22] accomplished the synthesis of "estrone-A" (87), one of the seven possible unnatural racemates in 1942. It took six more years before the first synthesis of natural estrone was achieved by Anner and Miescher [50] in 1948. Since then several syntheses [51–67] of estrone have been described in literature. By 1958, Johnson had completed the synthesis of all but one of the eight stereoisomers of estrone [54–59]. The eighth isomer 102, which should be the methyl ether of 96, was synthesized by the same group in 1968 [60].

3.2.1 Synthesis According to Bachmann

The starting material for the Bachmann synthesis [22] (Scheme 3-8) was β-m-anisylethyl bromide (76) which was condensed with sodiomalonic ester to give β-m-anisylethylmalonic ester (77). Condensation

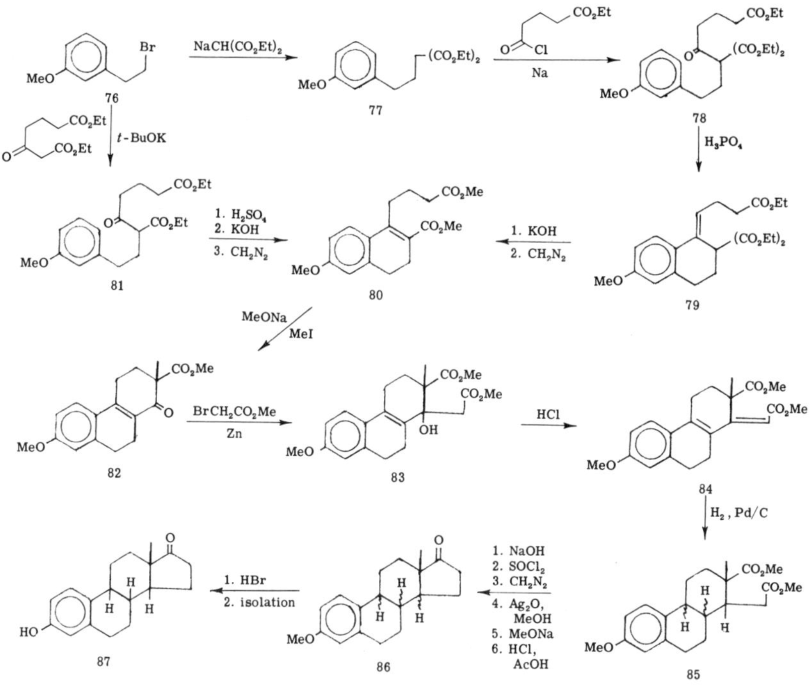

3.2 Estrone

Scheme 3-8

of the latter with the acid chloride of ethyl hydrogen glutarate afforded the keto triester **78**. Cyclization with orthophosphoric acid or sulfuric acid, followed by hydrolysis and esterification yielded the diester **80**. Later, Hunter and Hogg [68] also synthesized **80** by another route. The bromide **76** was condensed with the diethyl ester of β-oxopimelic acid to give the keto ester **81**. Successive cyclization, hydrolysis, and esterification afforded **80**.

The dimethyl ester **80** was next converted into the tricyclic keto ester **82**, by Dieckmann cyclization, followed by methylation. Reformatsky reaction afforded the crystalline hydroxy ester **83**, which on dehydration led to the $\Delta^{8,14}$-diester **84**. Catalytic hydrogenation of this compound afforded a mixture of stereoisomers **85**. Bachmann used this mixture for the construction of ring D, by following essentially the same procedure, developed by him for the synthesis of equilenin [1] (see Scheme 3-2). From the mixture of isomeric products in the final step, only one crystalline compound could be isolated in pure state. This product **87**, ("estrone A") was later shown to have the *cis-syn-cis* configuration by Johnson and co-workers [59].

3.2.2 Synthesis According to Anner and Miescher

Later, in a series of papers [50–52], Anner and Miescher announced the total synthesis of estrone and several of its stereoisomers (Scheme 3-9). The starting material was the so-called Robinson's ketone **90**, which had been synthesized from the diacid **88** [22,48,49]. Hydrogenation and esterification of this diacid gave the diester **89**. Dieckmann cyclization followed by angular methylation afforded the ketone **90**. The Ciba group was able to isolate three of the four theoretically possible racemates from the keto ester **90**. The major product was found to possess the "natural" configuration and was designated ketone A (**91**). Another isomer, ketone C (**92**), could be tranformed to ketone A (**91**) by treatment with alcoholic alkali.

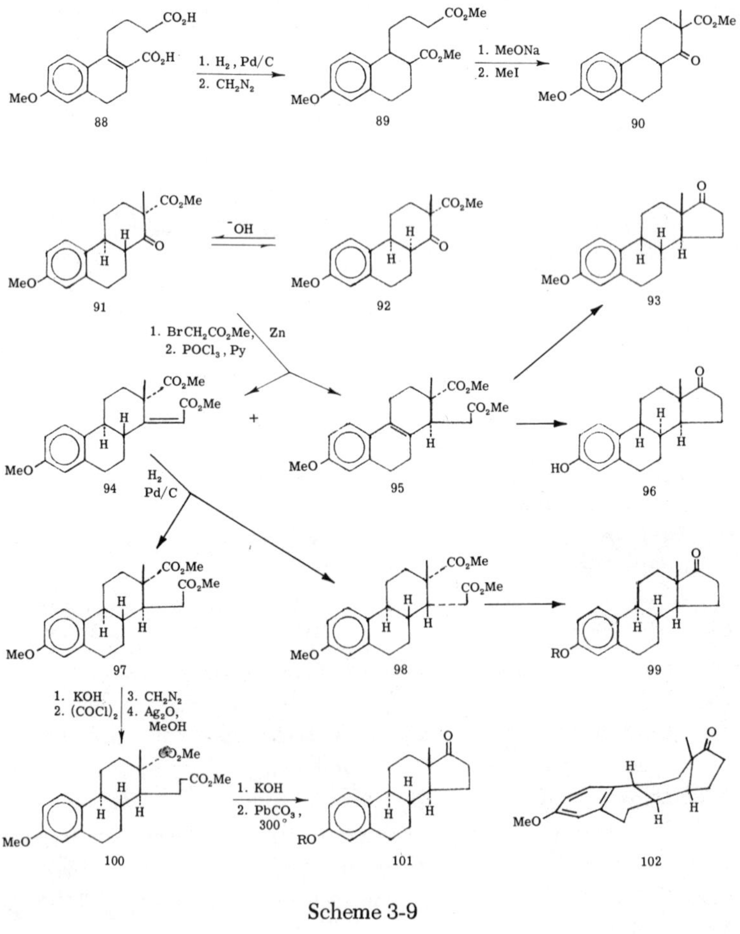

Scheme 3-9

3.2 Estrone

The Reformatsky reaction of the keto ester **91** followed by dehydration led to a mixture of isomeric unsaturated diesters **94** and **95**, in the ratio 3:1. The higher melting isomer **94** was hydrogenated to give another mixture of isomers **97** and **98**, which was separated by crystallization. The diester **97**, after selective hydrolysis, was submitted to Arndt-Eistert reaction. Subsequent alkaline hydrolysis and cyclization led to the tetracyclic methyl ether (**101**, R=Me). Demethylation with pyridine hydrochloride led to *dl*-estrone (**101**, R=H). Resolution was achieved by crystallization of the *l*-menthoxyacetates. The less soluble ester, after hydrolysis, afforded a product which was designated as "estrone-b" and was shown to be identical with the natural estrone.

Similar series of reactions on the diester **98** yielded another isomer of estrone, "estrone-a" (**99**). Anner and Miescher [52,69] also synthesized "estrone-d" (**93**) and "estrone-e" (**96**) starting from the lower-melting unsaturated diester **95**. The same isomers **93** and **96** could also be obtained by construction of ring D onto the keto ester **92**. The fifth isomer, "estrone-f," also synthesized by the Swiss workers, was later shown by Johnson [59] to have the *cis-syn-cis* configuration **87**, synthesized earlier by Bachmann [22]. The stereochemistry of these isomers was established in an equally brilliant series of studies by Johnson and co-workers [54–60], who completed the synthesis of all the eight possible isomers of estrone.

3.2.3 Sheehan's Synthesis of Estrone

In both the syntheses of estrone by Anner and Miescher [50–52] (Scheme 3-9) and by Johnson [56,57] (See Chapter 6, Section 6.1), a carbon atom is first added to the primary side chain of dimethyl marrianolate methyl ether (**97**) and subsequently removed after cyclization. In a series of papers, Sheehan and co-workers [61,67,70–73] described a novel direct cyclization procedure, using the acyloin condensation (Scheme 3-10). The diester **97** when subjected to acyloin condensation in a homogeneous medium, afforded stereospecifically the estrogenic steroid 16-keto-17β-estradiol-3-methyl ether (**103**) in excellent yield. Removal of the 16-keto group was effected via the intermediate thioketal acetate **105**, which on desulfurization led to the methyl ether **106** (R=Me). Demethylation of the latter yielded 17β-estradiol (**106**, R=H) in excellent yield, while oxidation with chromic acid gave estrone methyl ether (**101**, R=Me). Later, Sheehan and co-workers [61] developed a better method for the transformation of **103** into estrone. Reduction of the 16-oxo grup with sodium borohydride afforded the mixture of the

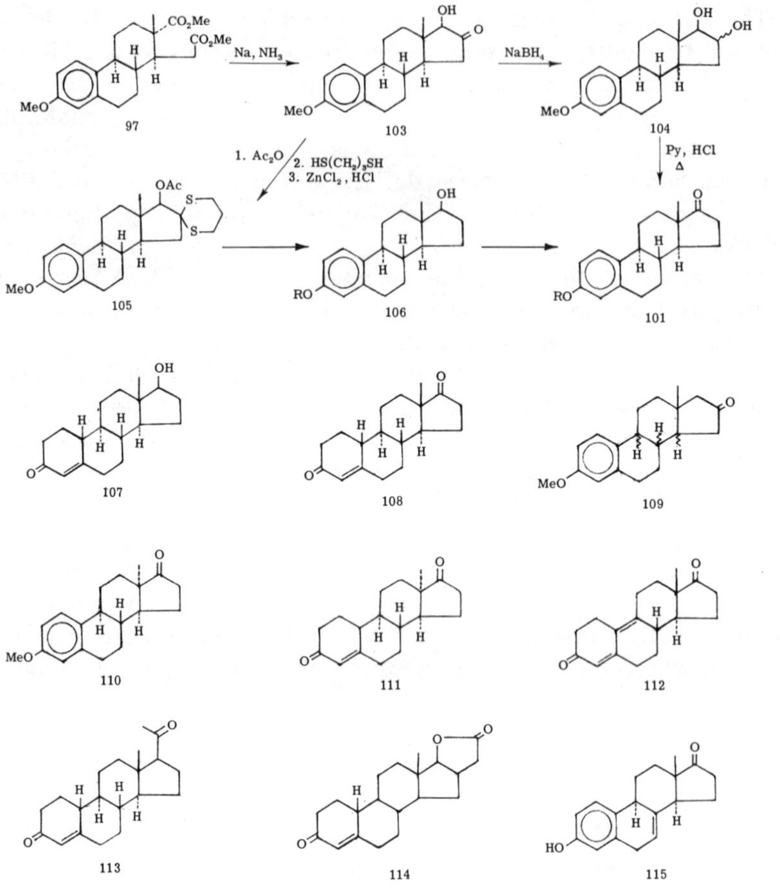

Scheme 3-10

epimeric glycols **104**. Dehydration of this mixture by heating with pyridine hydrochloride at 200–220° gave estrone (**101**, R=H) as the only product, in 90% overall yield. Further modifications of this method were later made by Kiprianov [74].

A large number of biologically important steroidal derivatives have been synthesized, starting with estrone (**101**, R=H) and estradiol (**106**, R=H). Conversion of the latter to 19-nortestosterone (**107**) by Birch and Mukherji [75,75a] has been marked as an important advance in synthetic organic chemistry. Reduction of 3-glyceryl ether of estradiol (**106**, R=glyceryl) with sodium or potassium and ethanol in liquid ammonia was followed by mild acid hydrolysis to give **107** in about 12%

overall yield from estradiol. Later, the syntheses of both 107 and 19-nor-4-androstene-3,17-dione (108) from estrone was achieved by Wilds and Nelson [76] utilizing a modified Birch reduction. In another series of studies, Wilds [33], following the early methods [6] developed for the synthesis of 3-hydroxy-16-equilenone (57, R=H) achieved the synthesis of 3-methoxy-1,3,5-estratrien-16-one (109). The configuration of the final compound, however, was not established.

Using estrone as the starting material, Crabbé and co-workers [77,78] synthesized several steroids with unnatural configuration. Irradiation of estrone 3-methyl ether (101, R=Me) with ultraviolet light afforded its 13α-isomer 110 [79], which was next converted into 13α-estr-4-ene-3,17-dione (111). Other examples, where estrone was used as the starting material, include estra-4,9-diene-3,17-dione (112) [80], 19-norprogesterone (113) [81], 17β-hydroxy-3-oxo-4-estren-16β-ylacetic acid lactone (114) [82], and equilin (115) [83].

3.3 Bisdehydrodoisynolic Acid

Bisdehydrodoisynolic acid (127, R=H), one of the most potent estrogens, was discovered by Miescher and co-workers [84,85] as a degradation product of equilenin. In a brilliant series of papers [86–94] they also described the total synthesis of this compound (Scheme 3-11). The keto ester 15, an intermediate in Bachmann's synthesis of equilenin (Scheme 3-2), was treated with ethylmagnesium bromide to form the hydroxy ester 117. Alternatively, the compound 117 could also be synthesized via the acetylene intermediate 116. Dehydration and alkaline hydrolysis afforded the unsaturated acid 118. Catalytic hydrogenation in the presence alkali led to the mixture of 14α- and 14β-isomers 119 and 120, in a ratio of 1:9.6. In a second synthesis [87,88], Anner and Miescher started with the bromide 121, which was alkylated with α-propionylpropionic ester to give the keto ester 122. Subsequent cyclization, dehydration, and alkaline hydrolysis afforded the unsaturated acid 118.

A third route to the synthesis of dl-bisdehydrodoisynolic acid was developed by Johnson [7,95,96], who utilized in the initial step the Stobbe condensation of dimethyl succinate with 6-methoxy-2-propionylnaphthalene (123). The product, upon catalytic reduction afforded the half-ester 124. The latter, then was cyclized by reaction of the corresponding acid chloride with stannic chloride to give the tricyclic keto ester 125. Subsequent hydrogenolysis of the keto group, followed by

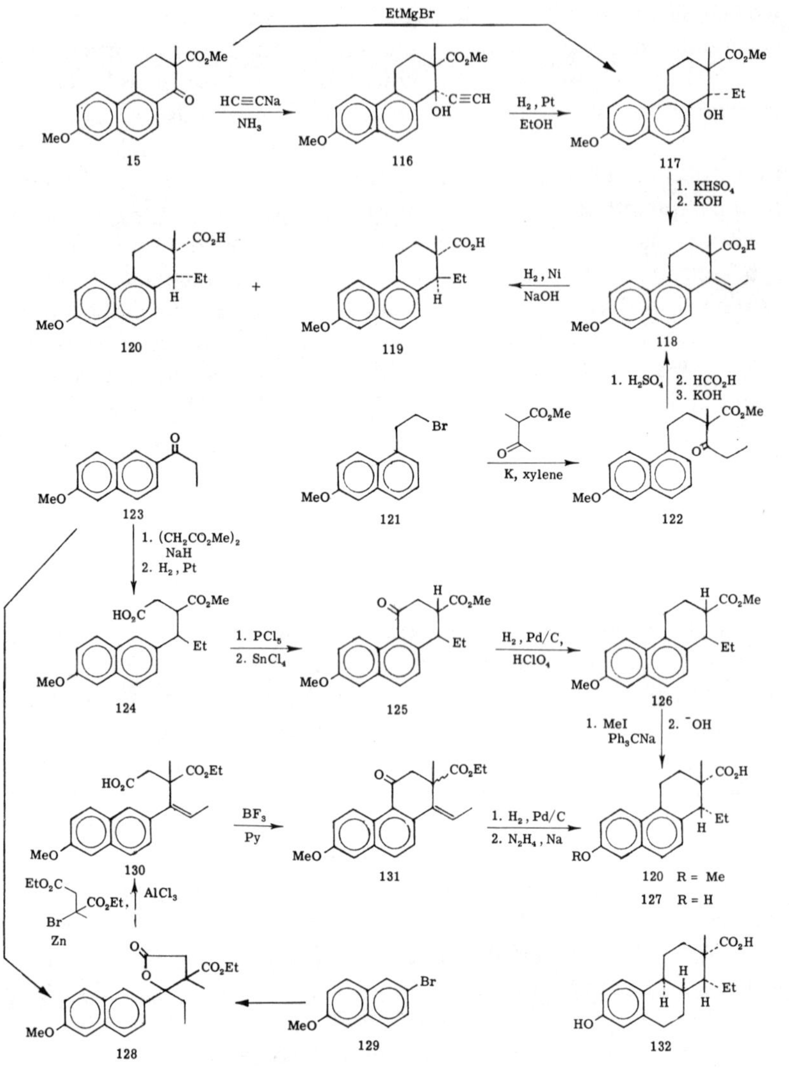

Scheme 3-11

angular methylation and hydrolysis led to *dl-cis*-bisdehydrodoisynolic acid methyl ether **120** (R=Me) in a yield of 64%, along with a small amount (7%) of the corresponding *trans* isomer.

Several other syntheses of bisdehydrodoisynolic acid have been described in the literature [97–103]. Gay and co-workers [97–100], starting with the ketone **123**, synthesized **127** (R=H) in poor yield. A better method was developed by Gastambide-Odier and co-workers [102] who started with the same bicyclic ketone 123. Reaction with α-bromo-α-methyl succinic ester and zinc yielded the lactone 128. Alternatively, the lactone could be synthesized by the Grignard reaction of 2-bromo-6-methoxynaphthalene (129) with diethyl α-methyl-α-propionyl succinate [104]. Treatment of the lactone 128 with aluminum chloride yielded the intermediate 130. The latter upon cyclization with boron trifluoride and pyridine yielded the unsaturated tricyclic ketone 131. Catalytic hydrogenation, followed by the removal of the 11-oxo group by Wolff-Kishner method led to cis-bisdehydrodoisynolic acid (127, R=H).

Several stereoisomers of dl-doisynolic acid, the degradation product resulting from the action of potassium hydroxide upon estradiol (106, R=H) [84] were synthesized by Anner and Miescher [105–107] and also by Hunter and Hogg [68,108]. They followed essentially similar methods developed for the synthesis of the bisdehydro analogs. The stereochemistry of doisynolic acid remained in doubt for a long time, until Crabbé and co-workers [109,110] established the absolute configuration of (+)-cis-doisynolic acid (132) by a stepwise conversion from estrone (101, R=H). They also synthesized [110] various cis-doisynolic, dehydro-cis-doisynolic, and bisdehydro-cis-doisynolic acid analogs.

3.4 18,19-Bisnorprogesterone and 19-Norpregnanes

The total synthesis of 18,19-bisnor-14α-hydroxyprogesterone (141) as well as of 18,19-bisnorprogesterone (142) was achieved by Nelson and Garland [111] in 1957 (Scheme 3-12). The key intermediate was 1-keto-7-methoxy-1,2,3,4,9,10-hexahydrophenanthrene (133), which could be prepared from the diester 80, by Dieckmann cyclization, and acid hydrolysis [22]. Alternatively, the ketone 133 could also be synthesized from the ester 7 by selective hydrogenation and cyclization [11,112].

The sequence of reactions for the construction of ring D was based on the method developed by Sarett and co-workers for the synthesis of cortisone [113,114] (See Chapter 9, Section 9.4). The unsaturated ketone 133, was alkylated via the hydroxymethylene derivative 134 to give the ketone 135. The latter on reduction with lithium in liquid ammonia afforded the ketone 136. The overall yield of 136 from the ketone 133 was 41%. In an alternative method, the ketone 133 was first reduced and then alkylated to give the same ketone 136 in an overall yield of 24%. Treatment of the ketone 136 with ethoxyacetylenemagnesium bromide followed by hydration with sulfuric acid in tetrahydrofuran yielded the

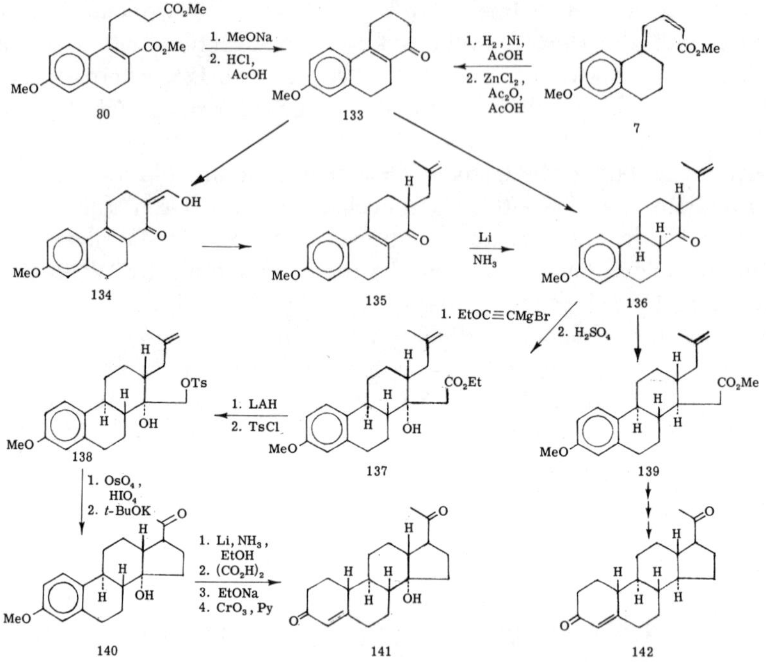

Scheme 3-12

14-hydroxy ester **137** as the major product (79%). In addition, about 9% of an oily mixture of unsaturated esters was obtained. Saponification of this mixture gave a noncrystalline mixture of acids, which was reduced with lithium in liquid ammonia. Esterification with diazomethane afforded the unsaturated ester **139**.

The hydroxy ester **137**, was reduced with lithium aluminum hydride and the resulting product was converted into the monotosylate **138**. Oxidation of the tosylate, followed by cleavage of the glycol with periodate and cyclization with base afforded 3-methoxy-17-acetyl-18-nor-1,3,5(10)-estratriene-14-ol (**140**). Birch reduction of the latter, followed by hydrolysis with oxalic acid, isomerization with sodium methoxide and oxidation with the chromium trioxide–pyridine complex led to 18,19-bisnor-14α-hydroxyprogesterone (**141**). Similar methods were used for the synthesis of 18,19-bisnorprogesterone (**142**) from the ester **139** [111]. The compound **142** was also prepared by Johns [115,116] and by Stork and co-workers [117] by partial synthesis.

The bicyclic ketone, 6-methoxy-2-tetralone (**145**) [118–122] was the starting material in the synthesis of 19-norpregnanes by Nagata and co-

3.4 18,19-Bisnorprogesterone and 19-Norpregnanes

workers [123–125] (Scheme 3-13). The tosylhydrazone [126] of 6-methoxy-1-tetralone (11) was treated with sodium glycolate and subsequently dehydrated with $KHSO_4$ to give 6-methoxy-3,4-dihydronaphthalene (144). Oxidation of the latter with peracetic acid, followed by treatment with hydrochloric acid led to the ketone 145. Following Robinson's method [127], ring C was constructed by the reaction of the ketone 145 with methyl vinyl ketone, in the presence of sodium methoxide. The resulting mixture of the isomeric ketones 146 and 147 was alkylated with 5-bromo-2-ethylenedioxypentane to form a mixture of the $\Delta^{8,9}$- and $\Delta^{8,14}$-isomers 148. The tricyclic ketone 147 had been previously synthesized by Nasipuri [128,129] by the Dieckmann cyclization of the ester 149.

The ketone 148, upon successive reduction, oxidation, acidification, and cyclization afforded the tetracyclic unsaturated methyl ketone 150. Hydrocyanation [123,130,131] of the latter with potassium cyanide, in presence of ammonium chloride and dimethylformamide yielded a mixture of the 13β-cyano derivative 151 and its 13α-epimer in a ratio 1:2.5. In later modifications [132–135] hydrocyanation of the ketone 150 was effected by treatment with hydrogen cyanide in presence of triethyl-

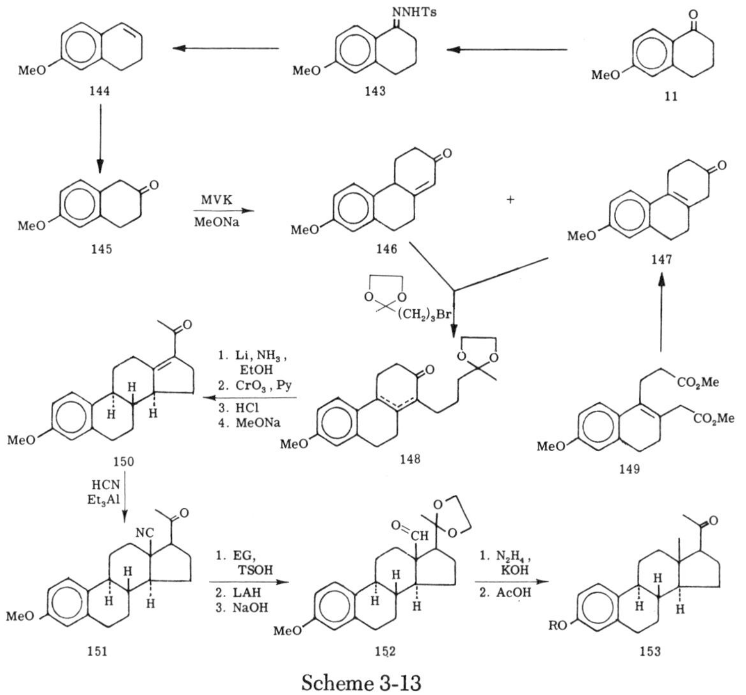

Scheme 3-13

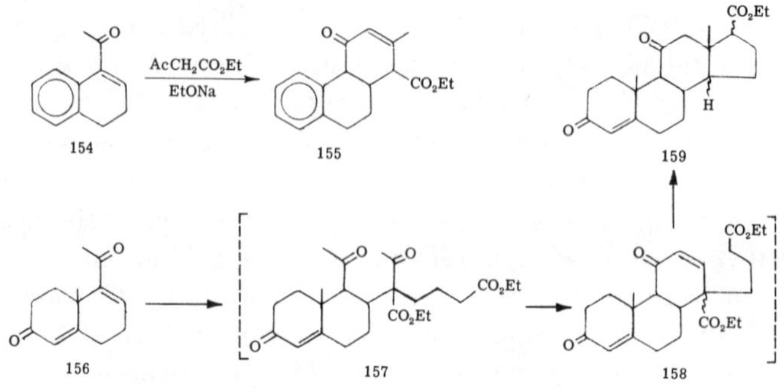

Scheme 3-14

aluminum, when the *trans-anti-trans* isomer **151** was obtained almost exclusively. After protection of the keto group as the ketal, the cyano compound **151** was converted into the 13-aldehyde **152** by successive reduction and alkaline hydrolysis. Huang-Minlon reduction of this aldehyde to the 13β-methyl derivative and subsequent removal of the ketal group led to a mixture of 3-hydroxy-19-norpregna-1,3,5(10)-trien-20-one **153** (R=H) and its methyl ether **153** (R=Me).

3.5 Miscellaneous Syntheses

As a part of the studies directed toward total synthesis of steroids, Mukharji [136–138] synthesized the intermediate tricyclic keto ester **155** (Scheme 3-14). The Michael condensation of ethyl sodioacetoacetate with 1-acetyl-3,4-dihydronaphthalene (**154**) gave the keto ester **155**. Later, Friedmann and Robinson [139,140] proposed an elegant route to the synthesis of 11-oxosteroids. Michael condensation of the unsaturated diketone **156** [139,141–143], with ethyl α-acetyladipate led to a mixture of compounds, from which a small amount of the tetracyclic keto ester **159** was isolated. Presumably, the reaction proceeded via the intermediates **157** and **158**. The stereochemistry of the products were not established. However, from the later studies by Barton [126] and Johnson [144] on other steroids, the product **159** could be assumed to have the unnatural CD *cis* configuration.

Thermal cyclization was used by Beslin and Conia [145] for the synthesis of 12-ketosteroids (Scheme 3-15). The tricyclic diester **160** was converted into the keto ester **161** in four steps. The latter was next transformed into the olefinic ketone **162** by a series of reactions, in which a three-carbon chain was added while the ketone group was protected as

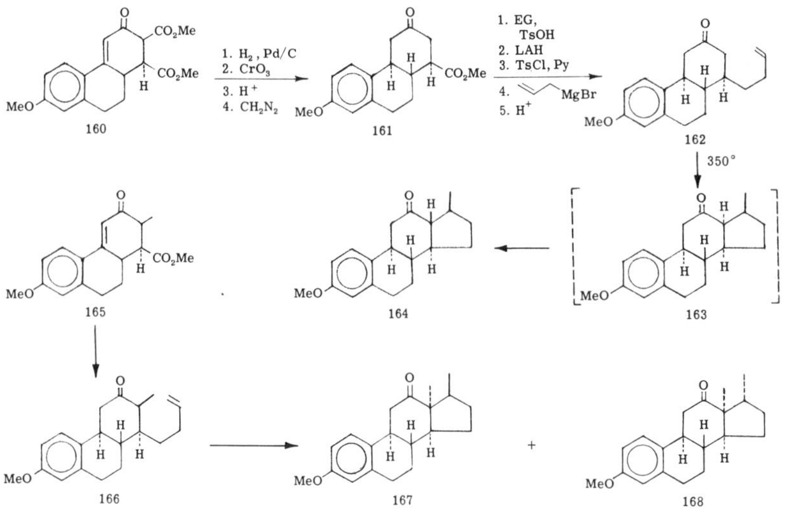

3.5 Miscellaneous Syntheses

Scheme 3-15

the ketal. Cyclization was effected by heating the ketone **162** at 350° C, when the 18-nor-17-methyl compound **164** was obtained in 69.5% yield. By similar methods, the keto ester **165** was transformed into the ketone **166**. Thermal cyclization of this compound, however, yielded a mixture of the 17α- and 17β-isomers **167** and **168** in which the CD junction had the *cis* stereochemistry.

In recent years, the synthesis of C-nor-D-homosteroid alkaloids has received considerable attention (see Chapter 12, Section 12.4). As a model compound for the synthesis of 17-acetyl-5α-etiojerva-12,14,16-trien-3β-ol (**177**), Brown and co-workers [146,147] (Scheme 3-16) synthesized benzo-[a]fluorenone-8 (**171**), starting from **169**. The total synthesis of the ketone **177** had been previously achieved by Johnson and co-workers [148]. Its conversion into the C-nor D-homosteroid alkaloid veratramine and other related alkaloids constituted, in formal sense, the total synthesis of these steroidal hormones [149,150].

The French group also developed different methods of annelation related to the construction of 5- and 6-membered cyclic ketones. By utilizing a method developed earlier by Sen and Mondal [151], they were able to synthesize the ketone **176**, starting from **175** (R=CHO) as well as from the keto ester **175** (R=CO$_2$Et). In another variation [152], the ring C could be constructed by the photocyclization of the mesylates **178** or **180** to give the ketone **179**.

The synthesis of the D-homo-C-nor derivative **174** was described by

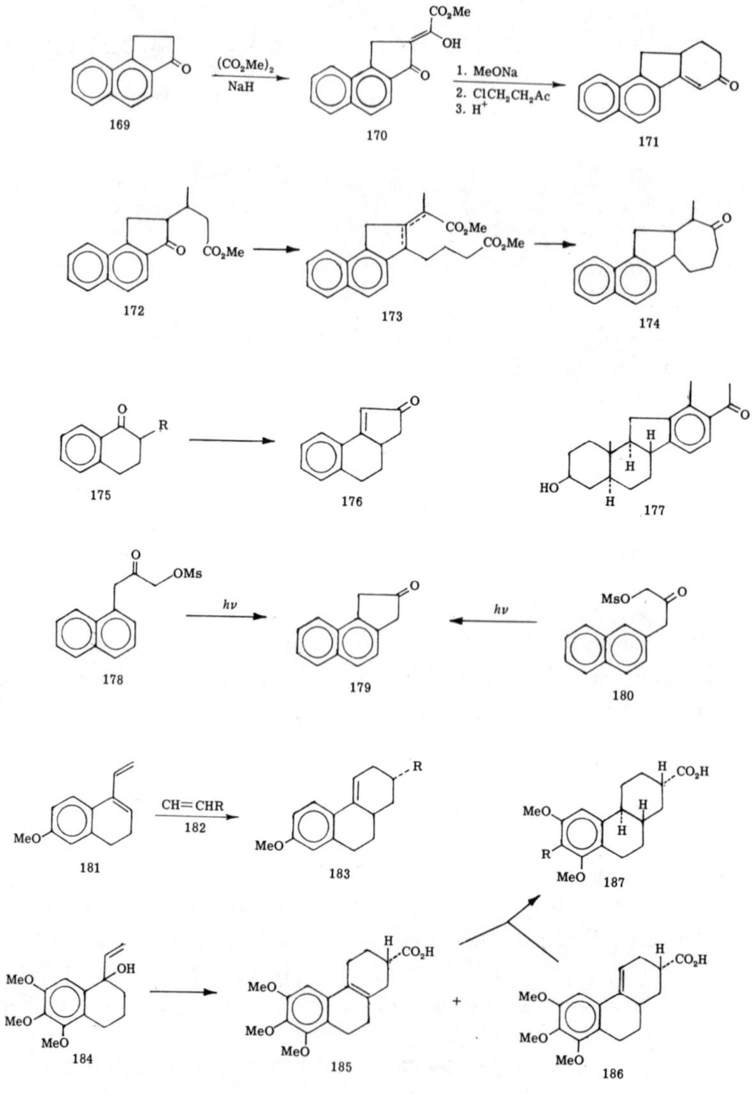

Scheme 3-16

Chatterjee [153,154]. The keto ester **172**, prepared from **169**, was subjected to Stobbe condensation with dimethyl glutarate. The resulting acidic material, on hydrolysis, decarboxylation, and subsequent esterification yielded the diester **173**, as a mixture of double bond isomers. Catalytic reduction, followed by base hydrolysis and pyrolysis led to the tetracyclic ketone **174**.

As a part of the program aimed at the synthesis of steroid-like compounds, Hajos and co-workers [155] studied the reaction of the diene 181 with methyl vinyl ketone (182, R=COCH$_3$). The product was identified as the endo addition product 183 (R=COCH$_3$). Similar treatment of 181 with ethyl acrylate (182, R=CO$_2$Et) followed by saponification led to 183 (R=CO$_2$H). Later, Rao and co-workers [156] studied the Diels-Alder addition of methyl acrylate (182, R=CO$_2$Me) to 3,4-dihydro-5,6,7-trimethoxy-1-vinylnaphthalene (generated *in situ* from 184). A mixture of two isomeric acids 185 and 186 was obtained. Both of these acids yielded the octahydrophenanthrene carboxylic acid 187 (R=OMe) on reduction with a limited quantity of sodium and liquid ammonia. On the other hand, when an excess of sodium and liquid ammonia was used, partial demethoxylation occurred, thus yielding the acid 187 (R=H).

3.6 Heterocyclic Steroids

During recent years, a large number of modified steroids containing nitrogen in the steroid skeleton has been synthesized. Evaluation of the biological activities of many of these compounds is far from complete. Nonetheless, synthetic efforts in this expanding area have often provided fascinating chemical problems and useful new reactions [157–164].

3.6.1 Oxasteroids

As a part of the work directed toward the total synthesis of heterocyclic steroids, Tilak and co-workers [165,166] synthesized *dl*-B-nor-6-oxaequilenin (190), the furano analog of equilenin (Scheme 3-17). The starting material was resorcinol, which was converted into the ketone 189 via γ-(2,4-dihydroxybenzoyl)-butyric acid 188. Following Johnson's route for the synthesis of equilenin (see Scheme 3-3), 7-methoxy-4-oxo-1,2,3,4-tetrahydrodibenzofuran (189) was next transformed into *dl*-B-nor-6-oxaequilenin (190). By analogy with Johnson's observation in the equilenin series, the CD rings in the final compound 190 were assumed to be *trans*-fused. The overall yield of 190 from the ketone 189 was 3.5%.

The synthetic routes to the 11-oxasteroids were studied by Kasturi and co-workers [167,168]. Following the method of Johnson for the synthesis of equilenin, these workers attempted the synthesis of 11-oxa analog of equilenin. The starting material was the nitrile 191, which was obtained by the cyanoethylation of 6-methoxy-1-naphthol. Cyclization with zinc chloride in presence of dry hydrogen chloride afforded the chromanone 192. The latter when treated with ethyl formate in the presence of sodium methoxide gave the corresponding hydroxymethylene derivative

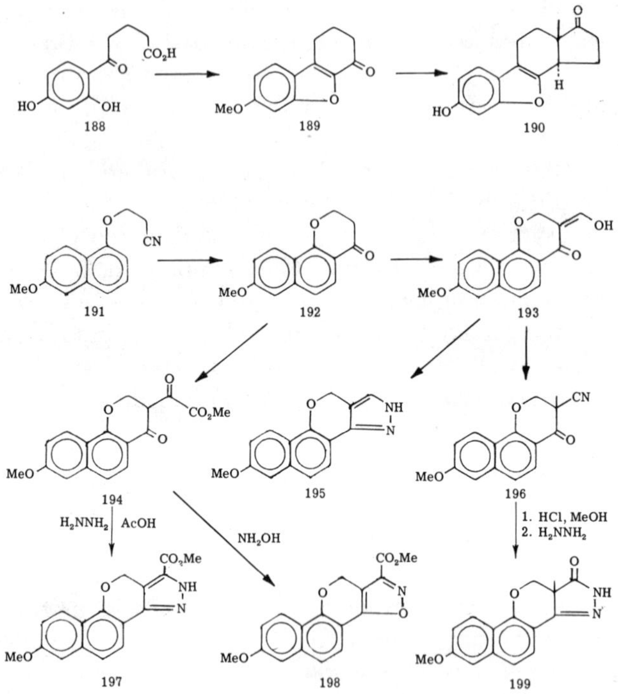

Scheme 3-17

193. This was next converted into 8-methoxy-3-cyano-3-methylbenz[h]-chroman-4-one (**196**) by following a modified procedure of Johnson's synthesis in the carbocyclic series. However, difficulties were encountered during the construction of the D ring. Stobbe condensation of the ketone **196** with dimethyl succinate failed to give the desired tetracyclic product.

Starting from the intermediate 8-methoxybenz[h]chroman-4-one **192**, several 11-oxa-15,16-diaza analogs of equilenin were synthesized [168]. Condensation of dimethyl oxalate with **192** in presence of sodium methoxide yielded the glyoxylate **194** in excellent yield. The latter on heating with hydrazine hydrate in acetic acid led to the 11-oxa-15,16-diaza derivative **197**. Also, the oxazole **198** was obtained from the glyoxalate **194**, by heating the latter with hydroxylamine hydrochloride in acetic acid. In the same manner, the hydroxymethylene derivative **193** was converted into the pyrazole analog of equilenin, **195**, by using hydrazine hydrate. Similarly, the cyanomethyl chromanone **196**, on methanolysis, followed by heating with hydrazine hydrate led to 11-oxa-15,16-diaza-14-dehydroequilenin methyl ether (**199**).

3.6 Heterocyclic Steroids

3.6.2 Thiasteroids

Early methods developed for the synthesis of steroids in the carbocyclic series have been frequently utilized for the preparation of thiasteroids. The synthesis of ring B thiophene analog of 3-deoxyisoequilenin was described by Mitra and Tilak [169–171] (Scheme 3-18) as well as by Collins and Brown [172]. The former group also synthesized thiophene analog of 3-deoxyequilenin **204** (R = H). Both groups of workers followed Johnson's route for the synthesis of equilenin (see Scheme 3-3). The starting material was 4-oxo-1,2,3,4-tetrahydrodibenzothiophene (**200**, R = H), which was converted into 14,15-dehydro-3-deoxy-B-nor-6-thiaequilenin (**201**). Catalytic hydrogenation yielded a mixture of two products. Tilak and co-workers isolated both the stereoisomers. The more stable, lower melting product was initially assigned the *trans* (13β, 14α) structure, while the higher melting product was assumed to have *cis* stereochemistry. However, Collins and Brown, following the same route, isolated the pure, lower melting isomer and assigned *cis* structure **203** (R = H) to it. The higher melting isomer could not be purified by them.

Later Tilak and co-workers [173] synthesized the compound **205** by an unambiguous route and also revised their earlier assignment of the structures of the hydrogenation products from **201**. Following Banerjee's modified procedure for the synthesis of equilenin (see Schme 3-4), the 14,15-dehydro compound **201** was reduced with sodium borohydride to yield the product **202**. Catalytic hydrogenation led to 6-thia-B-nor-

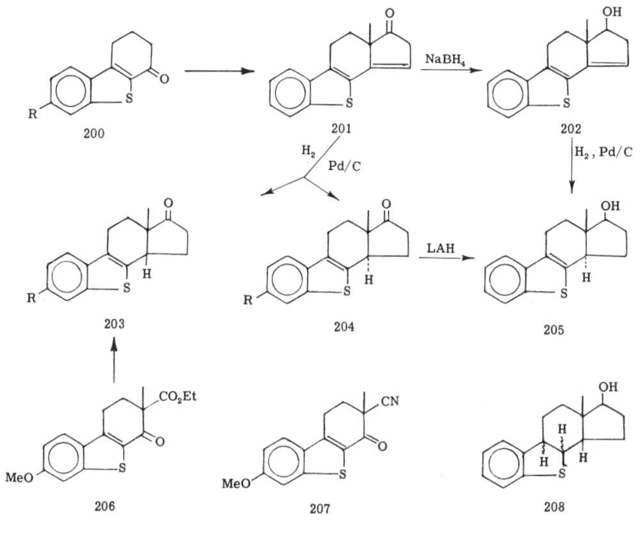

Scheme 3-18

1,3,5(10),8-estrapentaen-17β-ol (**205**), which was identical with the product obtained by the reduction of the higher melting stereoisomer derived from **201**. In view of the chemical and spectral evidence, the higher melting product from **201** was assigned to the 14α-isomer **204** (R=H) and the lower melting compound to the *cis* structure **203** (R=H). The latter compound has also been obtained from **200** (R=H) by following a modified Bachmann equilenin synthesis [173]. In another series of studies [174], the *cis* isomer **203** was used as the starting material for the synthesis of the thiophene analog of 3-deoxyisoestradiol (**208**) which was presumably obtained as a mixture of stereoisomers.

An unsuccessful attempt at the synthesis of B-nor-6-thiaequilenin following the Johnson route, has also been made [171]. The key intermediate **207** was synthesized from 7-methoxy-4-oxo-1,2,3,4-tetrahydrodibenzothiophene (**200**, R=OMe) in three steps in 54% yield. However, all attempts to build the ring D, by Stobbe condensation of **207** with diethyl or dimethyl succinate, failed. Later, following Bachmann's approach for the synthesis of equilenin (see Scheme 3-2), B-nor-6-thiaisoequilenin was synthesized via the intermediate **206** [175]. Following the analogy with Bachmann's synthesis, the CD-ring junction in the final product **203** (R=OMe) was assigned the *cis* stereochemistry. No chemical or spectroscopic evidence in support of this assignment was given, and obviously further work seems desirable. Later, the syntheses of *dl*-B-nor-6-thiaisoequilenin (**203**, R=OH) and *dl*-B-nor-6-thiaequilenin (**204**, R=OH) were achieved by Crenshaw and Luke [176], following a different route.

3.6.3 *Azasteroids*

A variety of new methods for the total synthesis of heterocyclic steroids has been developed by Huisman, Pandit, and co-workers. Elegant approaches involving enamine reactions were described by them for the synthesis of 13,14-diaza [177,178] and 11,13-diaza [179] steroidal systems (Scheme 3-19). The starting material for the 13,14-diaza-D-homosteroid **213**, was 6-methoxy-2-tetralone (**145**). Alkylation of its pyrrolidine enamine led to the γ-keto ester **209**. Treatment of **209** with perhydropyridazine in boiling xylene yielded the 13,14-diazasteroid **210**. Similarly, condensation of **209** with hydrazine hydrate gave the compound **211**. Reduction with lithium aluminum hydride led to the tricyclic hydrazine **212**, which was then converted into the 13,14-diaza-15,17-dioxosteroid **213**.

In another approach, Pandit, and co-workers [179] utilized the

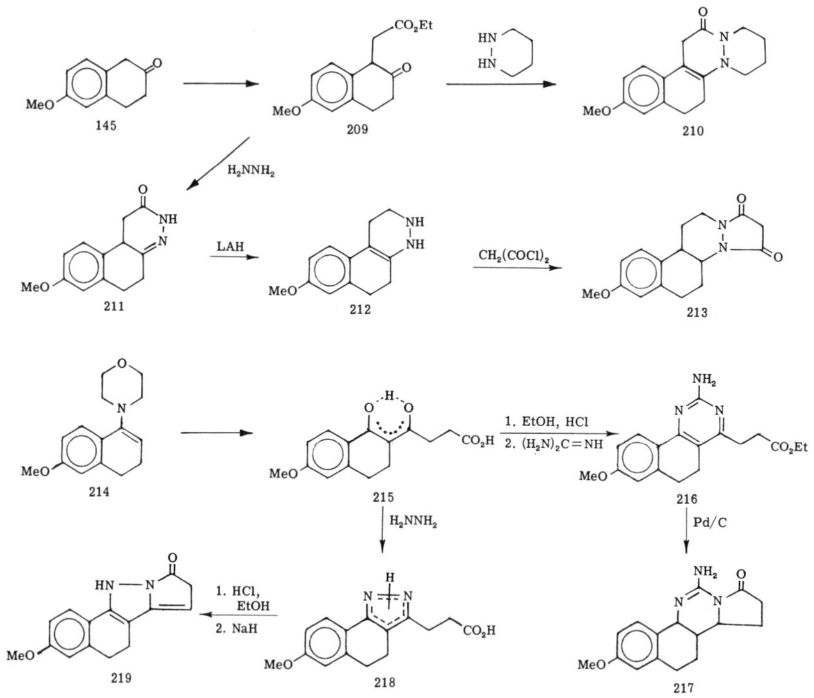

3.6 Heterocyclic Steroids

Scheme 3-19

morpholine enamine of 6-methoxy-1-tetralone (**214**) for the synthesis of 11,13-diazasteroids. Acylation of the enamine with the acid chloride of monomethyl succinate, followed by acid hydrolysis, afforded the diketo acid **215**. Esterification followed by heating with guanidine carbonate in 2-ethoxyethanol led to the tricyclic system **216**. Catalytic reduction of the latter in an acidic medium then yielded the 11,13-diazasteroid **217**. Finally, the location of the double bond and stereochemical assignments at C-8, C-9, and C-14 were determined by X-ray crystallographic studies. Treatment of the diketo acid **215** with hydrazine hydrate led to the formation of the hydrazine salt of **218**. The latter could be isolated in the zwitter ion form by careful adjustment of its aqueous solution to pH. 3.5. The acid **218** was next esterified and the resulting ester was refluxed with sodium hydride in dioxane. The crystalline product was assigned the C-nor steroidal structure **219**, on the basis of spectral data.

A versatile approach based on a reaction type that formally resembles a normal Diels-Alder cycloaddition between an imine and a diene, but belongs mechanistically to a different class [180], was used for the synthesis of several azasteroids [162,181,181a] (Scheme 3-20). Condensation of the ethyl biscarbamate **220** (R=R'=R"=Et) with the diene **221**, under the influence of boron trifluoride etherate, yielded the tricyclic ester **222** (R=R"=Et). Since difficulties were encountered during the hydrolysis of the carbamate group, later experiments were carried out with the corresponding benzyl biscarbamate **220** (R=R"=Bz, R'=Me), and the

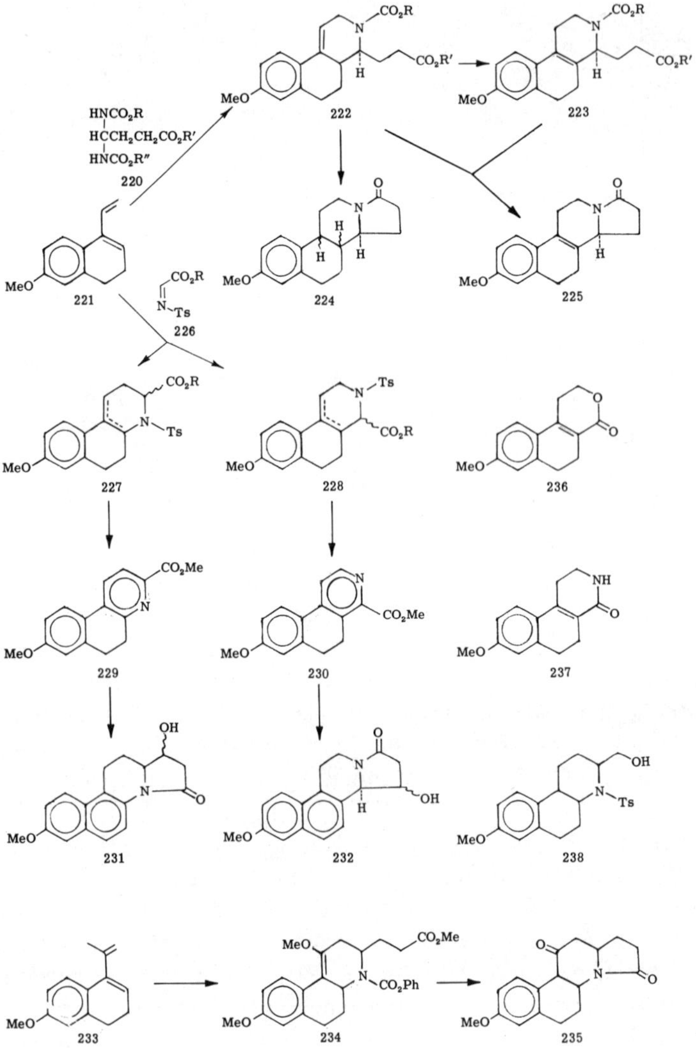

Scheme 3-20

3.6 Heterocyclic Steroids

crystalline adduct **222** (R = Bz, R' = Me) was isolated by chromatography. This latter compound could be isomerised to its 8,9-dehydro-isomer **223** (R = Bz, R' = Me) by treatment with acetic acid. Decarbobenzoxylation of either **222** (R = Bz, R' = Me) or **223** (R = Bz, R' = Me) went smoothly to give the intermediate γ-amino ester which on cyclization yielded the 8,9-dehydro-13-azaestrone (**225**). Also, the intermediate ester **222** (R = Bz, R' = Me) could be converted into the 13-azaestrone **224** by a combined hydrogenolysis of the benzyl ester and catalytic hydrogenation of the 9,11-double bond followed by cyclization.

In another series of studies [162], treatment of the imino ester **226** with the diene **221** led to a mixture of tricyclic isomeric adducts **227** and **228**. Alkaline hydrolysis yielded a mixture of dihydrobenzoquinoline acids, which were separated and esterified to give the methyl esters **229** and **230**. After dehydrogenation, **229** and **230** were converted into the 14-aza- and 13-azaequilenin systems **231** and **232**, respectively. This versatile method has also been applied to the synthesis of 11-oxygenated steroid analog **235** [162]. Condensation of the biscarbamate **220** (R = R'' = Bz, R' = Me) with the diene **233** yielded predominately the 14-azaisomer **234**. Acid-catalyzed cleavage of the benzyloxycarbonyl group led to the 14-azasteroidal system **235**. Further modifications of this above method led to the synthesis of the tricyclic intermediates **236**, **237** [182], and **238** [183].

The synthesis of the 13-aza compound **242** was accomplished by Kessar and co-workers [184] (Scheme 3-21). Condensation of the amine **239** (R = H) with β-carbomethoxypropionyl chloride yielded the amide **240** which was cyclized with phosphorous oxychloride. The basic material, obtained from the cyclization was hydrogenated and thermally cyclized to afford 13-aza-18-norequilenin (**242**). The same compound was also synthesized by Birch and Rao by a different route [185]. The amine **239** (R = H) was also used [186,187] for the synthesis of 13-aza-15-thia-18-norequilenin methyl ether (**244**). Reaction of **239** (R = H) with ethyl formate yielded the amide **239** (R = CHO), which on cyclization led to the dihydrobenz[f]isoquinoline (**243**). Reaction of the latter with mercaptoacetic acid furnished the thiazolidone **244**.

Several other approaches to azasteroids have been described. Attempts to synthesize B-nor-6-aza-6-methyl equilenin and the pyrrolo analog of 3-deoxyequilenin were unsuccessful [188,189]. Later the total syntheses of several 6-azaestrogens were described by Speckamp and co-workers [190]. The synthesis of 9-azasteroids was attempted by Jones and Wood [191–193] as well as by Schleigh and Popp [194.] Both groups of workers reported the synthesis of the tricyclic intermediate **245**. How-

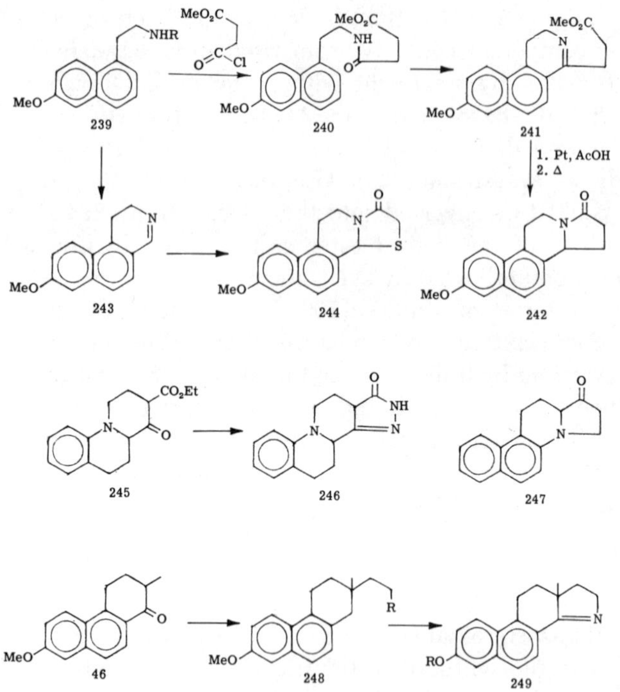

Scheme 3-21

ever, attempts to build up the ring D were not successful. Treatment of the keto ester **245** with hydrazine led to the pyrazolone **246**. The synthesis of 14-azasteroids was studied by Poirier and co-workers [195] as well as by Jones [196]. The former group achieved the synthesis of 3-deoxy-18-nor-14-azaequilenin (**247**). In another series of studies by Morgan and colleagues [23], the known tricyclic ketone **46** was condensed with acrylonitrile in t-butyl alcohol in the presence of potassium hydroxide to give the nitrile **248** (R=CN), which was hydrolyzed to the acid **248** (R=CO$_2$H). A modified Curtius rearrangement yielded the steroidal imine ether **249** (R=Me), which on heating with hydrobromic acid led to 15-azaequilenin derivative **249** (R=H). The same group of workers [9] described the synthesis of 15,16-diazasteroids. The synthesis of several other diazasteroid derivatives has been described in the literature. Examples of this group include 12,14- [197], 13,16- [198], 14,16-[198,199], and 15,16-diazasteroids [9,200].*

---

* 11-Amino-12,13,15,16-tetraaza-1,3,5(10),8,11,14,16-gonaheptaen, as well as other tetraazasteroids and a pentaazasteroid, have been synthesized by an ABC→ABCD route: B. Stanovnik, M. Tisler, and P. Skufca, *J. Org. Chem.* **33**, 2910 (1968).

References

1. W. E. Bachmann, W. Cole, and A. L. Wilds, *J. Amer. Chem. Soc.* **61**, 974 (1939).
2. W. E. Bachmann, W. Cole, and A. L. Wilds, *J. Amer. Chem. Soc.* **62**, 824 (1940).
3. A. Butenandt and G. Schramm, *Ber. Deut. Chem. Ges.* B **68**, 2083 (1935).
4. A. Cohen, J. W. Cook, C. L. Hewett, and A. Girard, *J. Chem. Soc., London* p. 653 (1934).
5. A. Cohen, J. W. Cook, and C. L. Hewett, *J. Chem. Soc., London* p. 445 (1935).
6. A. L. Wilds and W. J. Close, *J. Amer. Chem. Soc.* **69**, 3079 (1947).
7. W. S. Johnson and R. P. Graber, *J. Amer. Chem. Soc.* **72**, 925 (1950).
8. J. Hoch, *Bull. Soc. Chim. Fr.* p. 264 (1938).
9. J. G. Morgan, K. D. Berlin, N. N. Durham, and R. W. Chestnut, *J. Heterocycl. Chem.* **8**, 61 (1971).
10. G. Stork, *J. Amer. Chem. Soc.* **69**, 576 (1947).
11. G. Stork, *J. Amer. Chem. Soc.* **69**, 2936 (1947).
12. W. E. Bachmann and R. E. Holmen, *J. Amer. Chem. Soc.* **73**, 3660, (1951).
13. J. C. Bardhan, *Nature (London)* **134**, 217 (1934).
14. J. C. Bardhan, *J. Chem. Soc., London* p. 1848 (1936).
15. W. S. Johnson, J. W. Peterson, and C. D. Gutsche, *J. Amer. Chem. Soc.* **67**, 2274 (1945).
16. W. S. Johnson, J. W. Peterson, and C. D. Gutsche, *J. Amer. Chem. Soc.* **69**, 2942 (1947).
17. W. S. Johnson, C. D. Gutsche, R. Hirschmann, and V. L. Stromberg, *J. Amer. Chem. Soc.* **73**, 322 (1951).
18. D. K. Banerjee, S. Chatterjee, C. N. Pillai, and M. V. Bhatt, *J. Amer. Chem. Soc.* **78**, 3769 (1956).
19. D. K. Banerjee, B. Sugavanam, and G. Nadamuni, *Proc. Indian Acad. Sci., Sect. A.* **69**, 229 (1969).
20. D. K. Banerjee and G. Nadamuni, *Indian J. Chem.* **7**, 529 (1969).
21. W. S. Johnson and V. L. Stromberg, *J. Amer. Chem. Soc.* **72**, 505 (1950).
22. W. Bachmann, S. Kushner, and A. C. Stevenson, *J. Amer. Chem. Soc.* **64**, 974 (1942).
23. J. G. Morgan, K. D. Berlin, N. N. Durham, and R. W. Chesnut, *J. Org. Chem.* **36**, 1599 (1971).
24. G. Haberland and E. Blanke, *Ber. Deut. Chem. Ges.* B **70**, 169 (1937).
25. G. Haberland, *Ber. Deut. Chem. Ges.* B **69**, 1380 (1936).
26. G. Haberland, *Ber. Deut. Chem. Ges.* B **72**, 1215 (1939).
27. G. Haberland and E. Heinrich, *Ber. Deut. Chem. Ges.* B **72**, 1222 (1939).
28. G. Haberland, *Ber. Deut. Chem. Ges.* B **76**, 621 (1943).
29. C. Chang, *Acta Chim. Sinica* **21**, 190 (1955).
30. C. Chang, *Sci. Sinica* **4**, 547 (1955).
31. V. C. E. Burnop, G. H. Elliott, and R. P. Linstead, *J. Chem. Soc., London* p. 727 (1940).
32. A. L. Wilds and L. W. Beck, *J. Amer. Chem. Soc.* **66**, 1688 (1944).
33. A. L. Wilds and T. L. Johnson, *J. Amer. Chem. Soc.* **70**, 1166 (1948).

34. A. L. Wilds, J. A. Johnson, Jr., and R. E. Sutton, *J. Amer. Chem. Soc.* **72**, 5524 (1950).
35. A. L. Wilds, R. H. Zeitschel, R. E. Sutton, and J. A. Johnson, Jr., *J. Org. Chem.* **19**, 255 (1954).
36. A. L. Wilds, M. Harnik, R. Z. Shimizu, and D. A. Tyner, *J. Amer. Chem. Soc.* **88**, 799 (1966).
37. A. L Wilds and C H. Shunk, *J. Amer. Chem. Soc.* **70**, 2427 (1948).
38. B. Riegel and F. S. Prout, *J. Org. Chem.* **13**, 933 (1948).
39. A. L. Wilds, J. W. Ralls, W. C. Wildman, and K. E. McCaleb, *J. Amer. Chem. Soc.* **72**, 5794 (1950).
40. A. L. Wilds, J. W. Ralls, D. A. Tyner, R. Daniels, S. Kraychy, and M. Harnik, *J. Amer. Chem. Soc.* **75**, 4878 (1953).
41. P. T. Lansbury, E. J. Nienhouse, D. J. Scharf, and F. R. Hilfiker, *J. Amer. Chem. Soc.* **92**, 5649 (1970).
42. R. E. Juday and B. Bukwa, *J. Med. Chem.* **13**, 754 (1970).
43. O. R. Rodig, N. A. Nelson, E. M. Gross, M Harnik, and A L. Wilds, *Abstr. Pap., 131st Meet. Amer. Chem. Soc.* Pap. 32-0 (1957).
44. M. Harnik, *Isr. J. Chem.* **3**, 1 (1965).
45. M. Harnik and E. V. Jensen, *Isr. J. Chem.* **3**, 13 (1965).
46. M. Harnik and E. V. Jensen, *Isr. J. Chem.* **3**, 173 (1965).
47. M. Harnik, *Isr. J. Chem.* **3**, 183 (1965).
48. R. Robinson and J. Walker, *J. Chem. Soc., London* p. 747 (1936).
49. R. Robinson and J. Walker, *J. Chem. Soc., London* p. 183 (1938).
50. G. Anner and K. Miescher, *Helv. Chim. Acta* **31**, 2173 (1948).
51. G. Anner and K. Miescher, *Helv. Chim. Acta* **32**, 1957 (1949).
52. G. Anner and K. Miescher, *Helv. Chim. Acta* **33**, 1379 (1950).
53. S. N. Ananchenko, V. N. Leonov, A. V. Platonova, and I. V. Torgov, *Dokl. Akad. Nauk SSSR* **135**, 73 (1960).
54. W. S. Johnson, D. K. Banerjee, W. P. Schneider, and C. D. Gutsche, *J. Amer. Chem. Soc.* **72**, 1426 (1950).
55. W. S. Johnson, D. K. Banerjee, W. P. Schneider, C. D. Gutsche, W. E. Shelberg, and L. J. Chinn, *J. Amer. Chem. Soc.* **74**, 2832 (1952).
56. W. S. Johnson and R. G. Christiansen, *J. Amer. Chem. Soc.* **73**, 5511 (1951).
57. W. S. Johnson, R. G. Christiansen, and R. G. Ireland, *J. Amer. Chem. Soc.* **79**, 1995 (1957).
58. J. E. Cole, Jr., W. S. Johnson, P. A. Robins, and J. Walker, *Proc. Chem. Soc., London* p. 114 (1958).
59. W. S. Johnson, I. A. David, H. C. Dehm, R. J. Highet, E. W. Warnhoff, W. D. Wood, and E. T. Jones, *J. Amer. Chem. Soc.* **80**, 661 (1958).
60. W. S. Johnson, S. G. Boots, and E. R. Habicht, Jr., *J. Org. Chem.* **33**, 1754 (1968).
61. J. C. Sheehan, W. F. Erman, and P. A. Cruickshank, *J. Amer. Chem. Soc.* **79**, 147 (1957).
62. D. K. Banerjee and K. M. Sivanandaiah, *Tetrahedron Lett.* **5**, 20 (1960).
63. G. A. Hughes and H. Smith, *Chem. Ind. (London)* p. 1022 (1960).
64. I. V. Torgov, *Pure Appl. Chem.* **6**, 525 (1963).
65. D. J. Crispin and J. S. Whitehurst, *Proc. Chem. Soc., London* p. 22 (1963).
66. L. Velluz, G. Nominé, J. Mathieu, E. Toromanoff, D. Bertin, M. Vignau, and J. Tessier, *C. R. Acad. Sci.* **250**, 1510 (1960).

References

67. J. C. Sheehan, R. A. Coderre, and P. A. Cruickshank, *J. Amer. Chem. Soc.* **75**, 6231 (1953).
68. J. H. Hunter and J. A. Hogg, *J. Amer. Chem. Soc.* **71**, 1922 (1949).
69. G. Anner and K. Miescher, *Experientia* **4**, 25 (1948).
70. J. C. Sheehan and R. A. Coderre, *J. Amer. Chem. Soc.* **75**, 3997 (1953).
71. J. C. Sheehan, R. C. O'Neill, and M A. White, *J. Amer. Chem. Soc.* **72**, 3376 (1950).
72. J. C. Sheehan, R. C. Coderre, L. A. Cohen, and R. C. O'Neill, *J. Amer. Chem. Soc.* **74**, 6155 (1952).
73. J. C. Sheehan, U. S. Patent, 2,775,603 (1956); *Chem. Abstr.* **51**, 7446 (1957).
74. G. I. Kiprianov and L. M. Kutsenko, *Med. Prom. SSSR* **15**, 43 (1961).
75. A. J. Birch and S. M. Mukherji, *J. Chem. Soc., London* p. 2531 (1949).
75a. A. J. Birch, *J. Chem. Soc., London* p. 367 (1950).
76. A. L. Wilds and N. A. Nelson, *J Amer. Chem. Soc.* **75**, 5366 (1953).
77. P. Crabbé, A. Cruz, and J. Iriarte, *Chem. Ind. (London)* p. 1522 (1967).
78. J. Iriarte and P. Crabbé, *Chem. Commun.* p. 1110 (1972).
79. H. Wehrli and K. Schaffner, *Helv. Chim. Acta* **45**, 385 (1962).
80. J. P. Gesson, J. C. Jacquesy, and R. Jacquesy, *Tetrahedron Lett.* p. 4733 (1971).
81. A. M. Krubiner and E. P. Oliveto, *J. Org. Chem.* **31**, 24 (1966).
82. P. Kurath and W. Cole, *J. Org. Chem.* **26**, 4592 (1961).
83. J. A. Zderic, A. Bowers, H. Carpio, and C. Djerassi, *J. Amer. Chem. Soc.* **80**, 2596 (1958).
84. K. Miescher, *Helv. Chim. Acta* **27**, 1727 (1944).
85. J. Heer, J. R. Billeter, and K. Miescher, *Helv. Chim. Acta* **28**, 991 (1945).
86. J. Heer, J. R. Billeter, and K. Miescher, *Helv. Chim. Acta* **28**, 1342 (1945).
87. G. Anner and K. Miescher, *Helv. Chim. Acta* **29**, 586 (1946).
88. K. Miescher, *Experienta* **2**, 247 (1946).
89. J. R. Billeter and K. Miescher, *Helv. Chim. Acta* **29**, 859 (1946).
90. G. Anner, J. Heer and K. Miescher, *Helv. Chim. Acta* **29**, 1071 (1946).
91. J. Heer and K. Miescher, *Helv. Chim. Acta* **28**, 1506 (1945).
92. J. Heer and K. Miescher, *Helv. Chim. Acta* **30**, 777 (1947).
93. K. Miescher, *Chem. Rev.* **43**, 367 (1948).
94. K. Miescher, *Experientia* **5**, 1 (1949).
95. W. S. Johnson and R. P. Graber, *J. Amer. Chem. Soc.* **70**, 2612 (1948).
96. D. L. Turner, B. K. Bhattacharyya, R. P. Graber, and W. S. Johnson, *J. Amer. Chem. Soc.* **72**, 5654 (1950).
97. R. Gay, *C. R. Acad. Sci.* **242**, 3084 (1956).
98. A. Horeau and R. Gay, *Bull. Soc. Chim. Fr.* p. 581 (1958).
99. R. Gay, *Ann. Chim. (Paris)* [13] **4**, 995 (1959).
100. R. Gay and A. Horeau, *Tetrahedron* **7**, 90 (1959).
101. J. Lematre and A. Horeau, *C. R. Acad. Sci.* **251**, 257 (1960).
102. M. Gastambide-Odier, P. Carnero, J. Chevallier, B. Gastambide, M. J. Laroche, and A. Cottard, *Bull. Soc. Chim. Fr.* p. 1777 (1963).
103. W. R. J. Simpson, D. Babbe, J. A. Edwards, and J. H. Fried, *Tetrahedron Lett.* p. 3209 (1967).
104. H. Lapin, *Chimia* **18**, 141 (1964).
105. G. Anner and K. Miescher, *Experientia* **2**, 409 (1946).
106. G. Anner and K. Miescher, *Helv. Chim. Acta* **29**, 1889 (1946).
107. G. Anner and K. Miescher, *Helv. Chim. Acta* **30**, 1422 (1947).

108. J. H. Hunter and J. A. Hogg, *J. Amer. Chem. Soc.* **68**, 1676 (1946).
109. P. Crabbé, A. Cruz, and J. Iriarte, *Can. J. Chem.* **46**, 349 (1968).
110. P. Crabbé, in "Terpenoids and Steroids," Vol. 2, p. 393. Chem. Soc., London, 1972.
111. N. A. Nelson and R. B. Garland, *J. Amer. Chem. Soc.* **79**, 6313 (1957).
112. F. J. Villani, M. S. King, and D. Papa, *J. Org. Chem.* **18**, 1578 (1953).
113. L. H. Sarett, W. F. Johns, R. E. Beyler, R. M. Lukes, G. I. Poos, and G. E. Arth, *J. Amer. Chem. Soc.* **75**, 2112 (1953).
114. W. F. Johns, R. M. Lukes, and L. H. Sarett, *J. Amer. Chem. Soc.* **76**, 5026 (1954).
115. W. F. Johns, *J. Amer. Chem. Soc.* **80**, 6456 (1958).
116. W. F. Johns, *J. Org. Chem.* **28**, 1856 (1963).
117. G. Stork, H. N. Khastgir, and A. J. Solo, *J. Amer. Chem. Soc.* **80**, 6457 (1958).
118. R. Robinson and F. Weygand, *J. Chem. Soc., London* p. 386 (1941).
119. J. W. Cornforth, R. H. Cornforth, and R. Robinson, *J. Chem. Soc. London* p. 689 (1942).
120. W. S. Johnson, J. M. Anderson, and W. E. Shelberg, *J. Amer. Chem. Soc.* **66**, 218 (1944).
121. W. Nagata and T. Terasawa, *Chem. Pharm. Bull.* **9**, 267 (1961).
122. R. L. Kidwell and S. D. Darling, *Tetrahedron Lett.* p. 531 (1966).
123. W. Nagata, I. Kikkawa, and K. Takeda, *Chem. Pharm. Bull.* **9**, 79 (1961).
124. W. Nagata, S. Hirai, T. Terasawa, I. Kikkawa, and K. Takeda, *Chem. Pharm. Bull.* **9**, 756 (1961).
125. Shionogi & Co., British Patent 998,980 (1965); *Chem. Abst.* **63**, 10035 (1965).
126. D. H. R. Barton, A. S. Campos-Neves, and A. I. Scott, *J. Chem. Soc., London* p. 2698 (1957).
127. J. W. Cornforth and R. Robinson, *J. Chem. Soc., London* p. 1855 (1949).
128. D. Nasipuri, *Chem. Ind. (London)* p. 1389 (1956).
129. D. Nasipuri and K. K. Biswas, *J. Indian Chem. Soc.* **45**, 603, (1968).
130. W. Nagata, *Tetrahedron* **13**, 287 (1961).
131. W. L. Meyer and J. F. Wolfe, *J. Org. Chem.* **29**, 170 (1964)
132. W. Nagata, M. Yoshioka, and S. Hirai, *Tetrahedron Lett.* p. 461 (1962).
133. W. Nagata and M. Yoshioka, *Pro. Int. Congr. Horm. Steroids, 1966* Intl. Congr. Ser. No. 132, p. 327 (1967).
134. W. Nagata, *Nippon Kagaku Zasshi* **90**, 837 (1969).
135. W. Nagata, M. Yoshioka, and T. Terasawa, *J. Amer. Chem. Soc.* **94**, 4672 (1972).
136. P. C. Mukharji, *J. Indian Chem. Soc.* **24**, 91 (1947).
137. P. C. Mukharji, *J. Indian Chem. Soc.* **25**, 367 (1948).
138. P. C. Mukharji, *J. Indian Chem. Soc.* **25**, 373 (1948).
139. C. A. Friedmann and R. Robinson, *Chem. Ind. (London)* p. 777 (1951).
140. C. A. Friedmann and R. Robinson, *Chem. Ind. (London)* p. 1117 (1951).
141. S. Swaminathan and M. S. Newman, *Tetrahedron* **2**, 88 (1958).
142. I. N. Nazarov and I. A. Gurvich, *Zh. Obshch. Khim.* **25**, 956 (1955).
143. I. N. Nazarov and I. A. Gurvich, *Zh. Obshch. Khim.* **25**, 1723 (1955).
144. W. S. Johnson, S. Shulman, K. L. Williamson, and R. Pappo, *J. Org. Chem.* **27**, 2015 (1962).
145. P. Beslin and J.-M. Conia, *Bull. Soc. Chim. Fr.* p. 959 (1970).
146. E. Brown, M. Ragault, and J. Touet, *Bull. Soc. Chim. Fr.* p. 2195 (1971).

References

147. E. Brown, J. Touet, and M. Ragault, *Bull. Soc. Chim. Fr.* p. 212 (1972).
148. W. S. Johnson, J. M. Cox, D. W. Graham, and H. W. Whitlock, Jr., *J. Amer. Chem. Soc.* **89**, 4524 (1967).
149. T. Masamune, M. Takasugi, A. Murai, and K. Kobayashi, *J. Amer. Chem. Soc.* **89**, 4521 (1967).
150. W. S. Johnson, H. A. P. deJongh, C. E. Coverdale, J. W. Scott, and U. Burckhardt, *J. Amer. Chem. Soc.* **89**, 4523 (1967).
151. H. K. Sen and K. Mondal, *J. Indian Chem. Soc.* **5**, 609 (1928).
152. A. Tuinman, A. C. Ghosh, K. Schaffner, and O. Jeger, *Chimia* **24**, 27 (1970).
153. A. Chatterjee, S. Banerjee, A. K. Sarkar, and B. G. Hazra, *J. Chem. Soc., London* p. 661 (1971).
154. A. Chatterjee and S. Banerjee, *J. Indian Chem. Soc.* **41**, 643 (1964).
155. Z. G. Hajos, D. R. Parrish, and M. W. Goldberg, *J. Org. Chem.* **30**, 1213 (1965).
156. P. N. Rao, B. E. Edwards, and L. R. Axelrod, *J. Chem. Soc., London* p. 2863 (1971).
157. A. A. Akhrem and Yu. A. Titov, *Russ. Chem. Rev.* **36**, 311 (1967).
158. P. Morand and J. Lyall, *Chem. Rev.* **68**, 85 (1968).
159. B. D. Tilak, *J. Indian Chem. Soc.* **36**, 509 (1959).
160. H. O. Huisman, *Bull. Soc. Chim. Fr.* p. 13 (1968).
161. J. F. Bagli, *Can. J. Chem.* **40**, 2032 (1962).
162. H. O. Huisman, *Angew. Chem., Int. Ed. Engl.* **10**, 450 (1971).
163. V. N. Gogte, *J. Sci. Ind. Res.* **27**, 353 (1970).
164. H. O. Huisman, in "Steroids" (W. F. Johns, Ed.), Vol. 8, p. 235. Butterworth, London, 1973.
165. G. V. Bhide, N. L. Tikotkar, and B. D. Tilak, *Tetrahedron* **10**, 223 (1960).
166. G. V. Bhide, N. L. Tikotkar, and B. D. Tilak, *Chem. Ind. (London)* p. 1319 (1957).
167. T. R. Kasturi and A. Srinivasan, *Indian J Chem.* **8**, 482 (1970).
168. T. R. Kasturi and T. Arunachalam, *Indian J. Chem.* **8**, 203 (1970).
169. R. B. Mitra and B. D. Tilak, *J. Sci. Ind. Res., Sect. B* **14**, 132 (1955).
170. R. B. Mitra and B. D. Tilak, *J. Sci. Ind. Res., Sect. B* **15**, 497 (1956).
171. M. K. Bhattacharjee, R. B. Mitra, B. D. Tilak, and M. R. Venkiteswaren, *Tetrahedron* **10**, 215 (1960).
172. R. J. Collins and E. V. Brown, *J. Amer. Chem. Soc.* **79**, 1103 (1957).
173. B. D. Tilak, V. N. Gogte, A. S. Jhina, and G. R. N. Sastry, *Indian J. Chem.* **7**, 31 (1969).
174. R. B. Mitra and B. D. Tilak, *J. Sci. Ind. Res., Sect. B* **15**, 573, (1956).
175. B. D. Tilak and M. K. Bhattacharjee, *Indian J. Chem.* **7**, 36 (1969).
176. R. R. Crenshaw and G. M. Luke, *Tetrahedron Lett.* p. 4495 (1969).
177. U. K. Pandit, K. de Jonge, and H. O. Huisman, *Rec. Trav. Chim. Pays-Bas* **88**, 149 (1969).
178. E. R. deWaard, R. Neeter, U. K. Pandit, and H. O. Huisman, *Rec. Trav. Chim. Pays-Bas* **87**, 572 (1968).
179. U. K. Pandit, F. A. van der Vlugt, and A. C. van Dalen, *Tetrahedron Lett.* p. 3693 (1969).
180. C. K. Wilkins, *Tetrahedron Lett.* p. 4817 (1965).
181. W. N. Speckamp, R. J. P. Barends, A. J. de Gee, and H. O. Huisman, *Tetrahedron Lett.* p. 383 (1970).

181a. H. O. Huisman and W. N. Speckamp, U.S. Patent 3,597,436 (1971); *Chem. Abstr.* **75**, 141054f (1971).
182. R. J. P. Barends, W. N. Speckamp, and H. O. Huisman, *Tetrahedron Lett.* p. 5301 (1970).
183. P. P. M. Rijsenbrij, R. Loven, J. B. P. A. Wijnberg, W. N. Speckamp, and H. O. Huisman, *Tetrahedron Lett.* p. 1425 (1972).
184. S. V. Kessar, M. Singh, and A. Kumar, *Tetrahedron Lett.* p. 3245 (1965).
185. A. J. Birch and G. S. R. Subba Rao, *J. Chem. Soc., London* p. 3007 (1965).
186. S. V. Kessar and P. Jit, *Indian J Chem.* **7**, 735 (1969).
187. S. V. Kessar, P. Jit, K. P. Mundra, and A. K. Lumb, *J. Chem. Soc., C* p. 266 (1971).
188. G. V. Bhide, N. R. Pai, I. L. Tikotkar, and B. D. Tilak, *Tetrahedron* **4**, 420 (1958).
189. G. V. Bhide, N. L. Tikotkar, and B. D. Tilak, *Tetrahedron* **10**, 230 (1960).
190. W. N. Speckamp, H. de Koning, U. K. Pandit, and H. O. Huisman, *Tetrahedron* **21**, 2517 (1965).
191. G. Jones and J. Wood, *Tetrahedron* **21**, 2529 (1965).
192. G. Jones and J. Wood, *Tetrahedron* **21**, 2961 (1965).
193. J. D. Baty, G. Jones, and C. Moore, *J. Org. Chem* **34**, 3295 (1969).
194. W. R. Schleigh and F. D. Popp, *J. Chem. Soc., London* p. 760 (1966).
195. R. H. Poirier, R. D. Morin, F. Benington, and T. F. Page, *Abstr., 145th Meet., Amer. Chem. Soc.* p. 44Q (1963).
196. E. R. H. Jones, *J. Chem. Soc., London* p. 5907 (1964).
197. E. C. Taylor and Y. Shvo, *J. Org. Chem* **33**, 1719 (1968).
198. H. Zimmer and H. D. Benson, *Chimia* **26**, 131 (1972).
199. L. E. Katz and F. D. Popp, *J. Heterocycl. Chem.* **5**, 249 (1968).
200. J. Lematre and J. Soulier, *C. R. Acad. Sci.*, **265**, 199 (1967).

4
AB + D → ABD → ABCD

4.1 Introduction

ABD intermediates employed in the total synthesis of steroids belong to four different structural types (Scheme 4-1). In the first type, rings B and D are connected through an 8,14-bond and carbons 11 and 12 are attached to ring B. The synthesis of this type from fragment **1** and ring D is described in Sections 4.4 and 4.5. These same fragments are combined in Sections 4.2, 4.3, 4.4, and 4.6 to form the second type of ABD intermediate, in which rings B and D connect through carbons 11 and 12. This second type has been synthesized by six other routes, one of which, from **2**, is described in Section 4.3. The combination of fragments **3** and **4** is covered in Section 4.6, as is the ring closure of **5**. Closure of **6**, as well as combination of **7** and **8** are described in Section 4.7, while the joining of fragments **9** and nor-**4** is found in Section 4.4.

These same fragments combine differently in Section 4.8 to form a third type of ABD intermediate in which rings B and D are joined through the 8,14-bond and carbons 11 and 12 are attached to ring D. This intermediate is also prepared from **10**, as described in Section 4.7. The reaction of fragment **11** with ring D to form the fourth type of ABD intermediate is described in Section 4.8. This type has neither carbon 11 nor 12 attached to ABD, the parts of which are joined by an 8,14-linkage.

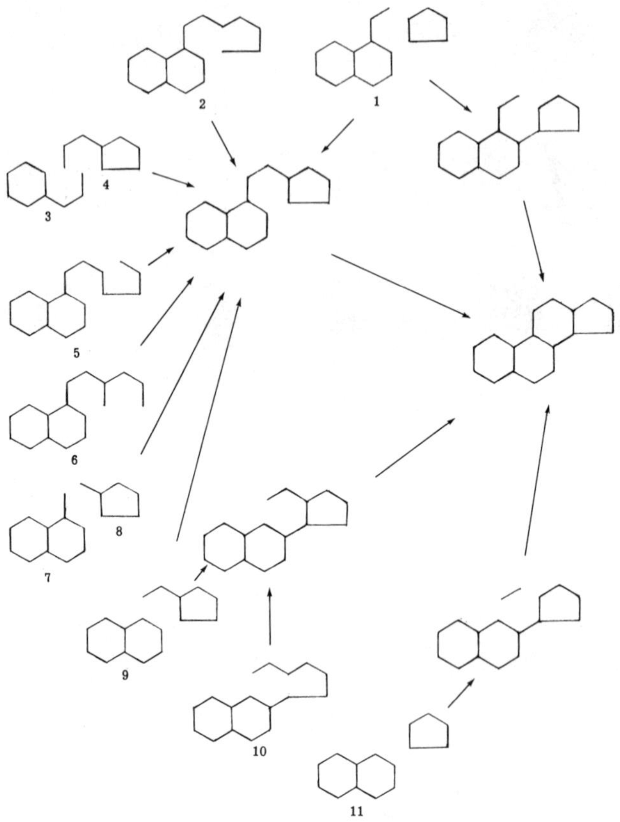

4 AB + D → ABD → ABCD

Scheme 4-1

Carbons 11 and 12 add to the ABD intermediate in a later stage of the synthesis.

4.2 The Torgov Synthesis

4.2.1 Estrone methyl ether

By far the most versatile synthesis of this group is that developed by I. V. Torgov and S. N. Ananchenko and their colleagues. It has been used by them and others to synthesize estrone, estradiol, equilenin, B- and D-homo analogs, B-nor analogs, 18-alkyl analogs, and 8α-isomers, as well as aza, thia, oxa and sila analogs. It is illustrated by the synthesis

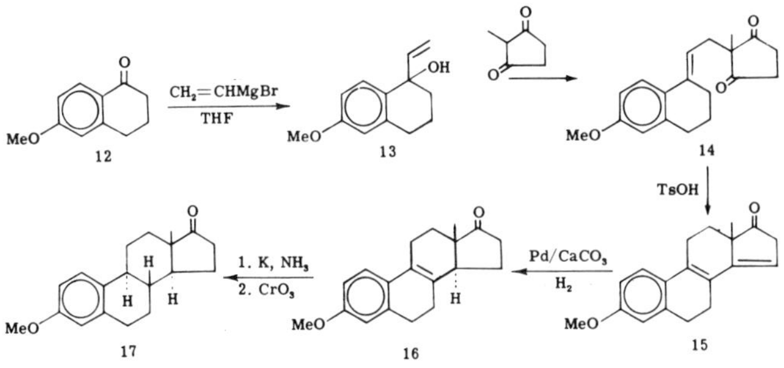

4.2 The Torgov Synthesis

Scheme 4-2

of the methyl ether (17) of estrone, as in Scheme 4-2 [1], which began with preparation of 6-methoxy-1-vinyl-1-tetralol (13) from 6-methoxy-1-tetralone (12) and vinylmagnesium bromide. Condensation of 2-methyl-1,3-cyclopentanedione with the vinyl alcohol 13 gave the ABD intermediate 14, which was ring closed under acid catalysis to the methyl ether (15) of 3-hydroxy-1,3,5(10),8,14-estrapentaen-17-one. Catalytic hydrogenation produced the tetraene 16, which was subjected to further reduction with potassium in liquid ammonia [2–5]. Some reduction of the 17-oxo group occurred simultaneously, so that the product mixture was oxidized with chromic oxide to give racemic estrone methyl ether (17). The success of this synthesis emerged out of four years of intensive investigations by the Russian group as well as by others. Four other groups published estrone syntheses very similar to Torgov's at essentially the same time [3,6–8].

4.2.2 Mechanism and stereochemistry

The condensation step between the vinyl alcohol 13 and the cyclic diketone was originally thought to be base catalyzed; benzyltrimethylammonium hydroxide (Triton B) [1,2,6,9–13], potassium hydroxide and potassium t-butoxide [10], triethylamine [14], and sodium bicarbonate [15] have been used. Recently, however, Kuo, Taub, and Wendler have shown that the condensation is acid catalyzed, the enolic form of the dione itself furnishing the acid to bring about the reaction [16; see also refs. 15 and 17]. Previous workers had always used less than molar amounts of base, so that sufficient unneutralized dione persisted to effect catalysis. Kuo and colleagues envision acid-catalyzed loss of OH from the vinyl alcohol 13 to produce the ion pair 18 (Scheme 4-3). Nucleophilic attack on C-12 by the dione anion then proceeds smoothly to the condensation product 14. They found that the dione and vinyl alcohol

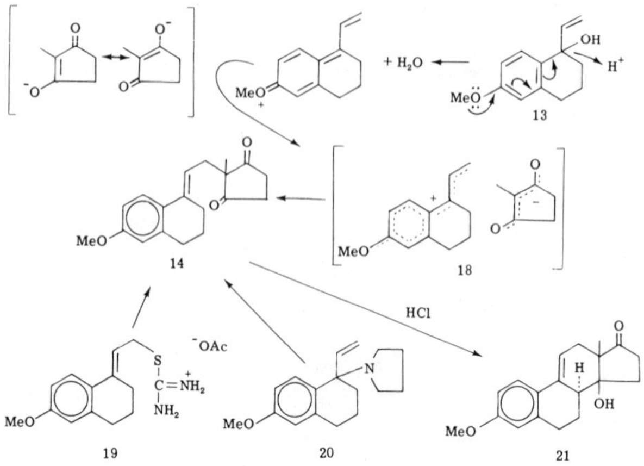

Scheme 4-3

reacted without catalyst in refluxing xylene and *t*-butyl alcohol to give a 70% yield of ABD intermediate 14. Furthermore, the isothiouronium salt 19 reacted with the dione in equeous media to give 14 in 90% yield, and this was thought to be an improvement in procedure because 19 reacts with diones that are unreactive toward 13. They found also that when methylcyclopentanedione and 13 were condensed in refluxing acetic acid and xylene, both condensation and cyclization to 15 were effected in 60% yield. Another variation in the Torgov synthesis employs the N-substituted pyrrolidine 20, which gives 14 in 80% yield [18]; piperidine and morpholine have been used in the same way [19].

In most applications of the Torgov synthesis the ABD intermediate is isolated, after which ring C is closed with the use of phosphorus pentoxide [7,14,20], *p*-toluensulfonic acid [1], formic acid [3], or concentrated hydrochloric acid [2,6]. Ring closure of 14 under mild acid conditions (dilute HCl) gave the 14-hydroxy intermediate 21, which was readily dehydrated to the pentaene 15 [14,21].

Compound 15 possesses only one asymmetric carbon, C-13; consequently, at this stage of the synthesis of estrone the product consisted of two enantiomers, 15 and 22 (Scheme 4-4). Reduction of the two nonaromatic double bonds created three more centers of asymmetry. if this were to occur in a nonstereospecific fashion, a total of sixteen isomers would be formed, a circumstance to be avoided ardently. Complete hydrogenation of the diene system in the presence of palladium-on-charcoal occurred stereospecifically, but gave 8α-estrone methyl ether (23, one enantiomer only shown) [1,3,4]. It was possible,

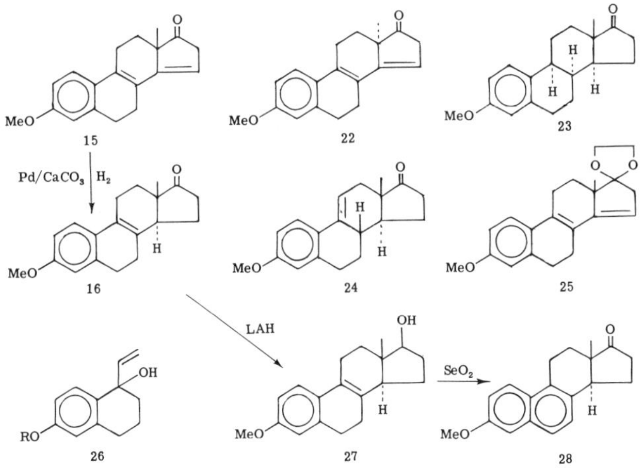

4.2 The Torgov Synthesis

Scheme 4-4

however, to reduce the D-ring double bond selectively with hydrogen and palladium-on-calcium carbonate to produce **16**. Chemical reduction (K,NH$_3$) of the 8-double bond of **16** has already been mentioned [2–5]. Other routes to estrone methyl ether include (a) isomerization of the 8-double bond to 9(11), as in **24**, and hydrogenation over palladium-on-calcium carbonate [2,4,22–25], and (b) protection of the 17-oxo group as a cyclic ketal **25**, followed by catalytic hydrogenation, lithium–ammonia reduction, and hydrolysis of the ketal linkage [1,5].

Although the Torgov synthesis is usually carried out with a 3-methoxyl group, it is possible to go directly to estrone by starting with the diol **26** (R=H); condensation, cyclization, hydrogenation, isomerization, and hydrogenation steps proceed in good yield [1,2,22,26–30]. Alternatively, the tetrahydropyranyl ether **26** (R=C$_5$H$_{10}$O) [2,31] or benzyl ether **26** (R=Bz) [31–33] may be condensed with the cyclic dione. The protective group is cleaved during the cyclization of ring-C step, which occurs in nearly quantitative yield. The condensation and cyclization steps occur normally also with dimethylamino or morpholino groups replacing RO– in **26**; the advantage here is that resolution of a tetracyclic intermediate may be carried out, and an oxygen function at C-3 ultimately regenerated [34].

Reduction of the 17-oxo group of **16** with lithium aluminum hydride or sodium borohydride gave the 17β-hydroxy intermediate **27** [1,2,5], which was further reduced to estradiol 3-methyl ether (structure not shown) with sodium and ammonia. The tetraene **27** was also converted

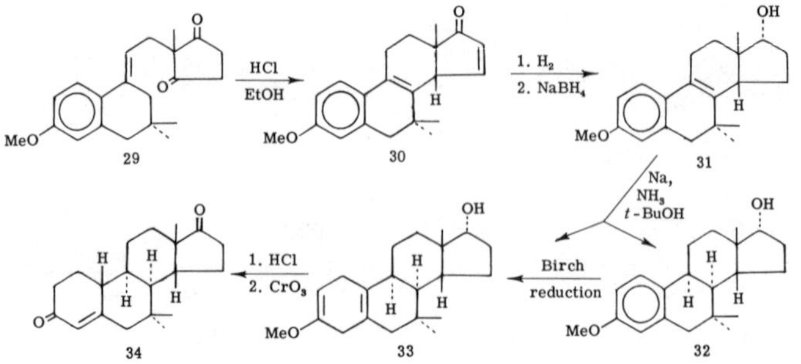

Scheme 4-5

to equilenin methyl ether (**28**) by selenium dioxide [1,2,4,14], chromic oxide [14], or with palladium-on-calcium carbonate [22]. Birch reduction of **16**, followed by acid hydrolysis, gave 19-nortestosterone [1,35,36].

4.2.3 Homo and Nor Variations

The dimethyl homolog of **13**, constructed so as to give rise to 7,7-dimethyl steroids, has been used in a Torgov synthesis [37], with some unexpected results. The corresponding ABD intermediate **29** (Scheme 4-5) was unstable and difficult to cyclize, but was convertible into the tetracyclic steroid **30** with ethanolic HCl. It differed from the usual Torgov product in that the ring-D double bond was at 15 rather than 14. The CD ring fusion was *cis*, which relieved strain that would otherwise be present caused by nonbonded interaction between the gem-dimethyl grouping and the C-15 hydrogen. Catalytic hydrogenation of **30**, followed by sodium borohydride reduction gave the 17α-hydroxy product **31**. Chemical reduction of **31** was difficult to stop at the dihydro stage, and mixtures of **32** and **33** were obtained. Hydrolysis of **33** was slow and reaction mixtures intractable, so they were oxidized to the 17-ketones, allowing separation of pure **34**. X-ray diffraction analysis indicated that the BC-ring juncture was *cis* and α.

By employing 2-methyl-1,3-cyclohexanedione (**35**, Scheme 4-6) in the condensation step of the Torgov synthesis, D-homo analogs were prepared by an identical sequence of reactions; in fact, conditions for the synthesis were worked out in this series before being applied to preparation of naturally occurring steroids. Thus, D-homoestrone methyl ether (**36**) was prepared via $\Delta^{8,14}$- and Δ^{8}-intermediates [9,10,20,38–41], and the corresponding 8α-isomer (not shown) was obtained by catalytic hydrogenation of the $\Delta^{8,14}$-intermediate [10,20,39]. The latter has been obtained in good yield recently via a 9α,14α-spiroether [42]. D-Homo-

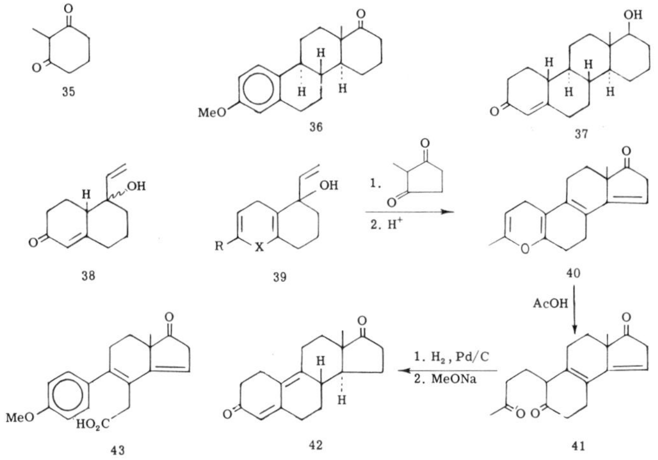

4.2 The Torgov Synthesis

Scheme 4-6

estrone methyl ether was converted to estrone methyl ether [20] by a ring-D contraction sequence devised by W. S. Johnson and colleagues [43]. Demethylation of the $\Delta^{8,14}$-intermediate with pyridine hydrochloride gave D-homoequilenin directly [10,44]. Birch reduction of the Δ^{8}-intermediate reduced the 8,9-double bond simultaneously giving, after the usual hydrolysis step, D-homo-19-nortestosterone (37) [45,46]. Similarly, Birch reduction of the corresponding 17-ethyleneketal led to D-homo-19-norandrostenedione [44]; reduction of the appropriate intermediate gave other D-homoandrostanes [10,45–57]. 19-Norsteroids of this type are not prepared directly via the Torgov synthesis simply by starting with an unsaturated ketone 38 in place of the aromatic vinyl alcohol. They are prepared, however, by Michael condensation between 6-oxo-1-vinyl-2,3,4,6,7,8-hexahydronaphthalene and cyclic 1,3-diones (discussed later in Section 4.4). The dihydro derivative of 13 (39, R=MeO, X=CH$_2$) did condense with 2-methyl-1,3-cyclohexanedione to give the customary ABD intermediate, although the latter did not cyclize [58]. Another route to nonaromatic A-ring steroids began with the heterocyclic vinyl alcohol 39 (R=Me, X=O) [59,60]. The normal steps of condensation and cyclization to the tetraene 40, were followed by opening of the A ring by heating in 50% acetic acid. Hydrogenation with palladium-on-carbon, and intramolecular condensation by sodium methoxide to close the A ring, gave 4,9-estradiene-3,17-dione (42). Other 4-oxasteroids are described in Section 4.2.4.

Other alkyl-substituted steroids have been synthesized by the Torgov method by utilization of the appropriate alkyl-substituted starting materials. For example, 12-homo-13 (Scheme 4-2) produced a 12α-methyl steroid [61], and 5,5-dimethyl-1,3-cyclohexanedione gave rise to 16,16-dimethyl-D-homo steroids [62].

The Torgov synthesis has been carried out with an open potential B ring, giving rise to the tricyclic intermediate 43 [63]. Resolution of this racemic acid, hydrogenation of the 14-double bond, potassium–ammonia reduction of the 8-double bond, ring closure (polyphosphoric acid), and reduction of the 6-keto group gave optically active estrone methyl ether (17). The condensation of 6-(m-methoxyphenyl)-1-hexen-3-one with 2-methylcyclopentane-1,3-dione gave a 78% yield of 2-[6-(m-methoxyphenyl)-3-oxo-1-hexyl]-2-methylcyclopentane-1,3-dione, an intermediate potentially cyclizable to estrone [64]. While topologically this would be an A+D →AD→ABCD synthesis, the condensation just described is a Torgov-like reaction.

Another simple variation was the employment of cyclopentanediones substituted at position 2 with phenyl [65] or alkyl groups larger than methyl: ethyl [8,30,32,66–69] n-propyl [30,68], isopropyl [68,69] n-butyl [68], isobutyl [68], and n-hexadecyl [68]. With the 13-phenyl derivative, Birch reduction of 44 (Scheme 4-7) brought about reduction of both aromatic rings, but Birch reduction of the 17-tetrahydropyranyl ether reduced only the A ring to give the 19-nortestosterone analog 45. The same device was used to protect the allyl group of 46 [70]. An alternative route to 13-allyl and -benzyl steroids employs alkylation of the thallium salt [71] of the ABD intermediate [72].

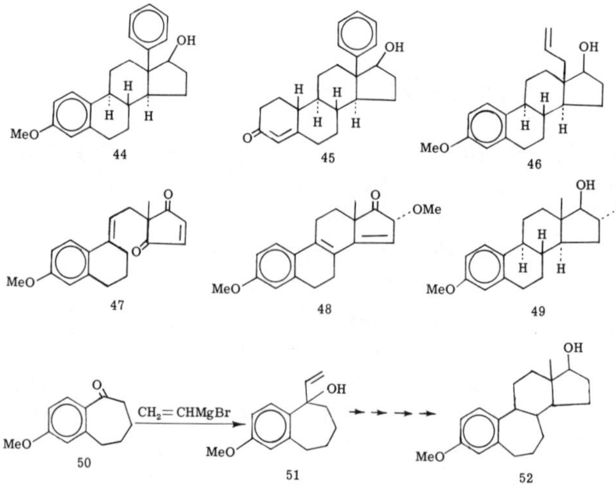

Scheme 4-7

4.2 The Torgov Synthesis

The Torgov synthesis, which usually produces 3-methoxy steroids, has been used also for various analogs including 4-methoxy [15], 2,3-dimethoxy [73], 2,4-dimethoxy [74] and 2,3,4-trimethoxy [75]. In the latter case, the 3-methoxyl was lost during reduction of the 8-double bond unless the proportion of sodium was held down to 2.5 gram atom equivalents. Condensation of the vinyl alcohol 13 with 4-hydroxy-2-methylcyclopentane-1,3-dione gave the pentaene 47; on cyclization in methanol it added methanol to give the triol dimethyl ether 48, which was reduced in the usual two stages to estriol 3,16-dimethyl ether (49) [76].

Normal Torgov syntheses have been carried out also with 5-chloro- and with 5-bromo-6-methoxy-1-vinyl-1-tetralol to give, after closure of ring C, the methyl ethers of 4-chloro- and 4-bromo-3-hydroxy-1,3,5(10),8,14-gonapentaen-17-one, respectively [77].

B-Homosteroids were obtained by utilizing the vinyl alcohol 51 prepared from vinylmagnesium bromide and 3-methoxybenzosuberone (50). Condensation with methylcyclopentanedione, cyclization, catalytic hydrogenation of the 17β-hydroxy derivative and lithium–ammonia reduction gave B-homoestradiol methyl ether (52) [78].*

The Torgov synthesis has been extended also to B-nor steroids (Scheme 4-8), beginning with the requisite indanone [79]. The allylic alcohol 53 was condensed with 2-methylcyclopentane-1,3-dione (with Triton B)

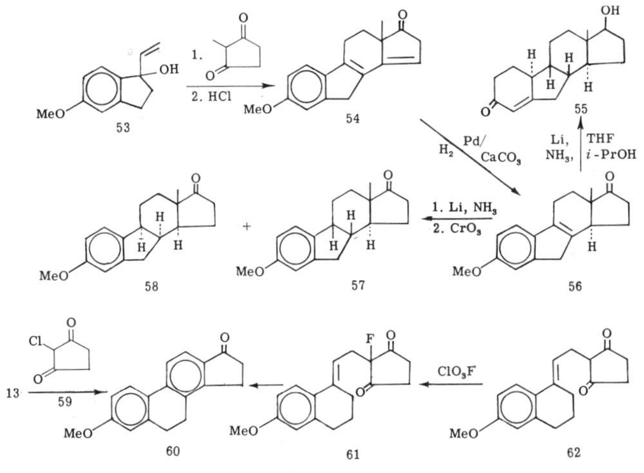

Scheme 4-8

* A-homosteroids have been synthesized by ring expansion reactions with dihalocarbenes [78a].

to give the ABD intermediate in 70% yield (based on the indanone). Cyclization to **54** in 92% yield was accomplished with methanolic HCl, and hydrogenation with palladium-on-calcium carbonate gave a 1:3 mixture of CD-*cis* and CD-*trans* isomers, from which only the latter, **56**, was isolated in pure form. Of the two general methods for reducing the 8,9-double bond in this type of intermediate to an 8β,9α-dihydro product, neither was successful. No isomeration to a $\Delta^{6(8)}$- or a $\Delta^{9(11)}$-isomer occurred under a variety of acidic conditions. Reduction of **56** with lithium and ammonia in tetrahydrofuran, followed by oxidation with Jones reagent, gave a 20:1 mixture of 3-methoxy-B-nor-9β-estra-1,3,5(10)-trien-17-one (**57**) and 3-methoxy-B-nor-8α-estra-1,3,5(10)-trien-17-one (**58**). Reductions with sodium and ammonia or calcium and ammonia occurred similarly. When isopropanol was added to the reaction mixture, a Birch reduction also took place resulting in 17β-hydroxy-B-nor-9β,10α-estra-4-en-3-one (**55**). Catalytic hydrogenation of **56** (palladium-on-carbon) gave a 1:8 mixture of **57** and **58**. Similar results were obtained by a German group except that the minor product of the lithium–ammonia reduction of the 8-double bond was shown to be the 8β,9α-dihydro derivative [80]. Reduction conditions were similar except for the presence of aniline in the latter case.

An aromatic C-ring steroid was synthesized by condensing the vinyl alcohol **13** with 2-chloro-1,3-cyclopentanedione (**59**) [81]. The same compound was prepared also by cyclization of the fluoro intermediate **61**, which arose by the action of ClO$_3$F on the ABD intermediate **62**.

4.2.4 Heterocyclics

The Torgov synthesis has proved fruitful for the preparation of heterocyclic steroids. Thus, beginning with the vinyl alcohol **63** (Scheme 4-9), the sequence of condensation (to **64**) and cyclization (to **65**) took place characteristically. The free phenolic hydroxy analog of **64** was prepared also, but would not cyclize [82].

Differences between carbocyclic steroids and N-containing analogs are even more pronounced in the 6-aza series. *N*-Arylsulfonyl pentaene intermediates with tosyl **67** [83,84] or benzenesulfonyl [85,86] substituents were synthesized in the usual fashion from the requisite allylic alcohols (e.g., **66**). Catalytic hydrogenation of the benzenesulfonyl analog gave the CD-*trans* tetraene as expected [86], but with **67** a mixture predominating in the 14β-isomer **69** resulted regardless of whether palladium-on-calcium carbonate, -strontium carbonate, or -charcoal was used as the catalyst [87]. When the hydrogenation was carried out on

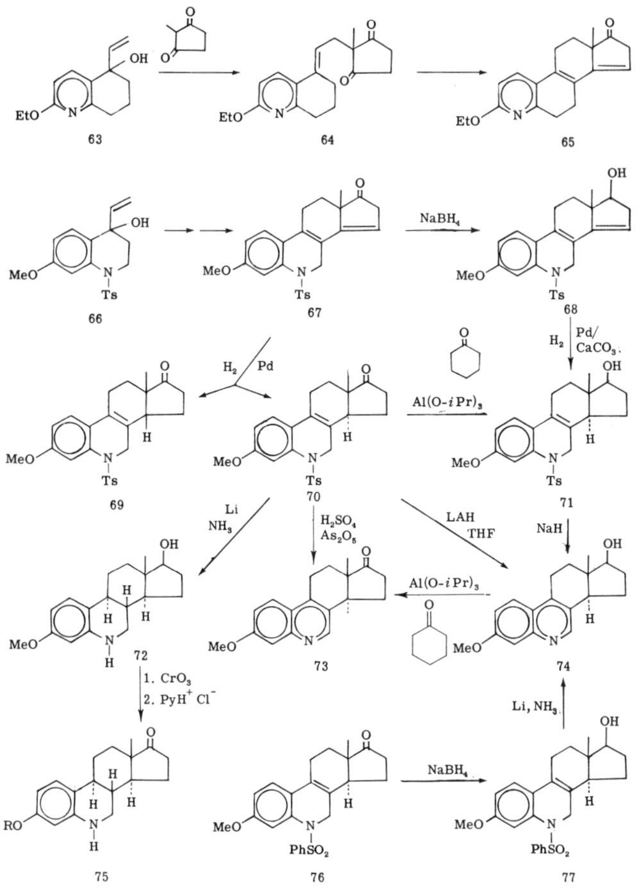

4.2 The Torgov Synthesis

Scheme 4-9

the 17β-hydroxy analog **68**, prepared from **67** by sodium borohydride reduction, the predominant product was the α-isomer. Oppenauer oxidation of **71** gave **70**, which was converted into the methyl ether (**73**) of 6-azaequilenin in low yield with sulfuric acid and arsenic pentoxide. A better route was via **74**, obtained by detosylation of **71** with sodium hydride or from **70** and lithium aluminum hydride in THF. In this series, lithium–ammonia reduction of the 8,9-double bond occurred normally, though in only 30% yield, to give the methyl ether (**72**) of 6-azaestradiol. In the benzenesulfonyl series, however, lithium–ammonia reduction of **77** (from **76** and sodium borohydride) gave **74** [86]. An alternative route in carbocyclic steroids is comprised of isomerization of the 8,9-double

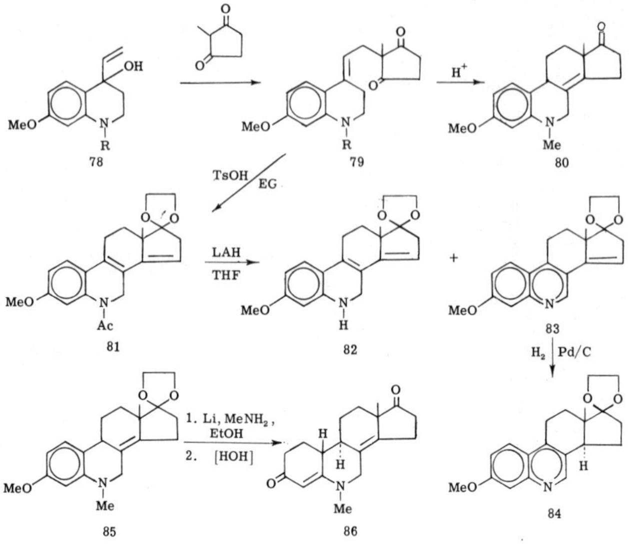

Scheme 4-10

bond to the 9(11)-position, followed by catalytic hydrogenation, but **77** would not isomerize under a wide variety of acidic conditions. 6-Azaestrone (**75**, R=H) was synthesized by oxidation of **72** and demethylation with pyridinium chloride [88]. The 14β-isomers of **72-75** were also prepared.

The N-methyl ABD intermediate, **79** (R=CH$_3$, Scheme 4-10), prepared in quantitative yield from the allylic alcohol **78**, cyclized atypically to give the $\Delta^{8(14)}$-steroid **80** [89]. The N-acetyl intermediate **79** (R=Ac), prepared from **78** (R=H) and 2-methyl-1,3-cyclopentanedione, followed by acetylation, cyclized normally under conditions of ketalization to give the pentaene **81** in 50% yield. Reduction of **81** with lithium aluminum hydride in THF gave a mixture of the deacetyl steroids **82** and **83**. Hydrogenation of **83** with palladium-on-charcoal gave a mixture of the 3-methyl ether, 17-ketal (**84**) of 6-azaequilenin and the corresponding 14β-isomer, separable by chromatography.

Birch reduction of **85** (17-ethyleneketal of **80**), which failed in ammonia, was successful in methylamine to give the dihydro product, which on hydrolysis gave the diketone **86** [90].

13-Azasteroids are available by replacing the diketone in the Torgov synthesis with succinimide (Scheme 4-11). Succinimide reacted slowly with the allylic alcohol **13** in methanol, and more rapidly as a melt

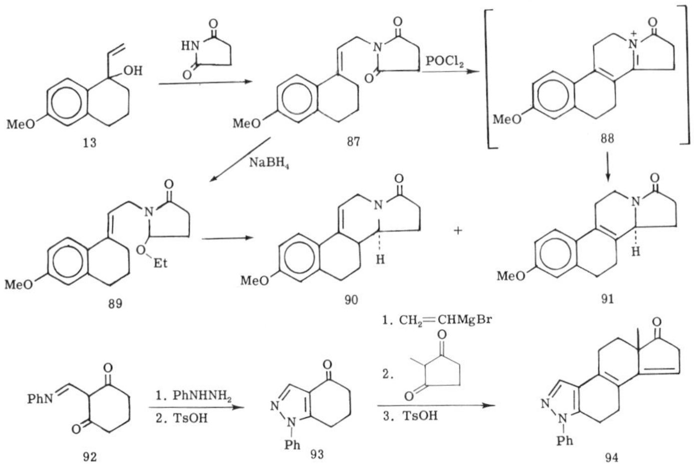

4.2 The Torgov Synthesis

Scheme 4-11

containing the potassium salt of succinimide to give the ABD intermediate **87** in 20–30% yield [91]. This intermediate could be cyclized in phosphorus oxychloride to give a tetracyclic steroid, presumably **88**, which was not purified, but was hydrogenated over rhodium on charcoal to the methyl ether (**91**) of 13-aza-8-dehydroestrone. Alternatively, **87** could be reduced with sodium borohydride in ethanol containing HCl to give a good yield of the ether **89** [92]. Cyclization of the latter gave an almost quantitative yield of a one-to-one mixture of **90** and **91**.

The Torgov synthesis worked equally well for steroids with a pyrazole A ring [93]. The starting allyl alcohol was prepared from the Schiff's base **92** which, with phenylhydrazine, gave the corresponding hydrazone. The latter was cyclized with p-toluenesulfonic acid to the pyrazole **93**; the customary Grignard reaction then gave an allylic alcohol whose condensation with 2-methyl-1,3-cyclopentanedione gave the expected ABD intermediate, which was cyclized to the diaza-A-nor steroid **94** with p-toluenesulfonic acid. The 3-p-tolyl analog also was synthesized. Sodium borohydride reduction of **94** gave the 17β-hydroxy derivative, and the 17-ethyleneketal of **94** was prepared from either **94** or its 8,14-seco precursor, ketalization and cyclization taking place simultaneously in the latter case.

A-nor-3-oxasteroids were synthesized from the allyl alcohol **95** (Scheme 4-12) [94]. It would not condense with 2-methyl-1,3-cyclopentanedione in the presence of either Triton B or diethylamine, but the corresponding isothiouronium salt **96** readily reacted to give the ABD

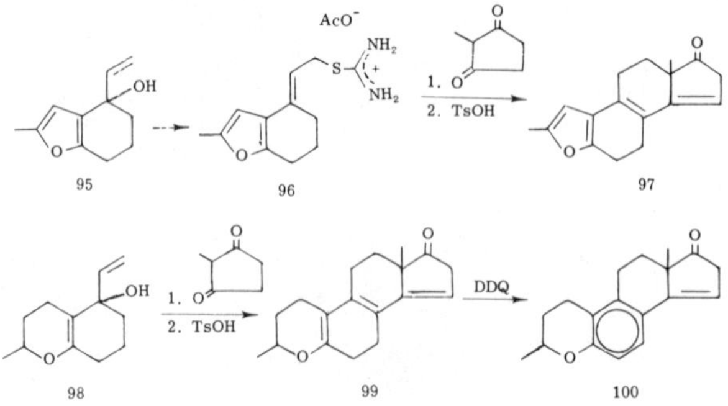

Scheme 4-12

intermediate in 44% yield. The latter was cyclized with p-toluenesulfonic acid, and the ketone **97** was quantitatively reduced with sodium borohydride to 2-methyl-3-oxa-A-nor-1,5(10),8,14-estratetraen-17β-ol.

The allyl alcohol **98** condensed with 2-methyl-1,3-cyclopentanedione to produce a 66% yield of the expected ABD intermediate, which readily cyclized to **99** under acid catalysis [95]. Dichlorodicyanobenzoquinone dehydrogenated **99** to the aromatic B-ring steroid **100**, which was converted to the corresponding 14α-dihydro derivative by sodium borohydride reduction of the 17-oxo group, catalytic hydrogenation, and oxidation with chromic oxide [95]. The 2-dehydro analog of **99** already has been discussed in Section 4.2.3. (see Scheme 4-6).

6-Oxasteroids were readily synthesized beginning with the allyl alcohol **101**, which was condensed with 2-methyl-1,3-cyclopentanedione in the presence of potassium hydrogen carbonate to the ABD intermediate (Scheme 4-13) [86]. Cyclization to **102**, hydrogenation, and isomerization proceeded normally to give the $\Delta^{9(11)}$-steroid **105**. Hydrogenation of the double bond, hydride reduction of the oxo group, and Birch reduction of the A ring gave the diene **108**. It was oxidized (Oppenauer) to the ketone, condensed with lithium acetylide, then hydrolyzed to the 6-oxa-19-nortestosterone derivative **107**. The analog **109** without the 17α-ethynyl group was prepared by hydrolysis of **108**. 8α-Isomers were obtained by exhaustive hydrogenation of **102** and sodium borohydride reduction of the ketone **103** [86]. C-18 homologs have been prepared also, by using 2-ethyl-1,3-cyclopentanedione in the condensation with **101** [96]. 7-Methyl homologs of **101**, **102**, **103**, **105**, and **106** were synthesized similarly [97], but it was necessary to reduce the 17-oxo group to hydroxyl to create a CD-*trans* ring fusion by hydrogenation.

Condensation of the isothiouronium salt **19** with 4-hydroxy-2-methyl-3-oxopentanoic acid lactone (**110**) in aqueous ethanol gave an 86% yield

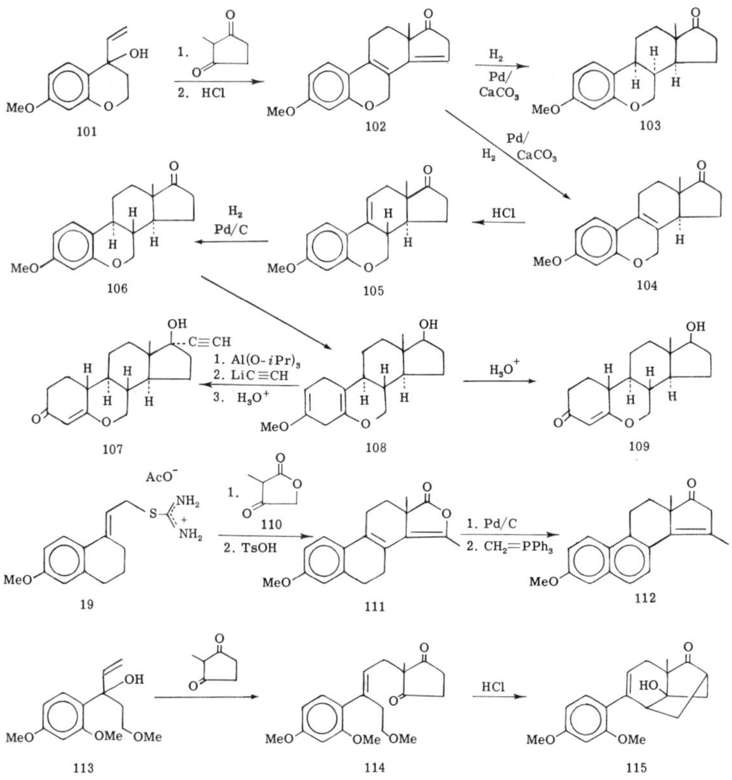

4.2 The Torgov Synthesis

Scheme 4-13

of the expected ABD intermediate. It was cyclized in 93% yield to the 16-oxasteroid **111**, which was dehydrogenated to the corresponding aromatic ring B analog (80% yield) [98]. The latter was converted to the carbocyclic steroid **112** in 60% yield by the action of methylenetriphenylphosphorane [99]. The D ring of **111** is an enol lactone; it was hydrolyzed with sodium hydroxide, and the resulting keto acid converted to *cis*-bisdehydrodoisynolic acid methyl ether, thereby confirming the carbon skeleton [98].

An attempt to synthesize a B-seco intermediate suitable for the preparation of heterocyclic steroids was initiated with the reaction of **113** with 2-methyl-1,3-cyclopentanedione. The condensation product **114** underwent bis-cyclization in refluxing ethanolic HCl, however, producing **115** in 20% yield [100].

Some aza-oxa-steroids have been synthsized beginning with condensation of the isothiouronium salt **19** and 2-acetylamino-1,3-cyclopentanedione (**116**, Scheme 4-14) [101]. Cyclization of the ABD intermediate **117** was effected by *p*-toluenesulfonic acid. Catalytic hydrogenation produced a CD-*trans* geometry as usual, and sodium borohydride reduction of the oxo group gave **119**. Exhaustive hydrogenation gave the 8α-steroid **120**, and potassium–liquid ammonia reduction (of **119**) with aniline as the proton source gave the normal BC-*trans* isomer **121**. Treatment of **121** with hydrochloride acid under conditions which might be expected to cause an N→O migration of the acetyl group gave instead the 2-oxazoline **123**. The corresponding dihydro derivative **122** was prepared by hydrolysis of the amide **121**, followed by condensation of the amino alcohol with acetaldehyde in refluxing benzene. Birch reduction of **121**, followed by the usual hydrolysis step, gave 13β-amino-17β-hydroxy-4-gonen-3-one (**124**).

The ABD intermediate **117** was found to be unstable to base in that it underwent cleavage of the D ring to **125** [101]. The structure of **125** was indicated by cyclization of its methyl ester (and accompanying deamination) to **126**, alkaline hydrolysis of the latter, and closure of ring

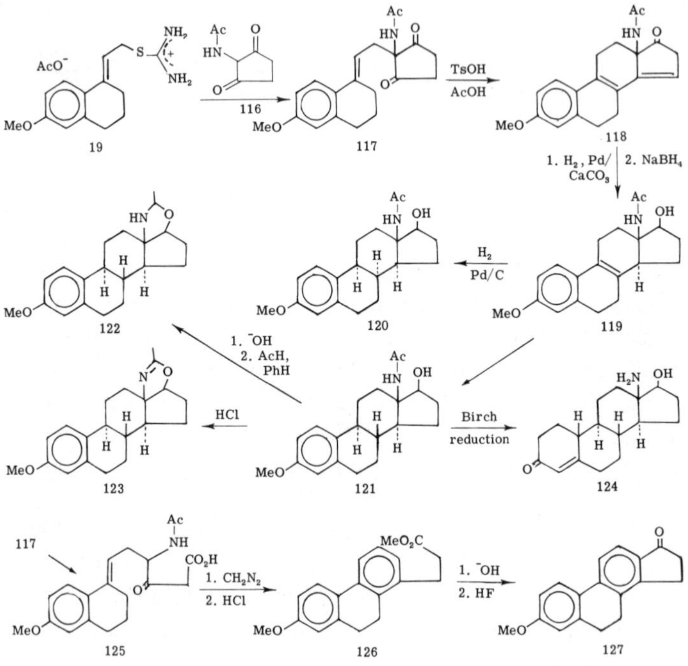

Scheme 4-14

4.2 The Torgov Synthesis

D to give the methyl ether (**127**) of 3-hydroxyl-1,3,5(10),8,11,13-gonahexaen-17-one, a known compound [81].

Thiasteroids, too, are available through the Torgov synthesis. The allylic alcohol **128** (Scheme 4-15) condensed with 2-methyl-1,3-cyclopentanedione to give a 55% yield of the expected ABD intermediate, which was cyclized to an unstable steroid **129** [102]. As in the case of the 6-azasteroid **67**, catalytic hydrogenation of the 14-double bond gave a mixture of CD-*trans* **132** and CD-*cis* **131** products, predominating in the latter. The problem was solved in the same way, namely, by hydride reduction to the 17β-hydroxy derivative, then catalytic hydrogenation in 90% yield to **130**, which gave **132** on Oppenauer oxidation. A more difficult problem was faced in reducing the 8-double bond in that attempted metal–ammonia reduction cleaved the B ring. Consequently, the tetraene **132** was hydrogenated catalytically producing a 1:1 mixture of the two BC-*cis* products **133** and **134**, which were separated by fractional crystallization [103]. Configurational assignments were based on NMR spectra. Dehydrogenation of **134** with 2,3-dichloro-5,6-dicyano-1,4-benzoquinone gave the $\Delta^{9(11)}$-olefin **135**, which was hydrogenated in 64% yield to the methyl ether (**136**) of 6-thiaestrone. The sulfones corresponding to **128**–

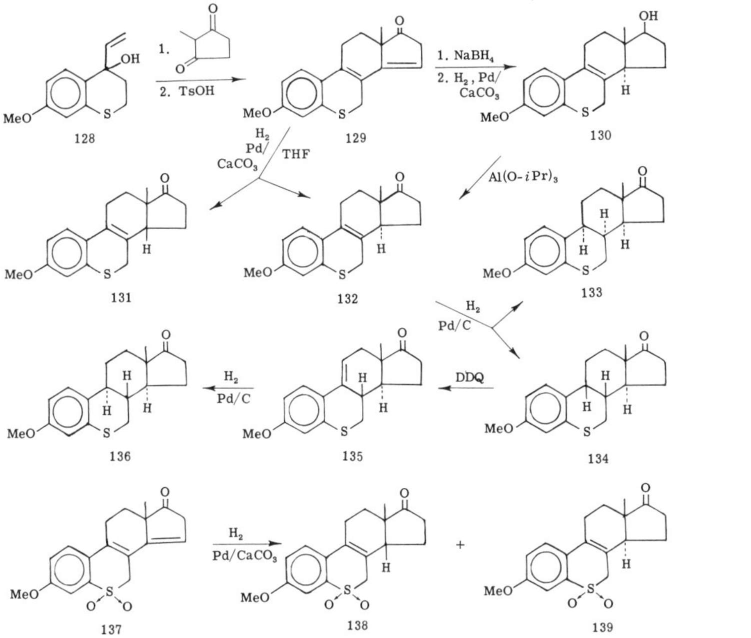

Scheme 4-15

131 were prepared by oxidation of the respective precursors with hydrogen peroxide. Neither the condensation nor the cyclization reactions took place with the sulfones (corresponding to 128 and the ABD intermediate, respectively), but catalytic hydrogenation of 137 gave a mixture of 14α-H and 14β-H products 139 and 138. 17α-Ethynyl and 17α-vinyl derivatives of 136 were synthesized by the usual methods [103]. See Konst et al. [72] for other 6-thiasteroids.

The thiazole A-ring steroids (Scheme 4-16, 144, R=C_6H_5, p-$CH_3OC_6H_4$, C_2H_5O) were synthesized beginning with condensation of the respective thioamides 140 with 2-bromocyclohexane-1,3-dione (141) [104]. The thiazole 142 was converted by the usual steps to the vinyl alcohol and thence to the isothiouronium salt 143. Condensation with 2-methylcyclopentane-1,3-dione in high yield and ring closure gave 144.

Condensation of the vinyl alcohol 128 (Scheme 4-15) with 2-methylcyclohexanedione led ultimately to the D-homo analog of 129 [105]. Similarly, condensation of 128 with succinimide or with 2-acetylaminocyclopentane-1,3-dione led to 145 or to 146, respectively [105].

Carboxyethyl dimethyl m-methoxyphenyl silane (147) was cyclized with phosphorus pentachloride, the customary condensation with vinyl Grignard reagent was carried out, and the resulting vinyl alcohol was converted to the isothiouronium salt 148. This was condensed and cyclized as usual to give the 6-silasteroid 149 [106].

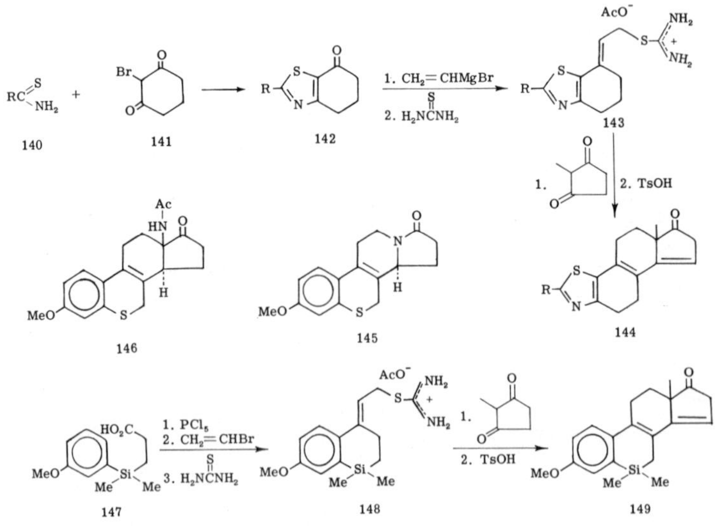

Scheme 4-16

4.3 Alkylation

In summary, the Torgov synthesis has proved to be a relatively simple and direct route to a wide variety of aromatic A-ring steroids. Reduced A-ring steroids are available from the former by known reduction methods. Δ^8-Intermediates in the Torgov synthesis have been converted to 11-oxygenated steroids [107], further extending the importance of this method.

4.3 Alkylation

In the Torgov synthesis, a cyclic diketone is alkylated with an allylic alcohol, allylic rearrangement taking place simultaneously. Alkylation with bromides and chlorides comprises another successful route to ABD intermediates. For example, alkylation of 2-carbethoxycyclopentanone or -cyclohexanone (Scheme 4-17, **151**, $n=1$ or 2, respectively) with bromoalkylnaphthalenes of the type **150** produced intermediates **152** [108–110]. Cyclization of **152** (R=CH$_3$O, $n=1$) with cold concentrated sulfuric acid, hydride reduction of the ester group to hydroxyl, tosylation of the latter, and reduction to a methyl group, gave 1,3,5(10),6,8,14-estrahexaen-3-ol methyl ether (**153**) [111]. Cyclization of **152** in hot sulfuric acid eliminated the carbethoxy group and aromatized the C ring, giving rise to **154** (R=H or CH$_3$O, $n=1$) [110,112,113]. D-Homo, 6-methoxy, and 6-, 15-, 16-, and 17-methyl analogs also have been prepared [88,110, 112,114,115].

The absence of a 17-oxo group in these products decreased the im-

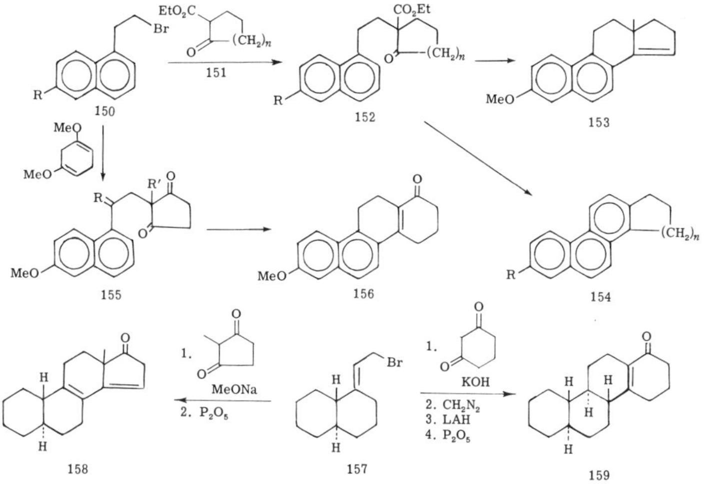

Scheme 4-17

portance of the method for the preparation of naturally occurring steroids. This handicap was eliminated by the use of diketones or their equivalents. Thus, alkylation of 1,5-dimethoxycyclohexa-1,4-diene (from Birch reduction of resorcinol dimethyl ether) with **150** (R=CH$_3$O), followed by acid hydrolysis, gave an 86% yield of the intermediate **155** (R=H$_2$, R'=H), which cyclized with polyphosphoric acid to the methyl ether (**156**) of 3-hydroxy-D-homo-1,3,5(10),6,8,13-estrahexaen-17a-one [116]. Alkylation of 2-methyl-1,3-cyclohexanedione with the bromoacetyl analog of **150** in the presence of potassium hydroxide gave **155** (R=O, R'=CH$_3$), which has not been cyclized [117]. Low yields of tetracyclic compounds have been obtained from the bromide **157** containing saturated A and B rings. Alkylation of 2-methyl-1,3-cyclopentanedione in the presence of sodium methoxide gave an ABD intermediate, which was cyclized to 5α-estra-8,14-dien-17-one (**158**) [117,118]. Alkylation of 1,3-cyclohexanedione, conversion to the enol ether with diazomethane, reduction with lithium aluminum hydride, and cyclization with phosphorus pentoxide gave D-homo-13-gonen-17a-one (**159**) [119,120]. Alkylation of 2-methyl-1,3-cyclopentanedione with an aromatic A-ring analog of **157** gave **14** (Scheme 4-2), which cyclized to **15** in 6% overall yield; O-alkylation is a competing reaction in alkylation of this diketone [121].

An 8-azasteroid was prepared via an alkylation reminiscent of the synthesis of **156** (above) [122]. Alkylation of 1,5-dimethoxycyclohexa-1,4-diene with the chloride **160** (Scheme 4-18) in liquid ammonia gave the ABD intermediate **161**. It was hydrolyzed and cyclized with 2 N hydrochloric acid, after which the 13-double bond was reduced with lithium aluminum hydride. This produced two racemic pairs of 8-aza-3-methoxy-D-homo-1,3,5(10)-gonatrien-17a-one (**162**), believed to be epimeric at C-13.

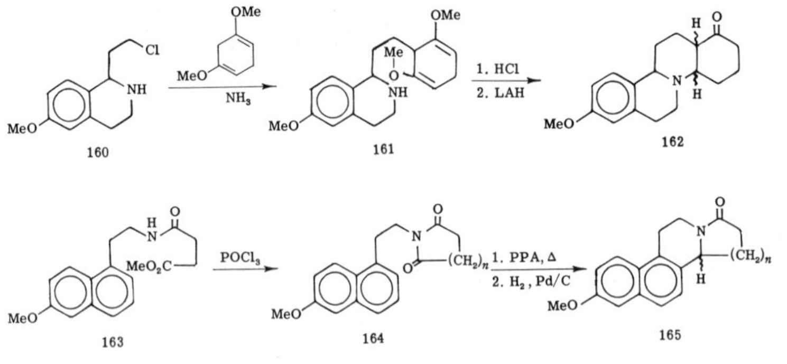

Scheme 4-18

4.4 Michael Addition

Alkylation of succinimide or glutarimide (as the potassium salts) with the bromide **150** (Scheme 4-17, R=CH$_3$O) produced **164** (n=1 or 2, respectively); the reaction took place only in dimethylsulfoxide. Cyclization was effected with polyphosphoric acid and the 14-double bond was reduced catalytically to the 13-azasteroid **165** [123]. The intermediate **164** (n=1) was alternatively synthesized from **163**, prepared from 1-(2-aminoethyl)-6-methoxynaphthalene and the acid chloride methyl ester of succinic acid [124]. Analogs of **164** (n=1), some of them with a saturated ring B, were reduced to the corresponding 14-hydroxy derivatives with sodium borohydride, then cyclized in high yield with p-toluenesulfonic acid in benezene [125].

4.4 Michael Addition

Another preparation of ABD intermediates that is formally similar to the Torgov synthesis is the Michael addition between the unsaturated ketone **166** and a cyclic diketone (Scheme 4-19) [58,126–129]. The trienone **166** can be prepared by hydrolysis and dehydration of **39** (Scheme 4-6, R=MeO, X=CH$_2$). With 2-methyl-1,3-cyclohexanedione the Michael addition using Triton B gave intermediate **167** (R=CH$_3$, n=2) which was cyclized with p-toluenesulfonic acid in benzene, while with diethylamine as catalyst, the reaction went directly to **168** (R=CH$_3$). Dehydration with p-toluenesulfonic acid in acetic acid gave D-homo-4,8(14),9-estratriene-3,17-dione (**169**, R=CH$_3$, n=2). Interestingly, when **166** was treated with 2-methyl-1,3-cyclopentanedione and diethylamine, addition, cyclization, and dehydration all took place and **169** (R=CH$_3$, n=1) was obtained directly [127,130]. The latter has been chloroethynylated in liquid ammonia to give **170** [130]. Conversely, it has been converted to

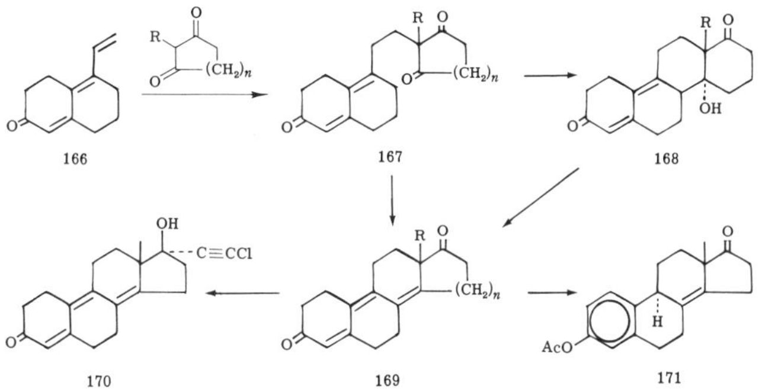

Scheme 4-19

the acetate (**171**) of 3-hydroxy-1,3,5(10),8(14)-estratetraen-17-one with isopropenyl acetate [131]. Analogs of **169** with R equal to ethyl, n-propyl, isopropyl, n-butyl, n-pentyl, allyl, trifluoromethyl, methoxymethyl, ethoxymethyl, n-propoxymethyl, n-butoxymethyl, n-pentoxymethyl, and difluoromethyl also have been prepared [132].

The Michael addition of 2-isopropylcyclopentanone to the acetyl compound **172** (Scheme 4-20) at room temperature gave the ABD intermediate **173**, while at reflux the cyclized product **174** was obtained directly [133]. An analog lacking the aromatic ring (and the methoxy) of **172** did not react under these conditions. An attempt to alkylate **174** at C-13 by the addition of methyl Grignard across the Δ^{12}-11-keto system failed; instead an 11-methyl-9(11),12-diene was formed by 1,2-addition and dehydration. Addition of 1,2-cyclohexanedione to **172** gave the D-homo steroid **175**, which on Birch reduction and hydrolysis produced 11α,17aβ-dihydroxy-D-homo-4-gonen-3-one (**176**) [134].

The Robinson group also carried out Michael additions in the other direction to produce ABD intermediates [135,136]. Addition of 6-methoxy-1-tetralone (**177**) to either 1-acetylcyclopentene or 1-acetylcyclohexene in ether and in the presence of sodium amide gave rise to the

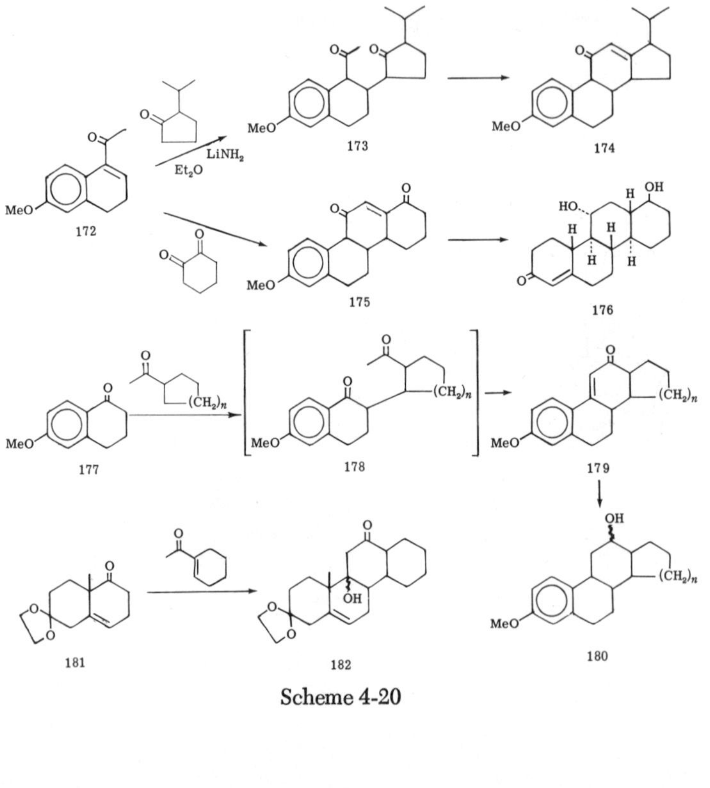

Scheme 4-20

4.4 Michael Addition

corresponding tricyclic intermediate (178). Normally 178 was not isolated, but was cyclized in the reaction medium to 179 (55% yield with $n=1$). The synthesis was admirable in its brevity, but suffered from lack of stereochemical control. Three asymmetric centers were created (at C-8, C-13, and C-14) in the formation of 178, which were still present in 179. When 179 was hydrogenated with palladium-on-strontium carbonate, reducing both double bond and ketone, the alcohol 180 ($n=2$) was obtained in three different forms, crystallized from petroleum ether [135,137]. Also, three different forms of 179 ($n=2$) have been isolated [137], two of which were recognized as isomerides. The synthesis was extended to 3-deoxy analogs [138,139].

When the Michael addition was carried out in liquid ammonia, nonaromatic ketones such as 181 reacted with 1-acetylcyclohexene to give, after cyclization, the D-homosteroid 182 [140]. The synthesis has been extended to oxa and aza analogs [141].

8-Azaestrone has been synthesized by a sequence of reactions, the first of which is the Michael addition of 2-methyl-1,3-cyclopentanedione to the isoquinoline derivative 183 (Scheme 4-21), and subsequent ring closure in the reaction medium. Hydrogenation of 184 (R=Me, $n=1$) occurred stereospecifically giving 185 (R=Me, $n=1$), which was demethylated to 8-azaestrone [142–144]. (A compound synthesized by a different route and originally assigned this same structure [145] was later assigned different stereochemistry on the basis of X-ray evidence [146] (see Chapter 6). 18-Homo (R=Et, n-Pr, $n=1$), D-homo (R=Me, $n=2$), and 2,3-dimethoxy analogs [147], as well as reduced A-ring and 17-acetyl analogs [148], were also prepared. The methyl ether of 3-hydroxy-6,8-diaza-D-homo-1,3,5(10),6,13-gonapentaen-17a-one has been prepared by a similar route [149].

An approach to 14-azasteroids began with the additon of 2-tetralone (186) to 2-vinylpyridine, with formation of the ABD intermediate 187 (R=H). Protection of the keto group as the ethylene ketal, catalytic hydrogenation of the pyridine ring, hydrolysis of the ketal, and ring closure gave the prototype 188 (R=H) [150]. *Trans* reduction of the 8-double bond was also accomplished, as was synthesis of a 187 intermediate with R equal to MeO.

Prototypes of 11,14-diazasteroids were synthesized beginning with the Michael addition of diethyl (2-naphthylamino)malonate to ethyl acrylate; under the conditions of the reaction the product 189 ring closed and decarboxylated to 5-carbethoxy-N-(2-naphthyl)-2-pyrrolidone [151]. The latter (as the free acid) was nitrated in the 1-position of the naphthalene ring giving (after reesterification) 190. On reduction of the nitro group, ring C closed spontaneously giving a 97% yield of 11,14-diaza-

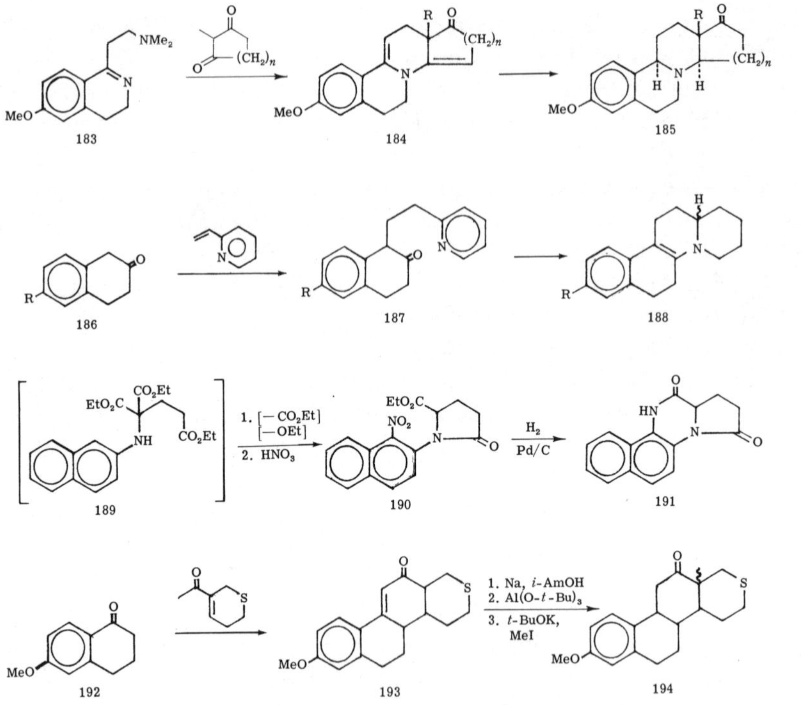

Scheme 4-21

1,3,5(10),6,8-gonapentaene-12,15-dione (**191**). Analogs with methyl or phenyl substituents on ring D also were synthesized.

A 17-thia-D-homo steroid with uncertain geometry was synthesized as part of a potential route to estrone [152]. Addition of 6-methoxy-1-tetralone (**192**) to 2-acetyl-4-thiacyclohexene and subsequent cyclization gave the 12-keto steroid **193**. Sodium and amyl alcohol reduced both the 9(11)-double bond and the 12-keto group, the resulting alcohol was oxidized with aluminum *t*-butoxide, and the product of alkylation, presumably 3-hydroxy-17-thia-D-homo-1,3,5(10)-estratrien-12-one methyl ether (**194**), was obtained by the action of methyl iodide and potassium *t*-butoxide on **193**.

4.5 Enamines

A series of 8-azasteroids has been synthesized via enamine ABD intermediates. For example, the condensation of the isoquinoline derivative **195** (Scheme 4-22, R=CO$_2$Et) with cyclopentanone furnished the

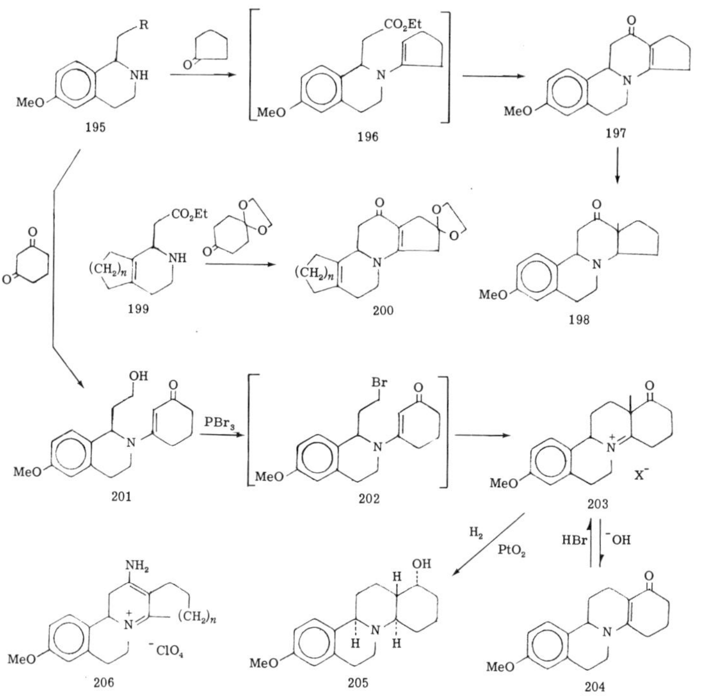

4.5 Enamines

Scheme 4-22

enamine **196**, which was cyclized to the 12-keto steroid **197**.* Methylation and catalytic hydrogenation produced the methyl ether (**198**) of 8-aza-3-hydroxy-1,3,5(10)-estratrien-12-one [147,153]. The 8-azasteroid **200** ($n=1,2$) was similarly synthesized by heating the amine ester **199** and the ketone in refluxing toluene [154].

17-Oxosteroids were synthesized beginning with the condensation of **195** (R=CH$_2$OH) with 1,3-cyclohexanedione [155]. The hydroxy enamine **201** reacted with phosphorus tribromide to give an oily bromide **202** which, on standing, cyclized and crystallized as **203** (X=Br). Alternatively, **202** could be converted with silver perchlorate to **203** (X=ClO$_4$), which was unstable and was treated with base to give the methyl ether (**204**) of 8-aza-3-hydroxy-D-homo-1,3,5(10),13-gonatetraen-17a-one. Catalytic hydrogenation of either salt **203** afforded the 17α-hydroxy derivative **205** [156].

---

* A 14-aza-11-oxosteroid has been synthesized by a similar sequence beginning with 6-methoxy-2-tetralone and ethyl 2-(3,3-ethylenedioxy-2-pyrrolidyl)acetate [152a].

The two cyclizations just discussed take place through the interaction of an enamine group with either an ester or bromide group. Enamine groups react also with nitriles, and this reaction, catalyzed by magnesium perchlorate, was used to close ring C, producing the 8-azasteroids **206** ($n=1,2,3$) [157].

The isoquinoline derivative **207** (Scheme 4-23, R=H or CH_3O) reacted with β-diketones in refluxing ethanol to give the 8-aza-12-ketosteroids **208** (R=CH_3O, R'=H_2 [158]; R=CH_3O, R'=O [159]) and **209** (R=CH_3O, R'=H_2, Z=NH [158]; R=CH_3O, R'=O, Z=$C(CH_3)_2$ [158]; R=H, R'=O, Z=$C(CH_3)_2$ [159]).

The enamine **210** was alkylated with ethyl bromoacetate, after which acid hydrolysis of the enamine group afforded **211**. The latter condensed with perhydropyridazine in refluxing xylene to give a 26% yield of the 13,14-diazasteroid **212** [160].

The enamine 1-pyrrolidinocyclopentene added to methyl 3,4-dihydro-1-naphthoate to give **213** in high yield. Hydrolysis of the enamine and ester groups gave **214**, which was converted to both aza-, **215** (Z=NH or NCH_2Ph), and oxasteroids, **215** (Z=O) [161].

4.6 Acylation

β-Keto esters (α-acetylsuccinic and α-acetylglutaric) were acylated with the acid chloride **216** (Scheme 4-24); the intermediates **217** ($n=1,2$) were hydrolyzed to the corresponding acids (with loss of Ac and CO_2Et groups), which were cyclized to the ABD intermediates **218**. Closure of ring C with phosphorus pentoxide gave the methyl ethers (**219**) of 3-hydroxy-1,3,5(10),13-gonatetraen-17-one (n=1) and 3-hydroxy-D-homo-1,3,5(10),13-gonatetraen-17a-one (n=2) [162-165]. α-Methyl homologs of **216** which would have provided estrone derivatives did not condense with α-acetylsuccinic ester. The angular methyl group was incorporated in the Friedel-Crafts acylation of 1,7-dimethoxynaphthalene with 2-carboxymethyl-2-methyl-3-oxocyclopentanecarboxylic anhydride (**220**). The unsymmetrical anhydride opened in the desired manner in only 4.25% yield to afford **221**, plus a 16% yield of the product of acylation by the other carboxyl. Cyclization of 11-deoxy-**221** (obtained by hydrogenolysis) was accompanied by inversion of configuration at C-17 [166].

Acylation of 1-naphthylamine with 2-carbethoxycyclopentanone at 180° gave **223** (R=H), along with some of the Schiff's base formed from **223** and 1-naphthylamine. Cyclization of **223** with concentrated sulfuric acid and reduction of the 13-double bond with sodium amalgam gave the

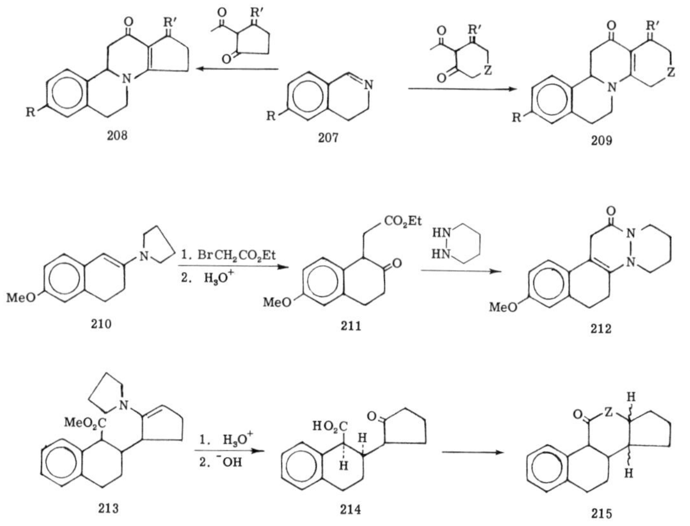

4.7 Aldol-Type Condensations

Scheme 4-23

11-azasteroid **224** (R=H) [167]. Analogs with R equal to CH₃O were prepared similarly. The reaction of **224** (R=H) with phosphorus oxychloride gave a 12-chloro derivative that was reduced with tin and hydrochloric acid to an 11-azasteroid with rings A, C, and D saturated.

Following the same initial approach, Popp and colleagues acylated 5-aminoisoquinoline (**225**) with 2-carboxycyclopentanone; the ABD intermediate was converted to its corresponding methiodide **226**. The A ring of the latter was hydrogenated in the presence of Adam's catalyst, and cyclization to the 3,11-diazasteroid **227** was effected in polyphosphoric acid [168].

3,5-Biscarbethoxy-1,2-cyclopentanedione condensed with 1,5-, 1,6-, and 1,7-naphthalenediol and ring closed to give intermediates **229**, which decarboxylated to the 2-hydroxy, 3-hydroxy, and 4-hydroxy isomeric 11-oxasteroids **230** [169].

4.7 Aldol-Type Condensations

The series of compounds **231–235** (Scheme 4-25) was synthesized where R=H [111] and where R=CH₃O [170]. The ketone **231** condensed with ethyl cyanoacetate, the intermediate product dehydrating to give the unsaturated ester **232**. The elements of HCN were added to the double bond, and the product was cyanoethylated with acrylonitrile. The cyano

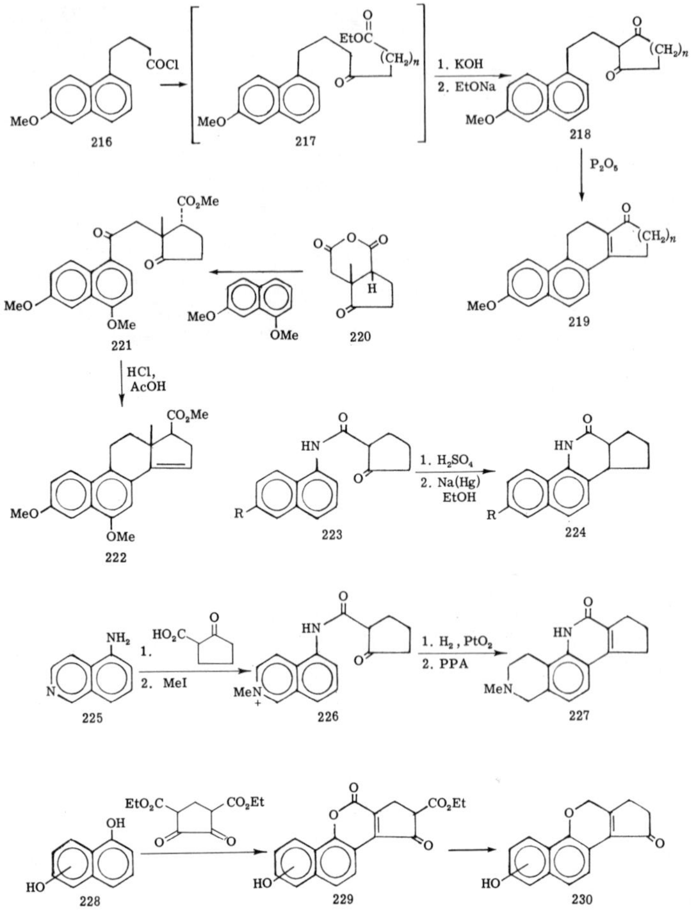

Scheme 4-24

groups and the ester group were hydrolized with hydrobromic and acetic acids, a carboxyl was lost by decarboxylation, and the resulting triacid was esterified with diazomethane. The resulting triester **233** was cyclized (NaOMe) to the ABD intermediate **234**, which was further cyclized (H[+]) to an acid which, on catalytic hydrogenation, gave the etiochola-1,3,5(10),6,8-pentaenoic acid **235** (R=H, CH$_3$O).

A synthesis developed by the Robinson group employs the aldol condensation between 2-acetyl-6-methoxynaphthalene (**236**) and furfural [171,172]. Hydrolysis of the furan ring of **237** gave 4,7-dioxo-7-(6-methoxy-2-naphthyl)heptanoic acid, which cyclized to **238** on treatment

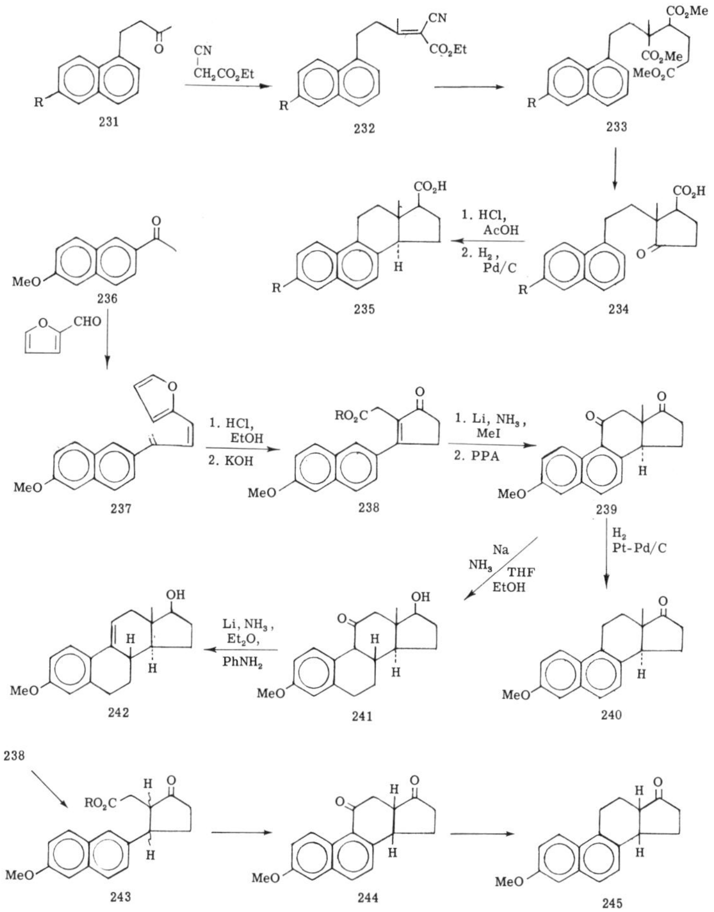

4.7 Aldol-Type Condensations

Scheme 4-25

with base. This intermediate has been converted into many steroid analogs by Robinson and others, but by far most of the reduced products have unnatural configurations at the ring junctures. Not until thirty years after the original work was **238** (R=H) converted into equilenin methyl ether [173]. The D-ring double bond was reduced with lithium and liquid ammonia, and methyl iodide was added to the reaction mixture to provide the angular methyl group. Two asymmetric centers were created in this step, and all four isomers were produced, racemic *cis* and racemic *trans* isomers at carbons 13 and 14. The mixture was not separated, but analysis by gas chromatography indicated the presence of 75%

trans and 25% *cis* product. The mixture was cyclized in 30% yield with polyphosphoric acid and the product was separated by chromatography on alumina. Eighty-five percent of the crude product was recovered as the *trans* isomer **239**, and 10% as the *cis* isomer. Each was selectively hydrogenated under acid conditions with a platinum–palladium-on-charcoal catalyst to remove the 11-oxo group. Both equilenin methyl ether (**240**) and the 14β-equilenin methyl ether were obtained (as racemic products).

The diketone **239** has been converted also into estrone methyl ether [174]. Its sodium–ammonia reduction in tetrahydrofuran (employing ethanol as the proton source) gave the tetraene **241**. Further reduction with lithium and ammonia gave a mixture from which the tetraene **242** was separated by thin layer chromatography. Hydrogenation of the 9(11)-double bond and oxidation of the 17-hydroxyl with Jones reagent led to the methyl ether of estrone. Some of the 9β-epimer was obtained as a by-product.

3-Hydroxy-14β-gona-1,3,5(10),6,8-pentaen-17-one methyl ether (**245**) was synthesized by catalytic hydrogenation of **238** (R=CH$_3$) (Pd/SrCo$_3$), hydrolysis to the free acid **243** (R=H), closure of ring C to give the 11,17-dione **244**, and hydrogenation (platinum-on-carbon) to remove the 11-oxygen [175]. The stereochemistry was unclear at the time, but is now believed [176] to constitute a *cis* hydrogenation of the 13-double bond followed by isomerization to a more stable *trans* configuration during hydrolysis to the free acid. Both *cis* and *trans* isomers **243** (R=H) were cyclized to the same 11,17-dione **244** [176], therefore *one* of them isomerized during cyclization. Because of the known greater stability of the *cis* hydrindane system, the cyclized product **244** was assigned *cis* geometry at the CD-ring juncture. Weidlich and Meyer-Delius confirmed [177] that hydrogenation of **238** under acid conditions gave a *cis* product and under alkaline conditions a *trans* product (though their melting points for the two acids differ from Robinson's). Comparable results were obtained with 3-deoxy analogs [175].

The acid **238** (R=H) was cyclized in nearly quantitative yield to **246** (Scheme 4-26) by being heated in acetic acid and acetic anhydride [171]. When the latter was hydrogenated, however, a mixture of products was obtained, the major portion of which had lost the 17-carbonyl oxygen [178].

An alternative approach to the reduction of these compounds was taken by opening the D ring. The 3,11-dimethyl ether obtained [135] from **246** was condensed with ethyl formate to give the 16-formyl derivative **247** [178]. Hydroxylamine and acetic acid formed the cyano intermediate **248**, which was not characterized, but was hydrolyzed to the

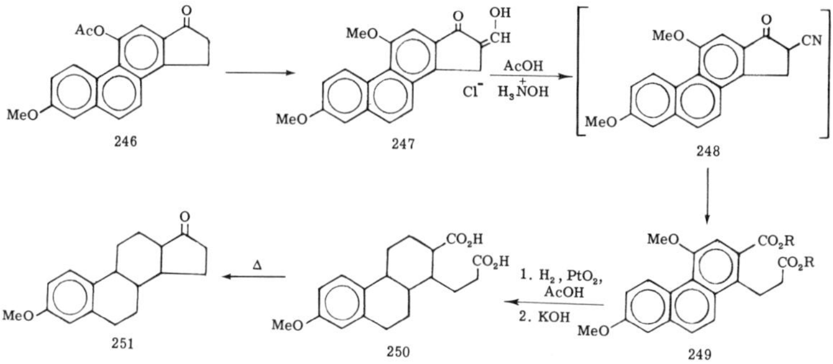

4.8 Organometallic Coupling

Scheme 4-26

diacid **249** (R = HO). Its dimethyl ester **249** (R = CH$_3$) was reduced with Adam's catalyst, and alkaline hydrolysis gave a mixture in which **250** was the major product. Pyrolysis of its lead salt gave an isomer **251** of 18-norestrone methyl ether probably having a *cis-syn-cis* geometry.

A 3-deoxy-D-homo analog was synthesized by a somewhat different route, but employing a ketone condensation in the formation of ring D of the ABD intermediate [179].

4.8 Organometallic Coupling

A number of steroids have been synthesized via ABD intermediates connected through an 8,14-bond that was formed with an organometallic reagent. One group of these (to be discussed later) was condensed with dienophiles to complete the construction of ring C. With the others, ring C was closed by an acylation reaction. Thus, the reaction of 6-methoxy-2-naphthyllithium (Scheme 4-27, **252**) with isopropyl 1-methyl-2-oxo-1-cyclopentanecarboxylate and subsequent dehydration gave the acid **253** [180]. Chain extension by the Arndt-Eistert method gave the next higher homolog **254**. The C ring was closed with phosphorus pentachloride, and catalytic hydrogenation of the 14-double bond gave the 11-keto steroid **255**. When the sequence was altered such that the hydrogenation was carried out before the Arndt-Eistert procedure, the 14β-epimer of **255** was obtained.

Condensation of **252** with the dimethyl ester of 2-carboxymethyl-2-methyl-3-oxocyclopentanecarboxylic acid, followed by ring closure with *p*-toluenesulfonic acid gave **256** [166,181]. By a series of reductions it was transformed into **257**, which was also synthesized from estrone [182].

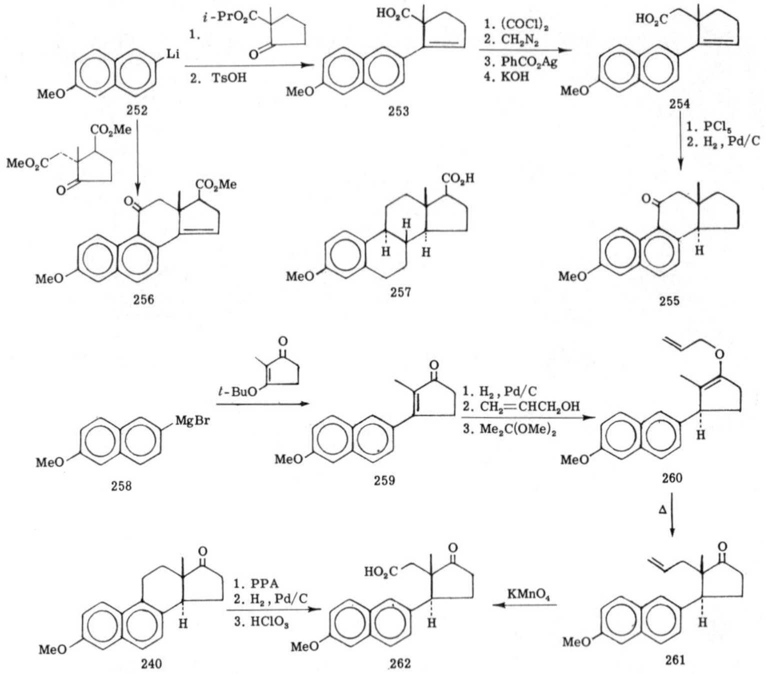

Scheme 4-27

In an attempt to incorporate a 17-oxygen, the condensation of **252** with 2-allyl-2-methyl-1,3-cyclopentanedione was attempted. Although the condensation took place, no ABD intermediate could be isolated because the potential D ring opened, producing a diketo product that could not be cyclized [183]. This route was abandoned, but the use of an allyl group to provide carbons 11 and 12 proved successful in the following example.

Reaction of the Grignard reagent **258** with the *t*-butyl enol ether of 2-methyl-1,3-cyclopentanedione gave ABD intermediate **259** [184,185]. Hydrogenation of the 13-double bond and formation of the allyl ether produced **260**, which underwent Claisen rearrangement by heating to give **261** in 80% yield. Permanganate oxidation to the acid **262**, cyclization with polyphosphoric acid, and hydrogenation in acid to remove the 11-oxygen afforded equilenin methyl ether (**240**).

The ABD intermediates **263** (Scheme 4-28, R=H,Me, n=1,2) have been prepared from the Grignard reagent **258** and the requisite cyclic ketone [186–188]. They undergo Diels-Alder addition of maleic anhydride forming **264** in yields ranging from 30 to 98% [189–191]. Normal

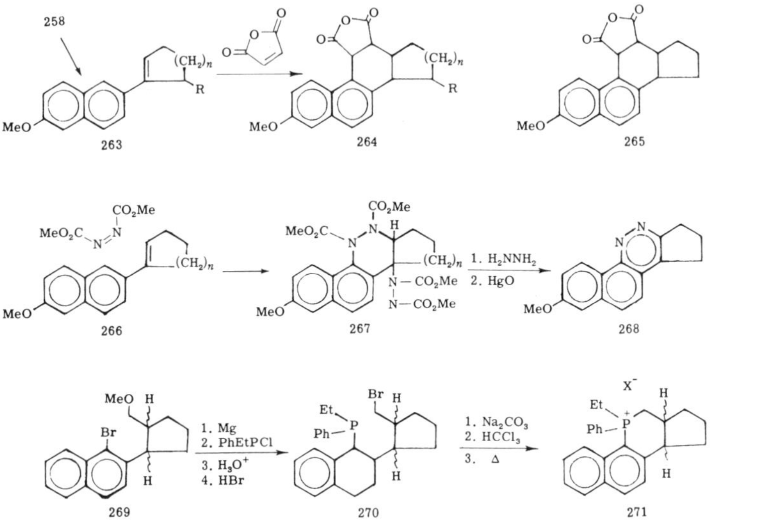

4.8 Organometallic Coupling

Scheme 4-28

adducts would have a $\Delta^{8(14)}$-double bond, but under the conditions of the reaction this isomerizes to Δ^8, reestablishing the naphthalene ring system. The location of the R group (at C-15) is the result of the greater stability of **263** compared to the tautomer that would condense to a 13-R adduct [192,193]. This "undersired" type of product was obtained also from **259**, which gave rise to **265** [176,194,195]. The condensation of **258** with enol ethers of 1,3-cyclohexanedione gave an ABD intermediate that added maleic anhydride, giving in 77% yield a 17a-keto steroid that underwent decarboxylation in quinoline to the methyl ether of 18-nor-D-homoequilenin [196]. The ABD intermediates have been prepared also by routes not involving organometallic reagents [176,178,197–201].

The ABD intermediates **266** (n=1,2), prepared from **258** and cyclopentanone or cyclohexanone, respectively, on standing in solutions of dimethyl azidodiformate gave 70% yields of **267** [202,203]. The latter were hydrolytically decarboxylated by heating with hydrazine hydrate, and the products oxidized with mercuric oxide to the 11,12-diazasteroids **268** (n=1,2).

An 11-phosphasteroid was prepared from **269**, which was synthesized from 2-naphthyllithium and 2-methoxymethylcyclopentanone. The reaction of ethylphenylphosphinous chloride with the Grignard reagent from **269** gave a phosphine that was converted to **270** with hydrobromic acid. An intramolecular displacement of Br by the ethylphenylphosphinous

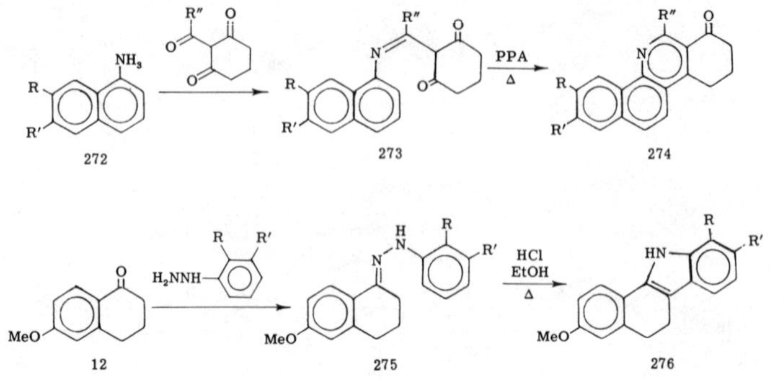

Scheme 4-29

group closed ring B to give the bromide **271** (X=Br). The corresponding perchlorate salt was easier to purify [204]. The CD ring fusion probably is *cis,* but the stereochemistry of the chiral phosphorus atom is unknown.

4.9 Miscellaneous Methods

Schiff bases **273** (Scheme 4-29), prepared in quantitative yield from α-naphthylamines **272** and 2-formyl- or 2-acetyl-1,3-cyclohexanedione, cyclized on heating in polyphosphoric acid to the 11-azasteroids **274** (R=R'=H, R''=Me; R=R'=R''=H; R=MeO, R'=R''=H; R=R''=H, R'=MeO; R=R'=MeO, R''=H) [205–207].

Phenylhydrazones **275** of 6-methoxytetralone (**12**) were cyclized in refluxing ethanolic HCl to give the 11-aza-C-nor-D-homosteroids **276** (R=H, R'=OMe; R=OMe, R'=H) [208]. 6,11-Diaza analogs were synthesized similarly.

References

1. S. N. Ananchenko and I. V. Torgov, *Tetrahedron Lett.* p. 1553 (1963).
2. G. H. Douglas, J. M. H. Graves, D. Hartley, G. A. Hughes, B. J. McLoughlin, J. Siddall, and H. Smith, *J. Chem. Soc., London* p. 5072 (1963).
3. T. B. Windholz, J. H. Fried, and A. A. Patchett, *J. Org. Chem.* **28**, 1092 (1963).
4. G. A. Hughes and H. Smith, *Chem. Ind. (London)* p. 1022 (1960).
5. K. K. Koshoev, S. N. Ananchenko, and I. V. Torgov, *Khim. Prir. Soedin.* [3], 172 (1965).
6. D. J. Crispin and J. S. Whitehurst, *Proc. Chem. Soc., London,* p. 22 (1963).
7. T. Miki, K. Hiraga, and T. Asako, *Proc. Chem. Soc., London,* p. 139 (1963).

8. H. Smith, G. A. Hughes, G. H. Douglas, D. Hartley, B. J. McLoughlin, J. B. Siddal, G. R. Wendt, G. C. Buzby, Jr., D. R. Herbst, K. W. Ledig, J. McMenamin, T. W. Pattison, J. Suida, J. Tokolics, R. A. Edgren, A. B. A. Jansen, B. Gadsby, D. H. R. Watson, and P. C. Phillips, *Experientia* **19**, 394 (1963).
9. S. N. Ananchenko and I. V. Torgov, *Dokl. Akad. Nauk SSSR* **127**, 553 (1959).
10. S. N. Ananchenko, V. E. Limanov, V. N. Leonov, V. M. Rzheznikov, and I. V. Torgov, *Tetrahedron* **18**, 1355 (1962).
11. T. B. Windholz, A. A. Patchett, and J. Fried, Belgian Patent 638,080 (1964); *Chem. Abstr.* **62**, 7846c (1965).
12. J. S. Whitehurst, French Patent 1,374,695 (1964); *Chem. Abstr.* **62**, 4093c (1965).
13. I. V. Torgov, S. N. Ananchenko, and A. V. Platonova, USSR Patent 157,056 (1963); *Chem. Abstr.* **60**, 5596 (1964).
14. T. Miki, K. Hiraga, and T. Asako, *Chem. Pharm. Bull* **13**, 1285 (1965).
15. D. P. Strike, T. Y. Jen, G. A. Hughes, G. H. Douglas, and H. Smith, *Steroids* **8**, 309 (1966).
16. C. H. Kuo, D. Taub, and N. L. Wendler, *Angew. Chem.* **77**, 1142 (1965); *J. Org. Chem.* **33**, 3126 (1970); French Patent 1,500,986 (1967).
17. A. V. Zakharychev, D. R. Lagidze, and S. N. Ananchenko, *Tetrahedron Lett.* p. 803 (1967).
18. U. K. Pandit, F. A. van der Vlugt, and A. C. van Dalen, *Tetrahedron Lett.* p. 3697 (1969).
19. U. K. Pandit and H. O. Huisman, German Patent 2,043,807 (1971); *Chem.* [3], 180 (1965).
20. S. N. Ananchenko, V. N. Leonov, A. V. Platonova, and I. V. Torgov, *Dolk. Akad. Nauk SSSR* **135**, 73 (1960).
21. A. V. Zakharychev, S. N. Ananchenko, and I V. Torgov, *Izv. Akad. Nauk SSSR* p. 1413 (1965); A. V. Zakharychev, I. Hora, S. N. Ananchenko, and I. V. Torgov, *Tetrahedron Lett.* p. 3585 (1966).
22. A. V. Zakharychev, S. N. Ananchenko, and I. V. Torgov, *Steroids* **4**, 31 (1964).
23. D. Banes and J. Carol, *J. Biol. Chem.* **204**, 509 (1953).
24. G. A. Hughes and H. Smith, *Proc. Chem. Soc., London* p. 74 (1960).
25. H. Smith, G. A. Hughes, and B. J. McLoughlin, *Experientia* **19**, 177 (1963).
26. W. E. Bachmann and N. L. Wendler, *J. Amer. Chem. Soc.* **68**, 2580 (1946).
27. A. V. Zakharychev, V. E. Limanov, S. N. Ananchenko, A. V. Platonova, and I. V. Torgov, *Izv. Akad. Nauk SSSR, Ser. Khim.* p. 1701 (1963).
28. A. V. Zakharychev, V. E. Limanov, S. N. Ananchenko, A. V. Platonova, and I. V. Torgov, *Izv. Akad. Nauk SSSR, Ser. Khim.* p. 1809 (1965).
29. A. V. Zakharychev, S. N. Ananchenko, and I. V. Torgov, *Tetrahedron Lett.* p. 171 (1964).
30. A. V. Zakharychev, D. R. Lagidze, S. N. Ananchenko, and I. V. Torgov, *Izv. Akad. Nauk SSSR, Ser. Khim.* p. 760 (1965).
31. H. Smith, Belgian Patent 632,348 (1963); *Chem. Abstr.* **61**, 1927c (1964).
32. H. Smith, G. A. Hughes, G. H. Douglas, and G. R. Wendt, *J. Chem. Soc., London* p. 4472 (1964).
33. H. Smith, Belgian Patent 632,349 (1963); *Chem. Abstr.* **61**, 3169b (1964).
34. R. Pappo, *Intra-Sci. Chem. Rep., Santa Monica, Cal.* **3**, 123 (1969).

35. K. K. Koshoev, S. N. Ananchenko, and I. V. Torgov, *Khim. Prir. Soedin.* [3], 180 (1965).
36. K. K. Koshoev, S. N. Ananchenko, A. V. Platonova, and I. V. Torgov, *Izv. Akad. Nauk SSSR, Ser. Khim.* p. 2058 (1963).
37. D. Lednicer, D. E. Emmert, C. G. Chidester, and D. J. Duchamp, *J. Org. Chem.* 36, 3260 (1971).
38. S. N. Ananchenko, I. V. Torgov, and V. N. Leonov, *Med. Prom. SSSR* 15, 38 (1961).
39. V. N. Leonov, S. N. Ananchenko, and I. V. Torgov, *Dokl. Akad. Nauk SSSR* 138, 384 (1961).
40. V. I. Zaretskii, N. S. Vul'fson, V. L. Sadovskaya, S. N. Ananchenko, and I. V. Torgov, *Dokl. Akad. Nauk SSSR* 158, 385 (1964).
41. I. V. Torgov, S. N. Ananchenko, A. V. Platonova, and V. N. Leonov, USSR Patent 136,368; (1960); *Chem. Abstr.* 59, 2908d (1963); T. A. Serebryakova, Z. N. Parnes, A. V. Zakharychev, S. N. Ananchenko, and I. V. Torgov, *Izv. Akad. Nauk. SSSR, Ser. Khim.* p. 725 (1969); A. V. Zakharychev, S. K. Kasymov, S. N. Ananchenko, and I. V. Torgov, *ibid.* p. 1790.
42. L. M. Kogan, V. E. Gulaya, B. Lakum, and I. V. Torgov, *Izv. Akad. Nauk SSSR, Ser. Khim.* p. 1356 (1970); V. E. Gulaya, L. M. Kogan, and I. V. Torgov, *ibid.* p. 1811.
43. W. S. Johnson, D. K. Banerjee, W. P. Schneider, C. D. Gutsche, W. E. Shelberg, and L. J. Chinn, *J. Amer. Chem. Soc.* 74, 2832 (1952).
44. S. N. Ananchenko, A. V. Platonova, V. N. Leonov, and I. V. Torgov, *Izv. Akad. Nauk SSSR, Otd. Khim. Nauk* p. 1074 (1961).
45. V. M. Rzheznikov, S. N. Ananchenko, and I. V. Torgov, *Izv. Akad. Nauk SSSR, Otd. Khim. Nauk* p. 465 (1962).
46. V. M. Rzheznikov, S. N. Ananchenko, and I. V. Torgov, *Khim. Prir. Soedin.* 1, 90 (1965).
47. S. N. Ananchenko, V. N. Leonov, V. I. Zaretskii, N. S. Vul'fson and I. V. Torgov, *Tetrahedron* 20, 1279 (1964).
48. S. N. Ananchenko, V. M. Rzheznikov, V. N. Leonov, and I. V. Torgov, *Izv. Akad. Nauk SSSR, Otd. Khim. Nauk* p. 1913 (1961).
49. V. N. Leonov, E. V. Shapkina, S. N. Ananchenko, and I. V. Torgov, *Izv. Akad. Nauk SSSR, Ser. Khim.* p. 375 (1964).
50. H. Smith, Belgian Patent 632,347 (1963); *Chem. Abstr.* 61, 1917a (1964); Belgian Patent 632,346 (1963); *Chem. Abstr.* 61, 4457d (1964); G. H. Douglas, J. M. H. Graves, D. Hartley, G. A. Hughes, B. J. McLoughlin, J. Siddall, and H. Smith, *J. Chem. Soc., London* p. 5072 (1963); H. Smith, G. A. Hughes, G. H. Douglas, G. R. Wendt, G. C. Buzby, Jr., R. A. Edgren, J. Fisher, T. Foell, B. Gadsby, J. McMenamin, T. W. Pattison, P. C. Phillips, R. Rees, J. Siddall, J. Suida, L. L. Smith, J. Tokolics, and D. H. P. Watson, *ibid.* p. 4472 (1964); D. Hartley and H. Smith, *ibid.* p. 4492.
51. A. L. Johnson, *J. Med. Chem.* 15, 360 (1972).
52. I. V. Torgov, S. N. Ananchenko, V. N. Leonov, and V. M. Rzheznikov, USSR Patent 147,589 (1962); *Chem. Abstr.* 58, 1516c (1963).
53. N. S. Vul'fson, I. V. Torgov, V. I. Zaretskii, V. N. Leonov, S. N. Ananchenko, and V. G. Zaikin, *Izv. Akad. Nauk SSSR, Ser. Khim.* p. 184 (1964).
54. V. I. Zaretskii, N. S. Vul'fson, V. G. Zaikin, S. N. Ananchenko, V. N. Leonov, and I. V. Torgov, *Tetrahedron* 21, 2469 (1965).

55. V. I. Zaretskii, N. S. Vul'fson, V. G. Zaikin, V. N. Leonov, S. N. Ananchenko, and I. V. Torgov, *Tetrahedron Lett.* p. 347 (1966).
56. V. M. Rzheznikov, S. N. Ananchenko, and I. V. Torgov, *Khim. Prir. Soedin.* 1, 7 (1965); T. D. Linni, A. V. Zakharychev, S. N. Ananchenko, and I. V. Torgov, *Izv. Akad. Nauk SSSR, Ser. Khim.* p. 1356 (1969).
57. W. F. Johns, *J. Org. Chem.* 31, 3780 (1966).
58. N. N. Gaidamovich and I. V. Torgov, *Izv. Akad. Nauk SSSR, Otd. Khim. Nauk* p. 1803 (1961).
59. D. Taub, French Patent 1,497,557 (1967); *Chem. Abstr.* 70, 4394z (1969).
60. D. Taub, U.S. Patent 3,454,600 (1969).
61. A. V. Zakharychev, D. R. Lagidze, S. N. Ananchenko, and I. V. Torgov, *Izv. Akad. Nauk SSSR, Ser. Khim.* p. 2332 (1968).
62. N. Al'Safar, A. V. Zakharychev, S. N. Ananchenko, and I. V. Torgov, *Izv. Akad. Nauk SSSR, Ser. Khim.* p. 2326 (1968); F. Kuok-kin, A. V. Zakharychev, S. N. Ananchenko, and I. V. Torgov, *ibid.* p. 2552.
63. L. Re, French Patent 1,489,526 (1967); *Chem. Abstr.* 69, 36358m (1968).
64. A. Horeau, L. Ménager, and H. Kagan, *Bull Soc. Chim. Fr.* [10] p. 3571 (1971).
65. T. B. Windholz, R. D. Brown, and A. A. Patchett, *Steroids* 6, 409 (1965); A. A. Patchett and T. B. Windholz, French Patent 1,493,563 (1967); *Chem. Abstr.* 69, 87328j (1968).
66. K. Hiraga, *Chem. Pharm. Bull* 13, 1289 (1965); *Chem. Abstr.* 64, 6711 (1966); K. Hiraga, T. Asako, and T. Miki, *Chem. Pharm. Bull.* 13, 1294 (1965).
67. C. Rufer, H. Kosmol, E. Schroder, K. Kieslich, and H. Gibian, *Justus Liebigs Ann. Chem.* 702, 141 (1967); C. Rufer, E. Schroder, and H. Gibian, *ibid.* 752, 1 (1971).
68. H. Smith, French Patent 1,341,755 (1963); *Chem. Abstr.* 60, 10753 (1964); French Patent A81,978 (1963); *Chem. Abstr.* 60, 10752 (1964); French Patent 1,394,701 (1965); *Chem. Abstr.* 63, 4371a (1965).
69. Takeda Chemical Industries Ltd., Japanese Patent 26,540 ('65) (1962); *Chem. Abstr.* 64, 8279g (1966).
70. K. Yoshioka, G. Goto, T. Asako, K. Hiraga, and T. Miki, *Chem. Commun.* p. 336 (1971).
71. E. C. Taylor, G. H. Hawkes, and A. McKillip, *J. Amer. Chem. Soc.* 90, 2421 (1968).
72. W. M. B. Konst, J. van Bruynsvoort, W. N. Speckamp, and H. O. Huisman, *Rec. Trav. Chim. Pays-Bas* 91, 869 (1972).
73. G. H. Douglas, C. R. Walk, and H. Smith, *J.Med. Chem.* 9, 27 (1966); G. A. Hughes and H. Smith, U.S. Patent 3,202,686 (1965); *Chem. Abstr.* 63, 18233 (1965).
74. P. N. Rao and L. R. Axelrod, *J. Chem. Soc., C* p. 2861 (1971).
75. P. N. Rao, E. J. Jacob, and L R Axelrod, *J. Chem. Soc., C* p. 2855 (1971).
76. K. Hiraga, T. Asako, and T. Miki, *Chem. Commun.* p. 1013 (1969).
77. H. Specht, D. Onken, and G. Adam, *Z Chem.* 10, 70 (1970).
78. E. Galantay and H. P. Weber, *Experientia* 25, 571 (1969).
78a. A. J. Birch, J. M. H. Graves, and J. B. Siddall, *J. Chem. Soc., London* p. 4234 (1963); A. J. Birch and G. S R Subba Rao, *Tetrahedron, Suppl.* 7, 391 (1966).
79. J. H. Burckhalter and F. C. Sciavolino, *J. Org. Chem.* 32, 3968 (1967).
80. H. Heidepriem, C. Rufer, H. Kosmol, E. Schroder, and K. Kieslich, *Justus Liebigs Ann. Chem.* 712, 155 (1968); C. Rufer, *Ibid.* 717, 228 (1968).

81. T. B. Windholz, B. Arison, R. D. Brown, and A. A. Patchett, *Tetrahedron Lett.* p. 3331 (1967).
82. M. A. T. Sluyter, U. K. Pandit, W. N. Speckamp, and H. O. Huisman, *Tetrahedron Lett.* p. 87 (1966).
83. W. N. Speckamp, U. K. Pandit, and H. O. Huisman, *Rec. Trav. Chim. Pays-Bas* **82**, 39 (1963).
84. H. O. Huisman, W. N. Speckamp, and U. K. Pandit, *Rec. Trav. Chim. Pays-Bas* **82**, 898 (1963).
85. S. V. Kessar, A. K. Lumb, and R. K. Mayor, *Indian J. Chem.* **5**, 18 (1967).
86. H. Smith, G. H. Douglas, an C. R. Walk *Experientia* **20**, 418 (1964).
87. W. N. Speckamp, H. de Koning, U. K. Pandit, and H. O. Huisman, *Tetrahedron* **21**, 2517 (1965).
88. G. A. R. Kon and F. C. J. Ruzicka, *J. Chem. Soc., London* p. 187 (1936).
89. W. N. Speckamp, J A. van Velthyusen, U. K. Pandit, and H. O. Huisman, *Tetrahedron* **24**, 5881 (1968).
90. W. N. Speckamp, J. A. van Velthyusen, M. A. Douw, U. K. Pandit, and H. O. Huisman, *Tetrahedron* **24**, 5893 (1968).
91. J. C. Hubert, W. N. Speckamp, and H. O. Huisman, *Tetrahedron Lett.* p. 1553 (1969); H. O. Huisman, U. S. Patent 3,597,436 (1969).
92. J. C. Hubert, W. Steege, W. N. Speckamp, and H. O. Huisman, *Syn. Commun.* **1**, 103 (1971).
93. G. Lehmann, H. Whelan, and G. Hilgetag, *Chem. Ber.* **100**, 2967 (1967); *Tetrahedron Lett.* p. 123 (1967).
94. G. Lehmann and B. Lucke, *Justus Liebigs Ann. Chem.* **727**, 88 (1969).
95. G. Sauer, U. Eder, and G.-A. Hoyer, *Chem. Ber.* **105**, 2358 (1972).
96. H. Smith, British Patent 1,069,844 (1967); *Chem. Abstr.* **68**, 114,835k (1968).
97. O. Dann, K-W. Hagedoen, and H. Hofmann, *Chem. Ber.* **104**, 3313 (1971).
98. W. R. J. Simpson, D. Babbe, J. A. Edwards, and J H. Fried, *Tetrahedron Lett.* p. 3209 (1967).
99. C. A. Henrick, E. Bohme, J. A. Edwards, and J. H. Fried, *J. Amer. Chem. Soc.* **90**, 5926 (1968).
100. T. R. Kasturi, R. Ramachandra, and K. M. Damodaran, *Tetrahedron Lett.* p. 5059 (1972).
101. D. B. R. Johnston, F .S. Waksmunski, T. B. Windholz, and A. A. Patchett, *Chimia* **22**, 84 (1968).
102. J. G. Westra, W. N. Speckamp, U. K. Pandit, and H. O. Huisman, *Tetrahedron Lett.* p. 2781 (1966).
103. W. N. Speckamp, J. G. Westra, and H. O. Huisman, *Tetrahedron* **26**, 2353 (1970).
104. G. Lehmann, H. Schick, B. Lucke, and G. Hilgetag, *Chem. Ber.* **101**, 787 (1968).
105. H. O. Huisman, *Angew. Chem., Int. Ed. Engl.* **10**, 450 (1971).
106. S. Barcza, U. S. Patent 3,637,782 (1972); *Chem. Abstr.* **76**, 154021p (1972).
107. H. Smith, *Proc. Int. Symp. Drug Res., 1967* p. 221 (1960).
108. R. A. Barnes and R. Miller, *J. Amer. Chem. Soc.* **82**, 4960 (1960).
109. J. C. Bardhan and S. C. Sengupta, *J. Chem. Soc., London* p. 2520 (1932).
110. G. A. R. Kon, *J. Chem. Soc., London* p. 1081 (1933).
111. L. Ehmann and K. Miescher, *Helv. Chim. Acta* **30**, 413 (1947); E. Buchta, H. Bayer, and K. Ruchlak, *Chem. Ber.* **93**, 1345 (1960).

References

112. A. Cohen, J. W. Cook, and C. L. Hewett, *J. Chem. Soc., London* p. 445 (1935).
113. L. Ruzicka, L. Ehmann, M. W. Goldberg, and H. Hosli, *Helv. Chim. Acta* **16**, 53 and 833 (1933).
114. D. J. C. Gamble and G. A. R. Kon, *J. Chem. Soc., London* p. 443 (1935).
115. S. H. Harper, G. A. R. Kon, and F. C. J. Ruzicka, *J. Chem. Soc., London* p. 124 (1934).
116. A. J. Birch and H. Smith, *J. Chem. Soc., London* p. 1882 (1951).
117. I. N. Nazarov, G. P. Verkholetova, S. N. Ananchenko, I. V. Torgov, and G. V. Aleksandrova, *Zh. Obshch. Khim.* **26**, 1482 (1956).
118. I. N. Nazarov, S. N. Ananchenko, and I. V. Torgov, *Zh. Obshsch. Khim.* **26**, 819 (1956).
119. V. I. Gunar and S. 1. Zav'yalov, *Izv. Akad. Nauk SSSR, Otd. Khim. Nauk* p. 527 (1962).
120. V. I. Gunar and S. I. Zav'yalov, *Izv. Akad. Nauk SSSR, Otd. Khim. Nauk* p. 38 (1963).
121. D. J. Crispin, A. E. Vanstone, and J. S. Whitehurst, *J. Chem. Soc., London* p. 10 (1970).
122. N. A. Nelson and Y. Tamura, *Can. J. Chem.* **43**, 1323 (1965).
123. A. J. Birch and G. S. R. Subba Rao, *J. Chem. Soc., London* p. 3007 (1965).
124. S. V. Kessar, M. Singh, and A. Kumar, *Tetrahedron Lett.* p. 3245 (1965).
125. J. C. Hubert, W. N. Speckamp, and H. O. Huisman, *Tetrahedron Lett.* p. 4493 (1972).
126. N. N. Gaidamovich and I. V. Torgov, *Izv. Akad. Nauk SSSR, Otd. Khim. Nauk* p. 1162 (1961).
127. N. N. Gaidamovich and I. V. Torgov, *Steroids* **4**, 729 (1964).
128. A. J. Birch, J. A. K. Quartey, and H. Smith, *J. Chem. Soc., London* p. 1768 (1952).
129. A. J. Birch, *Proc. Roy. Soc. N. S. W.* **83**, 245 (1949).
130. T. B. Windholz, J. H. Fried, H. Schwam, and A. A. Ratchett, *J. Amer. Chem. Soc.* **85**, 1707 (1963).
131. L. Re, D. B. R. Johnston, D. Taub, and T. B. Windholz, *Steroids* **8**, 365 (1966).
132. Merck and Co., Inc., Belgian Patent 638,079 (1964); *Chem. Abstr.* **62**, 11876d (1965).
133. A. J. Birch and R. Robinson, *J. Chem. Soc., London* p. 503 (1944).
134. A. J. Birch and J. A. K. Quartey, *Chem. Ind. (London)* p. 489 (1953).
135. W. S. Rapson and R. Robinson, *J. Chem. Soc., London* p. 1285 (1935).
136. J. R. Hawthorne and R. Robinson, *J. Chem. Soc., London* p. 763 (1936).
137. D. M. Crowfoot, W. S. Rapson, and R. Robinson, *J. Chem. Soc., London* p. 757 (1936).
138. D. A. Peak and R. Robinson, *J. Chem. Soc., London* p. 1581 (1937).
139. D. A. Peak and R. Robinson, *J. Chem. Soc., London* p. 759 (1936).
140. A. J. Birch and R. Robinson, *J. Chem. Soc., London* p. 501 (1943).
141. J. Sharp, Ph.D. Dissertation, Oxford University, 1954.
142. R. Clarkson, *J. Chem. Soc., London* p. 4900 (1965).
143. Imperial Chemical Industries, Ltd., Belgian Patent 633,066 (1963); *Chem. Abstr.* **60**, 8283e (1964).
144. I.C.I., Belgian Patents 633,213 and 633,214 (1963); *Chem. Abstr.* **60**, 15950g (1964).

145. R. I. Meltzer, D. Lustgarten, R. J. Stanaback, and R. E. Brown, *Tetrahedron Lett.* p. 1581 (1963).
146. J. N. Brown, R. L. R. Towns, and L. M. Trefonas, *J. Heterocycl. Chem.* **8**, 273 (1971); R. E. Brown, A. I. Meyers, L. M. Trefonas, R. L. R. Towns, and J. N. Brown, *ibid.* p. 279.
147. A. I. Meyers, G. G. Munoz, W. Sobotka, and K. Baburao, *Tetrahedron Lett.* p. 255 (1965).
148. Imperial Chemical Industries, Ltd., Belgian Patent 647,699 (1964); *Chem. Abstr.* **63**, 11664d (1965); Belgian Patent 647,700 (1964); *Chem. Abstr.* **63**, 11664b (1965); Belgian Patent 16,428 (1964); *Chem. Abstr.* **63**, 16428d (1965).
149. E. R. H. Jones, *J. Chem. Soc., London* p. 5911 (1964).
150. A. L. Logothetic, *J. Org. Chem.* **29**, 1834 (1964).
151. V. Nacci, G. Filacchioni, G. de Martino, R. Giuliano, and M. Artico, *Farmaco, Ed. Sci.* **27**, 548 (1972).
152. N. A. McGinnis and R. Robinson, *J. Chem. Soc., London* p. 404 (1941).
152a. R. J. Thill, Ph.D. Dissertation, University of Michigan, Ann Arbor, 1971.
153. W. Sobotka, W. N. Beverung, G. G. Munoz, J. C. Sircar, and A. I. Meyers, *J. Org. Chem.* **30**, 3667 (1965); A. H. Reine and A. I. Meyers, *ibid.* **35**, 554 (1970).
154. A. I. Meyers and W. N. Beverung, *Chem. Commun.* p. 877 (1968).
155. A. I. Meyers and J. C. Sircar, *Tetrahedron* **23**, 785 (1967).
156. J. C. Sircar and A. I. Meyers, *J. Org. Chem.* **32**, 1248 (1967).
157. A. I. Meyers and J. C. Sircar, *J. Org. Chem.* **32**, 1250 (1967).
158. M. von Strandtmann, M. P. Cohen, and J. Shavel, *J. Org. Chem.* **31**, 797 (1966)
159. A. A. Akhrem, A. M. Moiseenkov, V. A. Krivoruchko, F. A. Lakhvich, L. A. Saburova, and I. A. Poselenov, *Izv. Akad. Nauk SSSR, Ser. Khim.* p. 2338 (1969); *Chem. Abstr.* **72**, 43971 (1970); A. A. Akhrem, A. M. Moiseenkov, and A. I. Poselenov, *Dokl. Akad. Nauk SSSR* **203**, 95 (1972); *Chem. Abstr.* **77**, 5654r (1972); see also *Izv. Akad. Nauk SSSR, Ser. Khim.* p. 2078 (1972); *Chem. Abstr.* **78**, 30056a (1973).
160. U. K. Pandit, K. de Jonge, and H. O. Huisman, *Rec. Trav. Chim. Pays-Bas* **88**, 149 (1969).
161. U. K. Pandit and H. O. Huisman, *Tetrahedron Lett.* p. 3901 (1967).
162. C. K. Chuang, Y. L. Tien, and Y. T. Huang, *Ber. Deut. Chem. Ges.* B **70**, 858 (1937).
163. C. K. Chuang, Y. T. Huang, and C. M. Ma, *Ber. Deut. Chem. Ges.* B **72**, 713 (1939).
164. C. K. Chuang, C. M. Ma, Y. L. Tien, and Y. T. Huang, *Ber. Deut. Chem. Ges.* B **72**, 949 (1939).
165. R. Robinson and J. M. C. Thompson, *J. Chem. Soc., London* p. 1739 (1939).
166. M. Harnik, Y. Lederman, H. Frumkis, and N. Danieli, *Tetrahedron* **23**, 3183 (1967).
167. G. R. Clemo and L. K. Mishra, *J. Chem. Soc., London* p. 192 (1953).
168. F. D. Popp, W. R. Schleigh, P. M. Froelich, R. J. Dubois, and A. C. Casey, *J. Org. Chem.* **33**, 833 (1968).
169. N. P. Buu-Hoi and D. Lavit-Lamy, *Bull. Soc. Chim. Fr.* p. 773 (1962).
170. D. K. Banerjee, H. N. Khastgir, J. Dutta, E. J. Jacob, W. S. Johnson, C. F. Allen, B. K. Bhattacharyya, J. C. Collins, A. L. McCloskey, W. T. Tsatsos, W. A. Vredenburg, and K. L. Williamson, *Tetrahedron Lett.* p. 76 (1961).

References

171. R. Robinson, *J. Chem. Soc., London* p. 1390 (1938).
172. R. Robinson and W. M. Todd, *J. Chem. Soc., London* p. 1743 (1939).
173. A. J. Birch and G. S. R. Subba Rao, *Tetrahedron Lett.* p. 2763 (1967).
174. A. J. Birch and G. S. R. Subba Rao, *Aust. J. Chem.* 23, 547 (1970).
175. A. Koebner and R. Robinson, *J. Chem. Soc., London* p. 1994 (1938).
176. A. Koebner and R. Robinson, *J. Chem. Soc., London* p. 566 (1941).
177. H. A. Weidlich and M. Meyer-Delius, *Ber. Deut. Chem. Ges.* B 74, 1213 (1941).
178. R. Robinson and H. N. Rydon, *J. Chem. Soc., London* p. 1394 (1939).
179. D. Nasipuri, A. C. Chaudhuri, and J. Roy, *J. Chem. Soc., London* p. 2734 (1958).
180. G. Eglington, J. C. Nevenzel, A. I. Scott, and M. S. Newman, *J. Amer. Chem. Soc.* 78, 2331 (1956).
181. D. K. Banerjee, N. Mahishi, E. J. Jacob, G. Ramani, and D. Devaprabhakara, *Steroids* 16, 561 (1970).
182. D. K. Banerjee, E. J. Jacob, and N. Mahishi, *Steroids* 16, 733 (1970).
183. M. S. Newman and J. H. Mannhardt, *J. Org. Chem.* 26, 2113 (1961).
184. A. Horeau, E. Lorthioy, and J.-P. Guette, *C. R. Acad. Sci.* 269, 558 (1969).
185. A. Horeau and F. E. Lorthioy, French Patent 1,509,087 (1968); *Chem. Abstr.* 71, 30628u (1969).
186. W. E. Bachmann and M. C. Kloetzel, *J. Amer. Chem. Soc.* 60, 2204 (1938).
187. W. Bergmann and E. Bergmann, *J. Amer. Chem. Soc.* 62, 1699 (1940).
188. E. W. Gabisch, *J. Org. Chem.* 26, 4165 (1961).
189. E. Bergmann and F. Bergmann, *J. Amer. Chem. Soc.* 59, 1443 (1937).
190. L. H. Klemm, W. Hodes, and W. B. Schaap, *J. Org. Chem.* 19, 451 (1954).
191. V. R. Skvarchenko, L. Weng-Luan, and R. Ya. Levina, *Zh. Obshch. Khim.* 31, 2829 (1961).
192. L. H. Klemm and H. Ziffer, *J. Org. Chem.* 20, 182 (1955).
193. L. H. Klemm, H. Ziffer, J. W. Sprague, and W. Hodes, *J. Org. Chem.* 20, 190 (1955).
194. S. Carboni, *Gazz Chim. Ital.* 85, 1216 (1955).
195. S. Carboni and A. Marsili, *Gazz. Chim. Ital.* 89, 1717 (1959).
196. A. A. Akhrem, Yu. A. Titov, and I. N. Minaeva, *Izv. Akad. Nauk SSSR, Otd. Khim. Nauk.* p. 1164 (1961).
197. H. A. Weidlich and G. H. Daniels, *Ber. Deut. Chem. Ges.* B 72, 1590 (1939).
198. H. A. Weidlich and M. Meyer-Delius, *Ber. Deut. Chem. Ges.* B 72, 1941 (1939).
199. S. Carboni, F. Bottari, and A. Marsili, *Gazz. Chim. Ital.* 89, 2321 (1959).
200. W. E. Bachmann and R. D. Morin, *J. Amer. Chem. Soc.* 66, 553 (1944).
201. F. C. Novello and M. E. Christy, *J. Amer. Chem. Soc.* 75, 5431 (1953).
202. A. G. M. Willems, R. R. von Eck, U. K. Pandit, and H. O. Huisman, *Tetrahedron Lett.* p. 81 (1966).
203. A. G. M. Willems, R. R. van Eck, H. Nijhuis, U. K. Pandit, and H. O. Huisman, *Rec. Trav. Chim. Pays-Bas* 89, 885 (1970).
204. C. H. Chen and K. D. Berlin, *Phosphorus* 1, 49 (1971).
205. N. A. J. Rogers and H. Smith, *J. Chem. Soc., London* p. 341 (1955).
206. P. E. Cross and E. R. H. Jones, *J. Chem. Soc., London* p. 5916 (1964).
207. S. V. Kessar, I. Singh, and A. Kumar, *Tetrahedron Lett.* p. 2207 (1965).
208. P. E. Cross and E. R. H. Jones, *J. Chem. Soc., London* p. 5919 (1964).

5
AB + D → ABCD

5.1 Introduction

The Diels-Alder reaction is one of the major means of forming a six-membered ring; consequently, it is to be expected that it would be utilized in the total synthesis of steroids. It was encountered in Chapter 4, Section 4.8, where it added a C ring to an ABD intermediate. In the Sarett synthesis of cortisone, to be described in Chapter 9, a Diels-Alder reaction was used to build a BC intermediate. It was utilized to form the CD intermediate at the start of the Woodward synthesis to be described in Chapter 12. In the present chapter, a substituted AB moiety constitutes the diene, and ring D the dienophile. Ring C is formed when they unite.

Theoretically the dienophile could approach the diene in four different ways. As illustrated, an endo approach from the α-side would produce the $8\beta,13\beta,14\beta$-product:

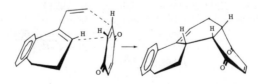

5.2 Carbocyclics

An endo approach from the β-side would produce the enantiomeric 8α,13α,14α-product:

An exo approach from the α-side would give rise to the 8β,13α,14α-product:

And, finally, an exo approach from the β-side would produce the enantiomeric 8α,13β,14β-product, which is not shown. It is well known, of course, that the favored approach for Diels-Alder additions is endo, presumably because the transition state is more stable as a result of maximal orbital overlap [1]. Consequently, the first two products shown constitute the racemic pair obtained in this reaction; exo products are not formed. The diene shown is asymmetric, and it is obvious that if the dienophile also were unsymmetrically substituted, there would be two possible orientations for each approach listed above. This additional complexity will be discussed in the next section.

There is a major difficulty inherent in this applicaiton of the Diels-Alder reaction to the synthesis of naturally occurring steroids: It is by the very nature of the addition that a *cis*-CD ring juncture is created. To obtain the desired *trans*-CD ring juncture, it is necessary to begin with a six-membered D ring, isomerize to the more stable *trans*-decalin system, and later contract to a 5-membered D ring. The 15-keto group is necessary for the isomerization, but its subsequent selective elimination can be achieved only by an indirect route.

5.2 Carbocyclics

The difficulties just described were surmounted with finesse in the synthesis of estrone (7, Scheme 5-1) [2], beginning with selective reduction of the 16-double bond with zinc and acetic acid, and selective reduction of the 17a-keto group to a mixture of epimers predominating

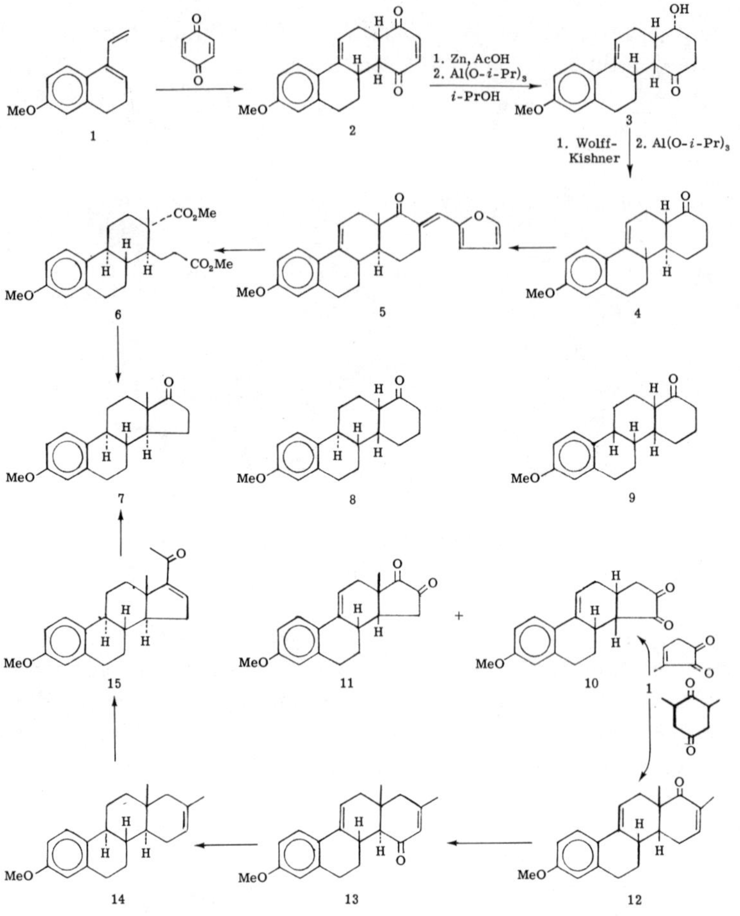

Scheme 5-1

in the 17aα-alcohol **3**. This second step was carried out for the purpose of protecting the 17a-ketone while the 15-ketone was removed. During Wolff-Kishner reduction of the 15-keto group, the 14β-hydrogen isomerized to α under the influence of the strong base. Oppenauer oxidation gave the 17a-ketone **4**. To prevent alkylation on carbon 16, **4** was converted to the furfurylidene derivative, which was then treated with methyl iodide and potassium *t*-butoxide. Both 13β-methyl **5** and 13α-methyl products were obtained, the former predominating as a result, in part, of the 9(11)double bond; the reader is referred to the work of W. S. Johnson described in Chapter 12, Section 12.2. Opening of the D ring with hydrogen peroxide, reduction of the double bond with sodium and ammonia, and esterification with diazomethane gave dimethyl homomarrianolate methyl ether (6). The overall yield up to this point was 5.3% of theoretical. As 6 had been converted previously by Anner and Miescher, as well as by Johnson and colleagues, to estrone in 80% yield by ring closure with sodium in ammonia, decarboxylation with lead carbonate, and demethylation with pyridinium chloride [3,4], this constituted a formal total synthesis of racemic estrone.

The methyl ethers (8 and 9, respectively) of 3-hydroxy-D-homo-14β-gona-1,3,5(10)-trien-17a-one and 3-hydroxy-D-homo-9β,14β-gona-1,3,5(10)-trien-17a-one, as well as their 14α-epimers also were prepared from 2 [2,5].

6-Methoxy-1-vinyl-3,4-dihydronaphthalene (1) reacted also with cyclic ketones already containing a methyl group; frequently two products were obtained, as was the case of 2-methyl-1-cyclopentene-3,4-dione, with the 14-methyl isomer (10) predominating [6–10]. Attempts to reverse this mode of addition have been largely unsuccessful until recently. Dickinson and colleagues have reported that, in contrast to the usual thermal reaction, boron trifluoride-catalyzed addition of 2,6-xyloquinone to 1 gave predominantly (69% yield) the methyl ether (12) of 3-hydroxy-17-methyl-D-homo-14β-estra-1,3,5(10),9(11),16-pentaene-15,17a-dione [11]. Isomerization of 12 to the CD-*trans* epimer with sodium bicarbonate, selective reduction of the 17a-keto group with lithium aluminum tri-*t*-butoxide, mesylation, and reduction of the mesylate with zinc in refluxing methanol gave the monoketone 13. Selective hydrogenation of the 9(11)-double bond, hydride reduction of the 15-keto group, acetylation of the mixture of epimeric alcohols, and reductive removal of the acetoxy groups with lithium and ammonia in tetrahydrofuran gave the olefin 14 (75% overall yield in the last three steps). Addition of osmium tetroxide to the 16-double bond, ring opening with lead tetraacetate, followed by ring closure of the keto aldehyde with aqueous HCl in tetrahydrofuran gave the methyl ether (15) of 3-hydroxy-1,3,5(10),16-pregnatetraen-20-one in 22% overall yield from 1. Beckmann rearrangement of the oxime of 15 gave a 40% yield of racemic estrone methyl ether (7). The dienophile 2,6-xyloquinone reacts similarly with a diene containing a saturated A ring.

Compounds such as 1, being essentially coplanar, lack asymmetry and form simple racemic mixtures in Diels-Alder reactions with symmetrical dienophiles. With asymmetrical dienophiles they form mixtures of positional isomers (such as 10 and 11). With alicyclic dienes, all of which in this work are asymmetric, additional possibilities exist. For example, the diene 16 (Scheme 5-2), prepared from 6-methoxy-9-methyl-Δ^6-1-decalone by ethynylation, acid hydrolysis of the vinyl ether linkage, hydrogenation

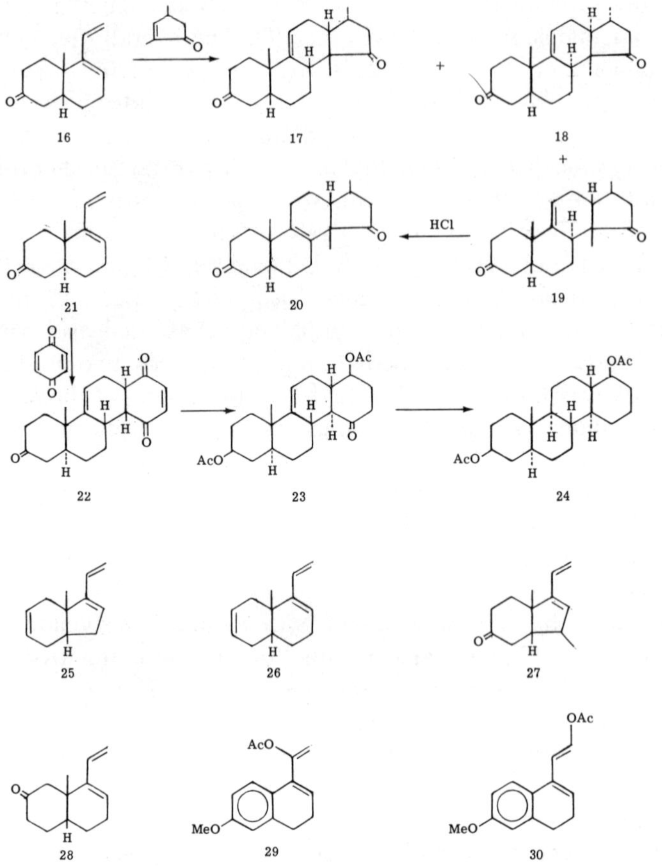

Scheme 5-2

of the ethynyl group to vinyl, and dehydration with potassium hydrogen sulfate, reacted with 2,5-dimethyl-1-cyclopenten-3-one to give predominantly **17** and small amounts of **18** and **19** [12]. Acid-catalyzed isomerization converted **17** and **19** into the same Δ^8-steroid, and **18** into the corresponding 13α,14α,17α-diasteriomer. As **17** comprised 95% of the Diels-Alder product, a high degree of steric control was evident.

The 5α-epimer of **16** was synthesized from 5(10)-dehydro-9-methyl-decalin-1,6-dione beginning with ethynylation at the 1-keto; the ethynyl group was hydrogenated, the 5(10)-double bond reduced (lithium–ammonia), and the t-alcohol dehydrated, giving **21**. The latter was treated with benzoquinone and two products were obtained, **22** and its 8α,13α,14α-isomer, the former predominating [13]. Several steps remi-

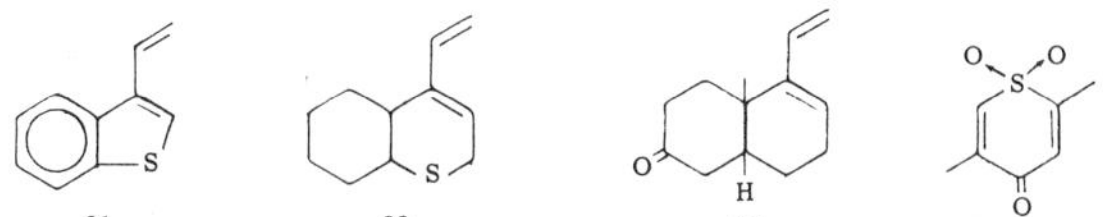

Scheme 5-3

niscent of the first synthesis of estrone by Diels-Alder condensation (Scheme 5-1) were carried out. The 16-double bond was reduced, the 14-hydrogen was isomerized with basic alumina, and the 3- and 17a-keto groups were reduced to hydroxyls, which were acetylated. The intermediate 23 was then converted into the known 24 [14] by addition of hypobromous acid to the 9(11)-double bond, oxidation to the 11,15-diketone, reductive removal of the 9α-bromine with zinc, and hydrogenolysis of the corresponding bisethylenedithioketal.

The diene 1 has been condensed with numerous other dienophiles [15–25,29], as has 16 [26–28,30]. A wide variety of dienes has been utilized, many of them by a Russian group associated with I. N. Nazarov. They include 25 [26,31–34], 26 [33,35], 27 [26,36,37], 28 [27], 29 [38], and 30 [39–43]. It must be emphasized that in many of these instances, as well as those in the next section, the stereochemistry of the products has not been proved in a definitive way, but merely is assumed on the basis of what is kown about the Diels-Alder reaction.

5.3 Thiasteroids

Steroids containing sulfur in the ring system have been synthesized both from thiadienes and from thiadienophiles (Scheme 5-3). Thus, both 31 [44] and 32 [45] have been employed in the synthesis of 6-thiasteroids. Furthermore, 25, 26, 27, and 33 have been condensed with 2,5-dimethyl-4-thiopyrone dioxide (34) [46].

References

1. J. G. Martin and R. K. Hill, *Chem. Rev.* 61, 537 (1961).
2. J. E. Cole, W. S. Johnson, P. A. Robins, and J Walker, *J. Chem. Soc., London* p. 244 (1962).
3. G. Anner and K. Miescher, *Helv. Chim. Acta* 31, 2173 (1948); 32, 1957 (1949); 33, 1379 (1950).
4. W. S. Johnson, D. K. Banerjee, W. P. Schneider, C. D. Gutsche, W. E. Shelberg, and L. J. Chinn, *J. Amer. Chem. Soc.* 74, 2832 (1952).
5. P. A. Robins and J. Walker, *J. Chem. Soc., London* p. 237 (1959).

6. E. Dane and J. Schmitt, *Justus Liebigs Ann. Chem.* **536**, 196 (1938).
7. G. Singh, *J. Amer. Chem. Soc.* **78**, 6109 (1956).
8. E. Dane and J. Schmitt, *Justus Liebigs Ann. Chem.* **537**, 246 (1939).
9. E. Dane, J. Schmitt, and C. Rautenstrauch, *Justus Liebigs Ann. Chem.* **532**, 29 (1937).
10. G. Stork and G. Singh, *Nature (London)* **165**, 816 (1950).
11. R. A. Dickinson, R. Kubela, G. A. MacAlpine, Z. Stojanac, and Z. Valenta, *Can. J. Chem.* **50**, 2377 (1972).
12. I. N. Nazarov and I. V. Torgov, *Izv. Akad. Nauk SSSR, Otd. Khim. Nauk* p. 901 (1953).
13. I. N. Nazarov and I. A. Gurvich, *Izv. Akad. Nauk SSSR, Otd. Khim. Nauk* p. 293 (1959).
14. I. A. Gurvich, T. V. Ilyukhina, and V. F. Kucherov, *Izv. Akad. Nauk SSSR, Otd. Khim. Nauk* p. 1706 (1960).
15. E. Dane and K. Eder, *Justus Liebigs Ann. Chem.* **539**, 207 (1939).
16. V. F. Kucherov and L. N. Ivanova, *Dokl. Akad. Nauk SSSR* **131**, 1077 (1960).
17. E. Dane, *Angew. Chem.* **52**, 655 (1939).
18. I. N. Nazarov and I. L. Kotlyarevskii, *Izv. Akad. Nauk SSSR Otd. Khim. Nauk* p. 1100 (1953).
19. E. Dane, O. Hoss, A. W. Bindseil, and J. Schmitt, *Justus Liebigs Ann. Chem.* **532**, 39 (1937).
20. E. Dane, O. Hoss, K. Eder, J. Schmitt, and O. Schon, *Justus Liebigs Ann. Chem.* **536**, 183 (1938).
21. P. A. Robins and J. Walker, *J. Chem. Soc., London* p. 3249 (1956).
22. I. N. Nazarov, I. V. Torgov, and G. P. Verkholetova, *Dokl. Akad. Nauk SSSR* **112**, 1067 (1957).
23. G. P. Verkholetova and I. V. Torgov, *Izv. Akad. Nauk SSSR, Otd. Khim. Nauk* p. 861 (1962).
24. I. I. Zaretskaya, T. I. Sorkina, and I. V. Torgov, *Izv. Akad. Nauk SSSR, Ser. Khim.* p. 1058 (1965).
25. I. V. Torgov and I. N. Nazarov, *Zh. Obshch. Khim.* **29**, 787 (1959).
26. I. N. Nazarov, L. D. Bergel'son, L. I. Shmonina, and L. N. Terekhova, *Izv. Akad. Nauk SSSR, Otd. Khim. Nauk* p. 439 (1949).
27. I. N. Nazarov, G. P. Verkholetova, I. V. Torgov, I. I. Zaretskaya, and S. N. Ananchenko, *Izv. Akad. Nauk SSSR, Otd. Khim. Nauk* p. 929 (1953).
28. I. N. Nazarov, I. I. Zaretskaya, G. P. Verkholetova, and I. V. Torgov, *Izv. Akad. Nauk SSSR, Otd. Khim. Nauk* p. 920 (1953).
29. E. Dane, U.S. Patent 2,230,233 (1941), *Chem. Abstr.* **35**, 3037[1] (1941).
30. I. N. Nazarov and L. D. Bergel'son, *Zh. Obshch. Khim.* **20**, 648 (1950).
31. I. N. Nazarov and M. S. Burmistrova, *Izv. Akad. Nauk SSSR, Otd Khim. Nauk* p. 56 (1954).
32. A. M. Gaddis and L. W. Butz, *J. Amer. Chem. Soc.* **69**, 1203 (1947).
33. I. N. Nazarov, L. I. Shmonina, and I. V. Torgov, *Izv. Akad. Nauk SSSR, Otd. Khim. Nauk* p. 1074 (1953).
34. I. N. Nazarov and L. I. Shmonina, *Zh. Obshch. Khim.* **20**, 876 (1950).
35. A. M. Gaddis and L. W. Butz, *J. Amer. Chem. Soc.* **69**, 1165 (1947).
36. I. N. Nazarov, V. F. Kucherov, and L. N. Terekhova, *Izv. Akad. Nauk SSSR, Otd. Khim. Nauk* p. 442 (1952).

References

37. I. N. Nazarov, L. N. Terekhova, and L. D. Bergel'son, *Zh. Obshch. Khim.* **20**, 661 (1950).
38. M. Mousseron, F. Winternitz, and J. Diaz, *C. R. Acad. Sci.* **250**, 3433 (1960).
39. F. Winternitz and J. Diaz, *Bull. Soc. Chim. Fr.* [0] p. 790 (1960).
40. T. I. Sorkina, I. I. Zaretskaya, and I. V. Torgov, *Dokl. Akad. Nauk SSSR* **129**, 345 (1959).
41. I. I. Zaretskaya, T. I. Sorkina, O. B. Tikhomirova, and I. V. Torgov, *Izv. Akad. Nauk SSSR, Ser. Khim.* p. 1051 (1965).
42. I. V. Torgov, T. I. Sorkina, and I. I. Zaretskaya, *Bull. Soc. Chim. Fr.* [0] p. 2063 (1964).
43. T. I. Sorkina, I. I. Zaretskaya, and I. V. Torgov, *Izv. Akad. Nauk SSSR, Ser. Khim.* p. 2021 (1964).
44. W. Davies and Q. N. Porter, *J. Chem. Soc., London* p. 4961 (1957).
45. I. N. Nazarov, A. I. Kuznetsova, and I. A Gurvich, *Zh. Obshch. Khim.* **22**, 982 (1952).
46. I. N. Nazarov, I. A. Gurvich, and I. A. Kuznetsova, *Izv. Akad. Nauk SSSR, Otd. Khim. Nauk* p. 1091 (1953).

6
A + C → AC → ABC → ABCD

With the exception of Johnson's estrone preparation (Scheme 6-2), this approach to total synthesis is of only minor significance. Historically a considerable effort has been directed toward the formation of ABC intermediates from AC fragments, but further work concerning elaboration to tetracyclic compounds has been minimal. Several of the ABC intermediates prepared (cf. Scheme 6-3) were identical, or very nearly so, to those obtained in a more straightforward manner via AB→ABC and BC→ABC routes, thus partially nullifying the impetus for extensive development of this synthetic method. This chapter is proportioned into two sections based on the type of AC intermediate utilized. In Section 6.1 the rings are bonded directly to each other via the potential C-9 and C-10 carbons, whereas Section 6.2 discusses the preparation of compounds possessing the AC rings linked by a two-carbon chain (the potential C-6 and C-7 carbons).

6.1 Synthesis from p-Anisylcyclohexanes

The primary effort in this section relates to Johnson's estrone synthesis and is remarkable for the high stereoselectivity of all the steps involved, with the exception of the introduction of the C-18 methyl group. The

6.1 Synthesis from p-Anisylcyclohexanes

key intermediate 9 has been produced by a number of workers by various routes outlined in Scheme 6-1. Johnson [1,2] and Turner [3,4] started with 1 (obtained in excellent yield from the Friedel-Crafts acylation of anisole with either glutaric anhydride or γ-carbomethoxybutyryl chloride) and, via a Stobbe condensation, hydrogenation, and esterification, isolated 6 in 41–47% yield. In fact, improvement in the yields of this and subsequent steps of the estrone synthesis has been reported [5]. Dieckmann cyclization produced intermediate 5 [1–6]), which could be isolated but was more conveniently methylated *in situ* to 9 (R = Me). This methylation procedure also provided the undesired 13α-methyl isomer; however, the natural 13β-epimer could be obtained in nearly 40% yield from 6 [2].

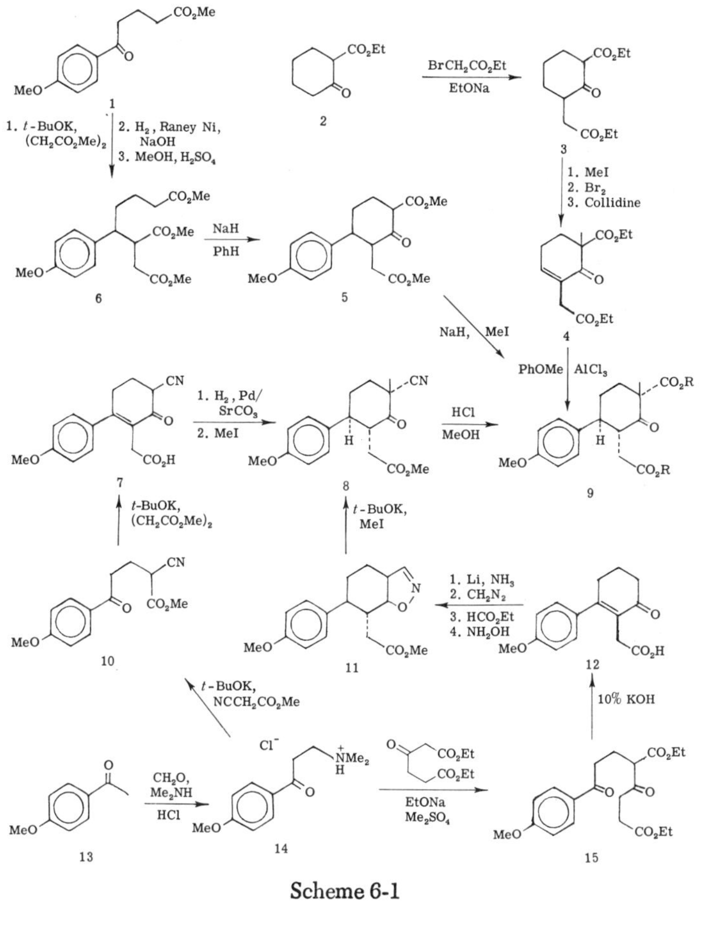

Scheme 6-1

An alternate method for the preparation of **9** employed as the last step the combination of the desired A and C rings [7–10]. Alkylation of **2** with ethyl bromoacetate gave **3**, the result of ethoxide-catalyzed rearrangement of the initially formed α,α-disubstituted cyclohexanone. It was found unnecessary to purify **3**; rather it was methylated, brominated, and dehydrobrominated to **4** in an overall yield of 40% based on **2** [10]. Friedel-Crafts alkylation of anisole afforded **9** (R=Et) in excellent yield (90%).

Two other, related procedures leading to **9** are also shown in Scheme 6-1 and start with the anisole derivative **14** (prepared via a Mannich reaction with p-methoxyacetophenone) [11–14]. Reaction with methyl cyanoacetate gave a 78% yield of **10** (note the similarity to **1**) which condensed with dimethyl succinate to provide **7** (44%). Hydrogenation and methylation gave a moderate (40–50%) yield of the cyano ester **8** [11,12]. This same precursor was prepared by condensing diethyl β-ketoadipate with **14** in the presence of dimethyl sulfate; the resulting **15** afforded **12** in the presence of 10% potassium hydroxide. The double bond was reduced with lithium–ammonia and esterification was effected with diazomethane. Introduction of the cyano and angular methyl groups was performed by Johnson's isoxazole method [15] involving hydroxymethylation and treatment with hydroxylamine to give **11**. Potassium t-butoxide with methyl iodide gave the previously prepared **8**, convertible to **9** with methanolic hydrochloric acid. This sequence provided **9** in an overall yield of 20% starting from **13**.

The synthesis of the methyl ethers of estrone and 14-isoestrone was completed as shown in Scheme 6-2 [2,5,6]. The Reformatsky reaction (methyl bromoacetate-zinc) of **9** gave a mixture (29:28%) of the 14α- and 14β-hydroxy derivatives, the former of which spontaneously lactonized (to the *cis* γ-lactone) whereas the latter could be converted into the

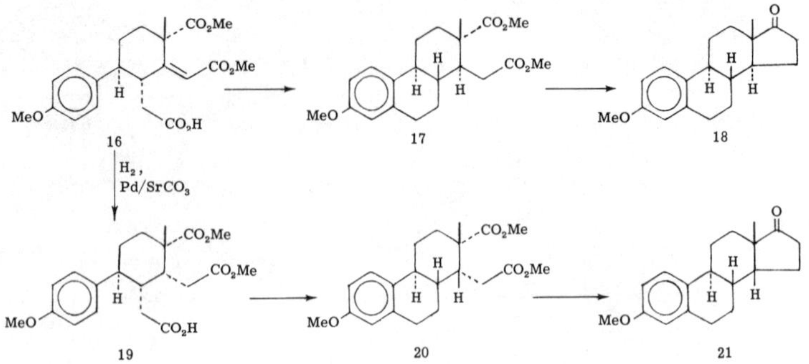

Scheme 6-2

6.2 Synthesis from C-5, C-8 Bridged Intermediates

analogous *trans* lactone by warming with formic acid. Both lactones were converted into **16** on treatment with hydroxide; for preparative purposes the lactones need not be separated nor purified. The crude reaction mixture was simply treated with formic acid and methanolic sodium hydroxide, and the dehydrated **16** was isolated in 39% yield [2]. Catalytic hydrogenation over 30% palladium-on-strontium carbonate gave a 79% yield of the unnatural 14β-isomer **19** and a 5% yield of the epimeric 14α-compound. Elaboration of the former to the methyl ether of 14-isoestrone (**21**) was accomplished by aluminum chloride catalyzed cyclization of the acid chloride and subsequent hydrogenolysis to **20** (79%). Formation of the D ring was similar to that described below for estrone methyl ether (**18**).

Because of the unfavorable stereochemical results found on hydrogenation of **16**, it was elected to perform cyclization first. To this end the acid chloride of **16** was treated with aluminum chloride and the resulting ABC intermediate was subjected to the action of hydrogen over palladium–charcoal–perchloric acid; simultaneous hydrogenation and hydrogenolysis afforded **17** in 71% yield. (This material was identical to that previously obtained by Anner and Miescher; Scheme 3-9, Chapter 3). Cyclization to **18** was effected in either of two ways. The first [2] involved chain homologation of the C-14 group via the Arndt-Eistert sequence, i.e., selective saponification, acid chloride formation, reaction with diazomethane, silver oxide and, finally, potassium hydroxide. The resulting diacid (*dl*-homomarrianolic acid) was pyrolyzed over lead carbonate to **18**. A better procedure [5] for the conversion of **17** into **18** utilized the Sheehan acyloin condensation (sodium–ammonia) of **17** followed by borohydride reduction and dehydration. The overall yield of estrone (demethylation of **18** was readily performed with pyridine hydrochloride) varied from 2–5%, depending on which steps were followed.

6.2 Synthesis from C-5, C-8 Bridged Intermediates

As in the preceding section, many ABC fragments have been prepared by this route, but completions to steroidal substances have been lacking. Two methods for preparing the dione **27** have been recorded and are outlined in Scheme 6-3. In the first, *m*-methoxyphenethyl bromide (**22**) was alkylated with the dimethyl ether of dihydroresorcinol to give, after hydrolysis, **27** (R = H) [16,17]. This alkylation procedure is obviously subject to considerable variation; and, historically, this potential has been exploited. Thus, the aryl group of **22** could possess a different sub-

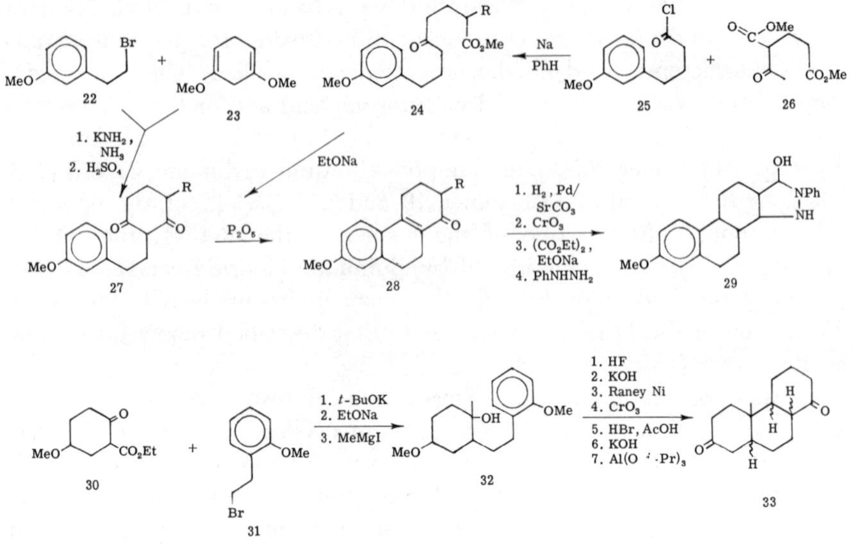

6 A + C → AC → ABC → ABCD

Scheme 6-3

stitution pattern; or various cyclohexane derivatives—primarily Hagemann's ester—may be used in place of **23** [18–21]. Alternatively, it has been possible to reverse the roles of the two reactants, i.e., an anionic form of **22** (the potassium salt of *m*-methoxyphenylacetylene) condensed with a ketonic analog of **23** [22–24]. These processes lead, of course, to ABC tricyclic intermediates potentially convertible to estrogens. Illustrated also in Scheme 6-3, and discussed below, is a second possibility in which the starting aromatic ring gives rise to the C ring of the tricyclic intermediates. Certainly each of the above variations carries with it its own uniqueness; however, since most were not completed (to steroid products), they have not been detailed here.

Early work by Robinson and Walker also led to **27** (R = H, Me) by a somewhat more circuitous route [25,26]. Using the sodium salt of **26** as the nucleophile, reaction with **25**, prepared via a Grignard reaction between 1-(*m*-methoxyphenyl)-3-iodopropane and methyl chloroformate, provided the keto ester **24** in poor (25%, R = H) yield. Cyclization to the dione **27** was effectively (ca. 80%) performed with sodium ethoxide; cyclodehydration gave **28** in good yield. This material has also been prepared by the AB→ABC route and has been converted into various estrogens (cf. Scheme 3-12, Chapter 3). The conversion of the unsubstituted material (R = H) into the novel 15,16-diazasteroid **29** by hydrogenation, oxidation, and condensation with diethyl oxalate and phenylhy-

6.2 Synthesis from C-5, C-8 Bridged Intermediates

drazine was also described [26]; the stereochemistry was undetermined.

Also outlined in Scheme 6-3 is a method, due primarily to Renfrow *et al.* [27–29] and Barnes *et. al* [30–36], for the synthesis of nonaromatic ABC fragments. Again, only a brief, general recount of the procedure is given. Potassium *t*-butoxide catalyzed alkylation of **30** (or derivatives thereof) with the bromide **31** provided, in 68% yield, the expected disubstituted ethane which was decarboxylated by heating with sodium ethoxide. These AC intermediates were also formed via a Grignard reaction of appropriate derivatives of **30** and **31** [31–33]. The angular methyl group was introduced to give **32**, which was readily cyclized (83%) with anhydrous hydrogen fluoride. This material was converted into **33** by first demethylating the aromatic oxygen and then hydrogenating with Raney nickel and oxidizing with chromic acid. This product (the 3-methoxy derivative of **33**) was treated with hydrogen bromide in acetic acid, and then potassium hydroxide to give the free alcohol which was oxidized to **33**. Several of the pure stereoisomers of **33** were separated and identified [35,36], and a correlation of the compounds prepared by this route and those prepared by the BC→ABC route [37–39] was established. Further work in this area produced a variety of substituted ABC intermediates but the final elaboration of the D ring was, for the most part, not undertaken. It was also shown that 11-oxygenated derivatives of **33** were preparable by starting with appropriately substituted bromides **31** [30,32,35].

Perhaps the most interesting synthetic work conforming to the A+C mode is that reported for the preparation of azasteroids and is presented in Scheme 6-4. Several nitrogen-containing ABC tricyclic intermediates have been prepared [40–42] (including **42** which was obtained in over 40% yield from *m*-methoxyphenethylamine and methyl coumalate [43]), but the synthesis discussed here has been brought to fruition [44,45]. Reaction of *m*-anisidine (**34**) with ethyl orthoformate gave **35**, which, when heated with cyclohexane-1,3-dione yielded **36**. Polyphosphoric acid cyclization gave the phenanthridine **39** which, in a manner quite similar to that employed by Johnson in his equilenin synthesis (isoxazole method [15]; cf. Scheme 3-3, Chapter 3) was converted into **38**. To effect decarboxylation without isomerization of the C-14 double bond, it was first necessary to reduce the 17-ketone to the 17β-ol, whereupon hydrolysis and loss of carbon dioxide readily occurred to form *dl*-6-aza-17-dihydroequilenin methyl ether (**37**; this material has also been prepared by the AB+D route, Scheme 4-9, Chapter 4). Further transformations led to **40** and **41**, the former prepared simply by stereospecific Raney nickel hydrogenation of **37**. The latter was obtained by first reacting **37**

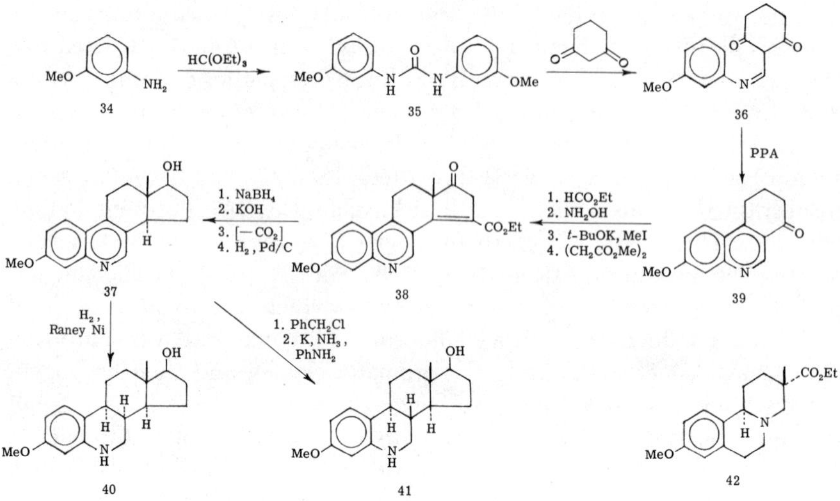

Scheme 6-4

with benzyl chloride and then reducing to the corresponding N-benzyl-6,7-dihydro material. Final reduction using potassium in liquid ammonia–aniline gave dl-6-azaestradiol 3-methyl ether (**41**).

References

1. W. S. Johnson, A. R. Jones, and W. P. Schneider, *J. Amer. Chem. Soc.* **72**, 2395 (1950).
2. W. S. Johnson, R. G. Christiansen, and R. E. Ireland, *J. Amer. Chem. Soc.* **79**, 1995 (1957).
3. D. L. Turner, *J. Amer. Chem. Soc.* **73**, 1284 (1951).
4. D. L. Turner, *J. Amer. Chem. Soc.* **73**, 3017 (1951).
5. G. I. Kipriyanov and L. M. Kutsenko, *Med. Prom. SSSR* **15**, 43 (1961).
6. W. S. Johnson and R. G. Christiansen, *J. Amer. Chem. Soc.* **73**, 5511 (1951).
7. B. K. Bhattacharyya and S. M. Mukherjee, *Sci. Cult.* **12**, 410 (1947).
8. B. K. Bhattacharyya and P. Sengupta, *Nature (London)* **170**, 710 (1952).
9. P. Sengupta and B. K. Bhattacharyya, *J. Indian Chem. Soc.* **31**, 337 (1954).
10. J. O. Jilek, V. Simak, and M. Protiva, *Collect. Czech. Chem. Commun.* **19**, 333 (1954).
11. W. S. Johnson, R. E. Ireland, and R. E. Tarney, (cited in) "Steroids," Fieser and Fieser, Reinhold Publ. Co., New York, N.Y., 1959, p. 500. 1959.
12. R. E. Tarney, *Diss. Abstr.* **21**, 2901 (1961).
13. D. K. Banerjee and K. M. Sivanandaiah, *Tetrahedron Lett.* p. 20 (1960).
14. D. K. Banerjee and K. M. Sivanandaiah, *J. Indian Chem. Soc.* **38**, 652 (1961).
15. W. S. Johnson, J. W. Petersen, and C. D. Gutsche, *J. Amer. Chem. Soc.* **69**, 2942 (1947).

References

16. A. J. Birch and H. Smith, *J. Chem. Soc., London* p. 1882 (1951).
17. A. J. Birch, H. Smith, and R. E. Thornton, *J. Chem. Soc., London* p. 1338 (1957).
18. J. C. Bardhan and S. C. Sengupta, *J. Chem. Soc., London* pp. 2520 and 2798 (1932).
19. J. A. Hogg, U. S. Patent 2,687,426 (1954); *Chem. Abstr.* **49**, 11717 (1955).
20. J. A. Hogg, *J. Amer. Chem. Soc.* **70**, 161 (1948).
21. J. A. Hogg, *J. Amer. Chem. Soc.* **71**, 1918 (1949).
22. M. Protiva, J. O. Jilek, E. Alderova, and L. Novak, *Experientia* **13**, 71 (1957).
23. E. Alderova, L. Novak, and M. Protiva, *Collect. Czech. Chem. Commun.* **23**, 681 (1958).
24. J. O. Jilek and M. Protiva, *Collect. Czech. Chem. Commun.* **23**, 692 (1958).
25. R. Robinson and J. Walker, *J. Chem. Soc., London* p. 192 (1936).
26. R. Robinson and J. Walker, *J. Chem. Soc., London* p. 747 (1936).
27. W. B. Renfrow, A. Renfrow, E. Shoun, and C. A. Sears, *J. Amer. Chem. Soc.* **73**, 317 (1951).
28. W. B. Renfrow and J. W. Cornforth, *J. Amer. Chem. Soc.* **75**, 1347 (1953).
29. W. B. Renfrow and A. Renfrow, *J. Amer. Chem. Soc.* **76**, 2268 (1954).
30. R. A. Barnes, H. P. Hirschler, and B. R Bluestein, *J. Amer. Chem. Soc.* **74**, 32 (1952).
31. R. A. Barnes and R. T. Gottesman, *J. Amer. Chem. Soc.* **74**, 35 (1952).
32. R. A. Barnes, H. P. Hirschler, and B. R. Bluestein, *J. Amer. Chem. Soc.* **74**, 4091 (1952).
33. R. A. Barnes and M. D. Konort, *J. Amer. Chem. Soc.* **75**, 303 (1953).
34. R. A. Barnes, *J. Amer. Chem. Soc.* **75**, 3004 (1953).
35. R. A. Barnes and A. H. Sherman, *J. Amer. Chem. Soc.* **75**, 3013 (1953).
36. R. A. Barnes and M. T. Beachem, *J. Amer. Chem. Soc.* **77**, 5388 (1955).
37. J. W. Cornforth and R. Robinson, *J. Chem. Soc., London* p. 676 (1946).
38. J. W. Cornforth and R. Robinson, *J. Chem. Soc., London* p. 1855 (1949).
39. A. L. Wilds, J. W. Ralls, W. C. Wildman, and K. E. McCaleb, *J. Amer. Chem. Soc.* **72**, 5794 (1950).
40. R. H. Wiley, N. R. Smith, and L. H. Knabeschuh, *J. Amer. Chem. Soc.* **75**, 4482 (1953).
41. N. A. Nelson and Y. Tamura, *Can. J. Chem.* **43**, 1323 (1965).
42. G. V. Bhide, N. L. Tikotkar, and B. D. Tilak, *Tetrahedron* **10**, 230 (1960).
43. N. A. Nelson and J. M. Schuck, *Can. J. Chem.* **43**, 1527 (1965).
44. J. H. Burckhalter and H. Watanabe, *Abstr. 143rd Nat. Meet., Amer. Chem. Soc.* p. 14a (1963).
45. J. H. Burckhalter and H. Watanabe, *Abst., Int. Symp. Chem. Natur. Prod., 1964* p. 253 (1964).

7
A + CD → ACD → ABCD

This chapter is divided into six portions according to the type of reaction used in the condensation of the A fragment with the CD portion. The first section concerns the reaction between an anionic A ring with a keto function in the CD portion and discusses, for the most part, Johnson's first estrone synthesis. In Section 7.2 the process is reversed such that an anion formed from the CD portion reacts with A; Section 7.3 discusses the use of the Wittig reaction in steroid total synthesis and includes some recent, promising work by a Russian group as well as brief mention of Inhoffen's classic work in the vitamin D field (a detailed presentation of the latter, as well as some recent work by Lythgoe on the synthesis of calciferol, has been omitted because the products are, for the present purposes, not considered true steroids). Section 7.4 discusses the recent, intriguing use of CD enamines in steroid total synthesis, and Section 7.5 presents the preparation of some thiasteroids. The last section describes some brief, miscellaneous schemes which have been reported in, as yet, little detail.

7.1 Condensation of Ring-A Anionic Fragments with CD Ketones

The first steroid total synthesis of Johnson resulted in the preparation of seven of the eight possible racemic stereoisomers of estrone and was

7.1 Condensation of Ring-A Anionic Fragments with CD Ketones 143

exceedingly important in correlating and assigning configurations to the many estrone isomers previously reported by others. The basic plan is outlined in Scheme 7-1 [1–3].

m-Methoxyacetylene (prepared from *m*-methoxyacetophenone) condensed with decalin-1,5-dione (cf. Chapter 12, Section 12.1; in this instance it made little difference whether **2** was pure *cis* or pure *trans* or a mixture) in the presence of potassium *t*-butoxide to give the inter-

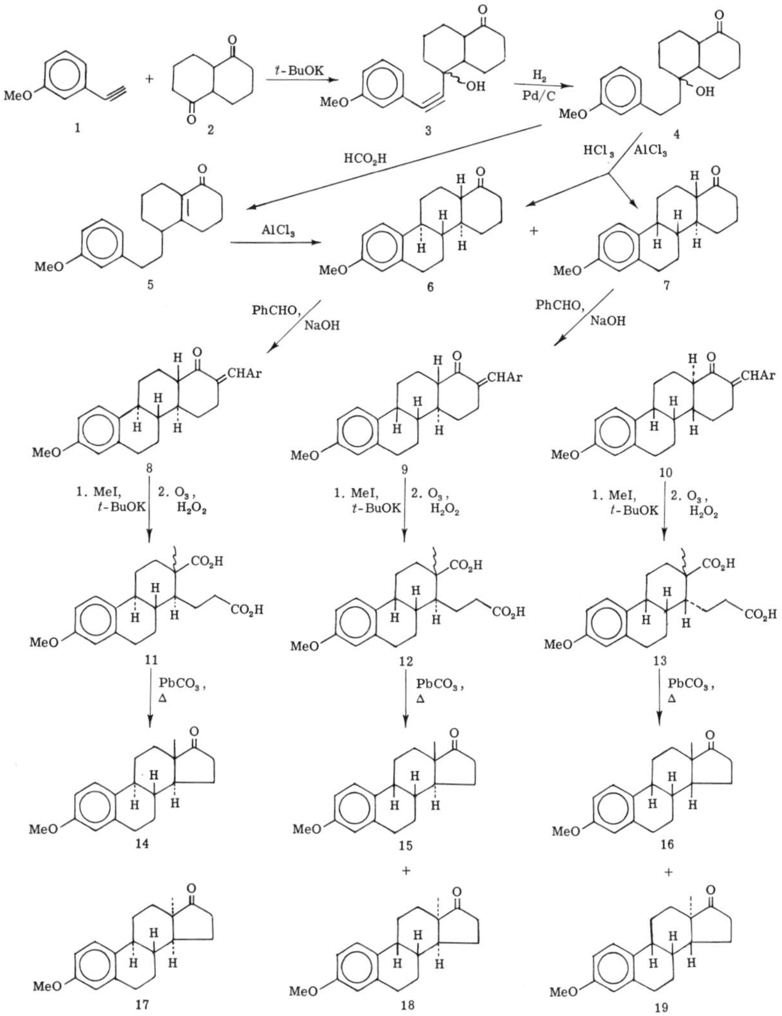

Scheme 7-1

mediate acetylene **3**[*]. The product (50–75%) could be separated into two isomeric forms, presumably epimeric at C-8, and possessing CD *trans*; however, it was more convenient to hydrogenate the acetylene over palladium–on–charcoal to give the epimeric mixture **4** (ca. 90%). At this stage two different routes were followed. The mixture of saturated alcohols could be dehydrated with formic acid to provide the racemic α,β-unsaturated ketone **5** in good yield (as high as 70% from **2**). This material was, in turn, cyclized in the presence of aluminum chloride to a mixture of **6** and **7**. Alternatively, **6** and **7** could be produced directly from **4** by aluminum chloride catalyzed cyclization with or without added hydrochloric acid. The results obtained in this cyclization varied appreciably with the conditions employed. In the direct conversion of **4**, the major product, almost to the exclusion of the second, was the *cis-anti-trans* **7** (isolated in overall yield of 25% from **2**). However, if dehydration to **5** was first performed, cyclization gave appreciable quantities of the natural *trans-anti-trans* **6**. Initially this latter sequence gave a 2:1 mixture of **7**:**6** [3,5], but later experiments [6] showed that **6** (and its demethylated analog) could be made the only readily isolable product (ca. 10%). It was felt that the purity of the aluminum chloride was a major factor in determining the outcome of the reaction.

A third configurational isomer in this series, **10** (Ar = 2-furanyl), was prepared in approximately 25% yield from **7**. Thus, **7** was converted to the enol acetate, brominated and dehydrobrominated to give 13,14-dehydro-**7** (40%). Catalytic hydrogenation provided the *cis-syn-trans* isomer, presumably as a result of *cis-β* hydrogenation followed by base-catalyzed epimerization at C-13; the material was condensed with furfuraldehyde to provide **10** [5]. The analogous benzylidene derivatives **8** and **9** (Ar = phenyl) were also readily prepared. Angular methylation of **8**, **9**, and **10** by standard methods provided, in each case, a mixture of C-13α-methyl and C-13β-methyl derivatives. With **8**, the α-methyl compound predominated by 50:17%; for **9**, the yields were 56:13%; but for **10**, the ratio was reversed (7:78%). This latter, apparent anomaly was explained by noting the increased steric hindrance to α-approach of methyl iodide to the anion of **10**. In practice, each of these mixtures was separated at this point and oxidized to the dibasic acids **11**, **12**, and **13**. Cyclization by pyrolysis over lead carbonate gave the estrone isomers (as their methyl ethers) **14–19**. Representative yields in these steps were

---

[*] Lythgoe and his co-workers have recently employed a similar reaction in their extensive researches in the total synthesis of calciferol and its relatives. They utilized lithium acetylides rather than the current potassium salts [4].

7.1 Condensation of Ring-A Anionic Fragments with CD Ketones

8→11 (51%); 11→14 and 17 (46–82%). More specifically, *dl*-estrone methyl ether (14) was prepared in an overall yield of 0.27% based on starting decalin-1,5-dione (2); the yield of lumiestrone (17) approached 2%. As *dl*-14-isoestrone had been prepared by an alternate approach (A+C), Johnson and his associates accounted for all but one of the eight possible racemic isomers of estrone (the products in Scheme 7-1 were demethylated by heating with pyridine hydrochloride) and, although several of the steps in this synthetic approach lacked the excellent yields he and others later obtained in alternate total syntheses, this work provided a much needed, sound stereochemical basis for steroid chemists to build on.

This preparative sequence was further utilized for obtaining *dl*-18-norestrone (Scheme 7-2) [7]. For this synthesis, the *trans-anti-trans* 8 (Ar = 2-furanyl) was subjected to ring contraction without first undergoing angular methylation. Ozonolysis followed by periodic acid decomposition and esterification gave 20 in over 80% yield. This *dl*-18-norhomomarrianolic acid diester readily underwent Dieckmann cyclization to 21; however, unsatisfactory results were obtained on attempted acid hydrolysis and decarboxylation. The conversion to 22 (56%) and 23 (28%) was conveniently effected by heating to 186° in aqueous triethylene glycol. Both materials were then demethylated with pyridine hydrochloride.

Scheme 7-2 also illustrates one further conversion in this series. In this instance, the D-homo-18-norestrone ether 6 was converted to *dl*-18,19-

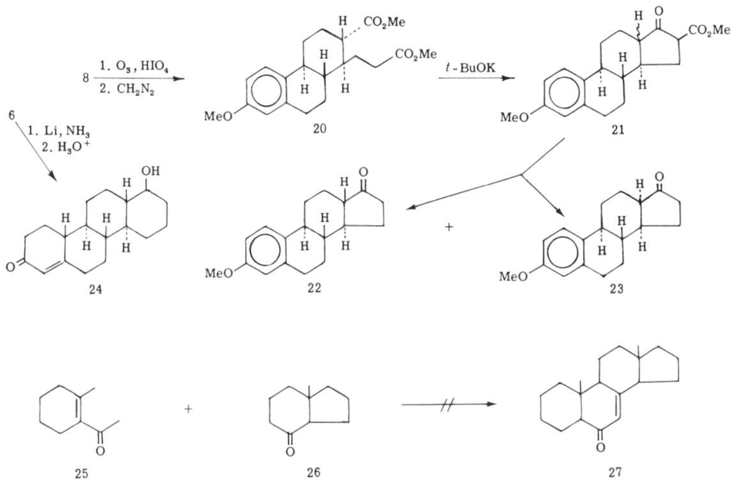

Scheme 7-2

bisnor-D-homotestosterone **24** in a 61% yield by Birch reduction and hydrolysis [8].

Before leaving this section, mention should be made of a few attempts to apply the Robinson-Rapson synthesis (cf. Chapter 4, Section 4.7) to the A+CD plan. Bagchi and Banerjee [9] proposed **27** to result from the isopropoxide catalyzed condensation of **25** and **26**. Additionally, Dimroth [10] believed he obtained an analogous D-homo product on starting with the decalone relative of **26**. Potentially a very good method for total synthesis, it was soon shown to be invalid. The actual product obtained was, in fact, not **27** but instead the corresponding 9,10-secosteroid [11]; the reaction thus gave the product expected from a normal aldol condensation. Presumably the C-10 methyl group obviates the possible Robinson-Rapson reaction that occurs in its absence [12].

7.2 Alkylation of an Anionic CD Fragment

This particular route to totally synthetic steroids has been employed, in variously modified forms, by an exceptionally large number of workers, but a correspondingly large number and variety of steroids have not resulted. The general route was utilized by Birch et al. [13,14] in their preparation of D-homosteroids. Scheme 7-3 indicates the preparation of **31** starting from the tetralone **28** and *m*-methoxyphenethyl bromide (**29**). Subsequent cyclization with polyphosphoric acid at 70° gave approxi-

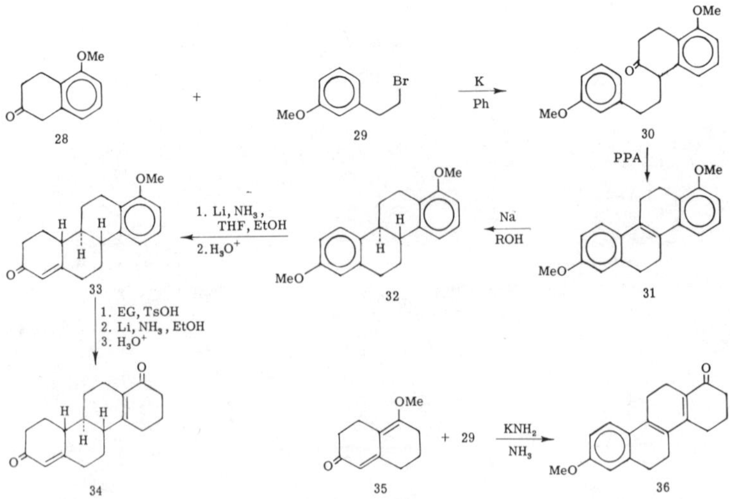

Scheme 7-3

7.2 Alkylation of an Anionic CD Fragment

mately 50% of the hydrochrysene derivative. The same condensation was also carried out less successfully using potassium t-butoxide as the base in place of metallic potassium [15]. Reduction of **31** with sodium in refluxing n-butanol provided a mixture (ca. 50%) of **32** and its *cis* isomer in a 4:1 ratio. Further reduction by the Birch method was performed in a stepwise manner such that **33** could be isolated (56%). Conversion to the ketal and treatment with 300 g-atoms of lithium (instead of the above used 60 g-atoms) gave, after hydrolysis, the D-homobisnor steroid **34**, the yield being less than 10%. Further conversion to the 13-dehydro-17aβ-alcohol also proceeded poorly [16]. By employing an alternate CD fragment (**35**), this same format provided an unsatisfactory yield of **36** (17%) [17].

The soundness of the general alkylation procedure having been proved (albeit, in less than outstanding yields), the principle was then applied in a pleasingly short synthesis of estrone [18–22]. Crispin and Whitehurst made use of the indane derivative **37**, shown in Scheme 7-4, as the tetrahydropyranyl ether [23,24]. The potassium enolate thereof was alkylated with **38** to give, after hydrolysis and oxidation, the dione **39**. The tosylate of **38** (X=OTs) was a more effective alkylating reagent than the bromide (32% versus 17%), the major competing reaction in both cases being elimination to m-methoxystyrene. Smith, Hughes, and McLoughlin [21,22] reported the corresponding ketone (**37**, OR=ketone) could be alkylated with **38** (X=Br); attempts to repeat this experiment were largely unsuccessful [20], although the tosylate did give some product (15%).

Interestingly, the dione, **39** could be converted to isoequilenin methyl ether (**40**, 25%) by refluxing with p-toluenesulfonic acid in benzene or, alternatively, into 3-methoxy-1,3,5(10),8,14-estrapentaene-17-one (**41**; 51%)

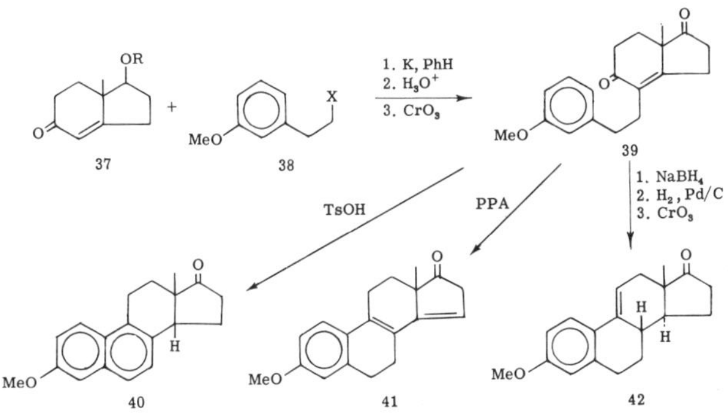

Scheme 7-4

by warming with phosphorus pentoxide and phosphoric acid. This latter material had previously been prepared by an alternate route and converted to estrone (cf. Chapters 4 and 8). If the dione **39** was reduced to the 17β-ol with borohydride, hydrogenated over palladium-on-charcoal, and then reoxidized, acid-catalyzed cyclization produced a 60% yield of 9(11)-dehydroestrone methyl ether (**42**). The latter, on metal–ammonia reduction followed by oxidation, gave the methyl ether of racemic estrone. Additionally, the analogous series of reactions starting with $\Delta^{5,10}$-9-methyloctalin-1,6-dione in place of **37** gave rise to various D-homosteroids.

The major drawback to the above sequence would appear to be the poor yields obtained in the alkylation step. An in depth study of this reaction has been reported [20] and some of the controlling factors delineated. And as will be seen below, various modifications have led to definite improvements.

Two groups have reported the total synthesis of 8β-substituted steroids by the CD + A route. Scheme 7-5 depicts a very nice synthesis of such B-nor-D-homo- and D-homoestranes. The starting material **43** was readily prepared in 48% yield starting from 2-methylcyclohexane-1,3-dione and ethyl vinyl ketone [25]. The very stable anion of **43** was generated with sodium hydride in dimethoxyethane and alkylated in good yield to provide **44** (84%) or **49** (85%) [26,27]. Use of the phenacyl derivative rather than a phenethyl compound (which was tried but gave no product) circumvented one of the major problems previously encountered in this type of synthesis, i.e., the elimination of HX to form a styrene derivative (*vide supra*). Of course, the latter possibility is obviated with *m*-methoxybenzyl chloride. Both halides, as well as acrylonitrile [25], added stereospecifically from the α-side to provide only the β-methyl orientation (analogous results are recorded in Scheme 7-6). The blocking group was then removed from **44** with base in aqueous ethanol, and subsequent acid hydrolysis of the ketal provided **45** in 93% yield. Similar basic treatment of **49** was more complex and two products, **50** (18%) and **51** (62%), were produced. The phenacyl carbonyl group was thought to participate in formation of an intermediate cyclic ketol which was resistant to cleavage; however, more extensive treatment with aqueous base converted **50** into **51** [28]. Hydrogenolysis of **51** over 5% palladium hydroxide-on-carbon in acetic acid solution followed by further acid treatment (to remove the 17a-ketal) gave a nearly quantitative yield of **53**. Less vigorous reducing conditions (5% palladium-on-carbon; methanol solvent) gave the tricyclic furan derivative **52** (R = H and R = Me) which could also be converted to **53**.

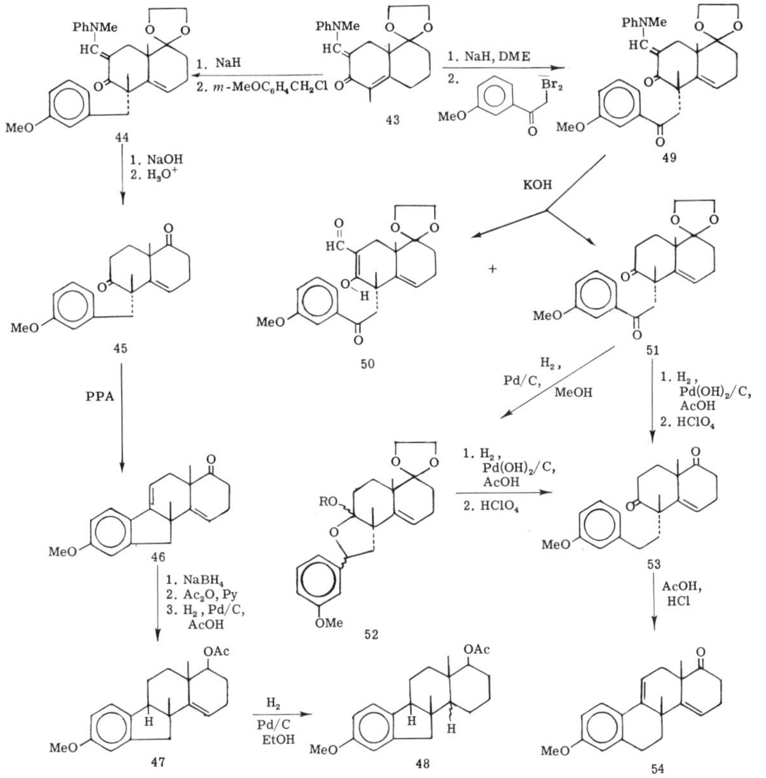

7.2 Alkylation of an Anionic CD Fragment

Scheme 7-5

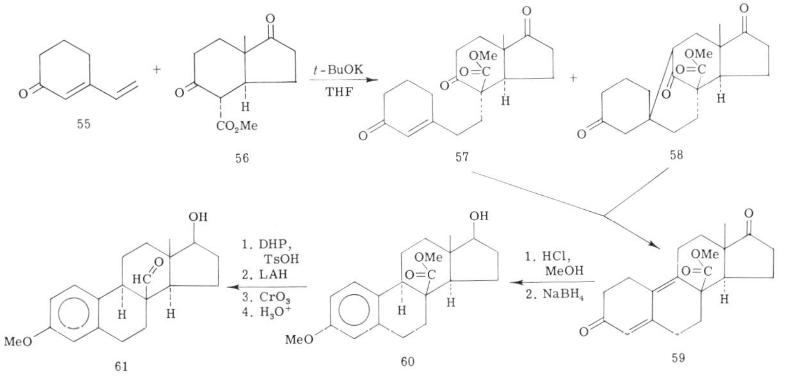

Scheme 7-6

The steroid precursors **45** and **53** were cyclized to **46** and **54**, respectively, in excellent yields. Interestingly, polyphosphoric acid was the reagent of choice for performing this conversion in the B-nor series, while the same reagent was much inferior to acetic acid-hydrochloric acid in the latter case. Overall, **46** and **54** were realized in yields of 25–30% starting from 2-methylcyclohexane-1,3-dione.

Also of note in this synthesis was the catalytic reduction of the 17aβ-acetate of **46** (prepared by borohydride reduction and acetylation). Thus, hydrogenation over 5% palladium-on-charcoal for a short time in acetic acid gave, stereospecifically, the 9β material **47** in 87% yield. This β-face hydrogenation occurred in spite of the adjacent 8β-methyl substituent. Further reduction under more forcing conditions led to **48** as a mixture of the 14α (22%) and 14β (49%; isolated as the free 17aβ-ol) isomers.

Another, closely related, synthesis of 8β-substituted steroids has been reported by Sakai and Amemiya [29]. Michael addition of the anion of **56** to the dienone **55** (Scheme 7-6) gave an 81% yield (based on **56** consumed) of what proved to be a mixture of **57** and **58**. The proportions of the two depended on conditions employed, and it was felt that **57** was the progenitor of **58**, as the latter was the sole isolated product obtained on prolonged reaction time. For preparative purposes, however, it was a moot point, since both **57** and **58** individually or together, gave the same result. The conjugated diene **59** was afforded (70%) from the mixture on treatment with potassium carbonate in methanol. It was then possible to effect A-ring aromatization by refluxing with 10% hydrochloric acid in methanol; borohydride reduction gave 8β-carbomethoxy-estra-3,17-diol 3-methyl ether (**60**) in good yield. Final conversion to the 8β-formyl derivative (**61**) was accomplished by hydride reduction of the tetrahydropyranyl ether of **60**, Collins' oxidation, and hydrolysis. Overall, a yield of about 10% **61** was realized from the mixture of **57** and **58**.

One final report on total synthesis by this route deserves mention and is outlined in Scheme 7-7. The uniqueness of this approach lies in the use of 6-vinyl-2-picoline (**62**) as the masked equivalent of two moles of methyl vinyl ketone. Building on the known conversion of the pyridine ring system into cyclohexenones by a sequence involving Birch reduction to dihydropyridines and subsequent hydrolysis [30], Danishefsky *et al.* [31–34] were able to prepare the D-homosteroid **65** in a minimum number of steps. The ketal ketone **63** underwent *t*-pentoxide catalyzed alkylation with the vinyl picoline **62** to provide, after hydrolysis, a 70% yield of **64**. Reduction, hydrogenation, and ketalization provided (28%) the steroid precursor **67**. The disappointing yield of this transformation was partially compensated by the excellent (90%) conversion of **67** to **66**

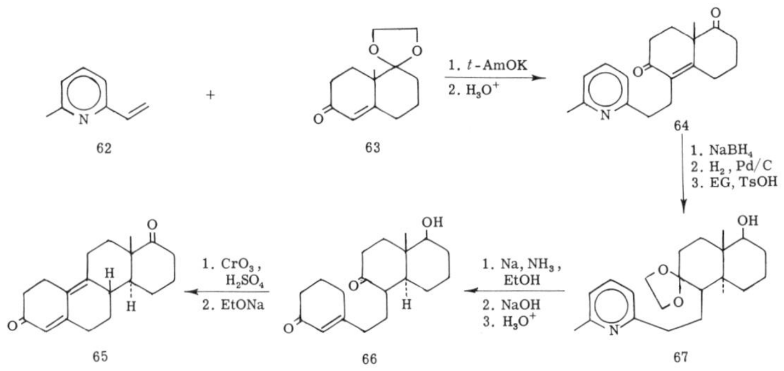

7.3 The Wittig Reaction in Total Synthesis

Scheme 7-7

by sodium–ammonia reduction, hydrolytic cyclization with base, and deketalization. The final cyclization was accomplished, after oxidation to the 17a-ketone, with sodium ethoxide in ethanol. The overall yield of **65** (which had previously been isomerized to D-homoestrone [22]) from the CD fragment **63** was 8%. Interestingly, the shortcoming of this route lies not in the novel pyridine-to-cyclohexenone conversion but, instead, in the reduction–hydrogenation–ketalization sequence (**64→67**).

7.3 The Wittig Reaction in Total Synthesis

Since its introduction nearly twenty years ago, the Wittig reaction has found extensive use in modifying steroids and their functional groups, but little effort has been devoted to its utilization in construction of the steroid skeleton. Inhoffen [35,36], in his studies on vitamin D derivatives, and Lythgoe [37,38], in analogous work on the synthesis of calciferol and its derivatives, briefly studied the use of appropriate phosphoranes in preparing secosteroids; however, they found other modes of attack more satisfactory. Interest in the reaction, as applied to steroid total synthesis, has been revived by a Russian group, and their work, culminating in the total synthesis of dl-estrone methyl ether, is presented in Scheme 7-8.

The starting material (**68**; R = Me, Et) was obtained in virtually quantitative yield by hydrogenation of the corresponding 8(14)-olefin over 10% palladium-on-calcium carbonate (cf. Chapter 12, Section 12.1) [39,40]. To prepare the necessary aldehyde **69**, the two ketone groups were protected as the ethylene ketals and the ester was reduced to the primary alcohol. Oxidation with chromic anhydride and pyridine provided crystalline **69** (R = Me, 62%; R = Et, 32%). The phosphonium salt, formed from *m*-methoxybenzyl chloride and triphenylphosphine, was converted to the

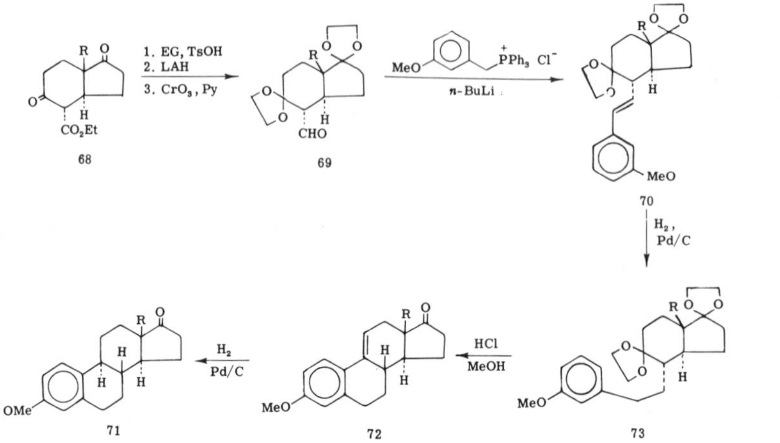

152   7 A+CD → ACD → ABCD

Scheme 7-8

ylid with n-butyllithium and combined with the aldehyde to yield the pure trans **70** (R=Me, 65%, R=Et, 46%); none of the cis isomer was detected. This olefin was hydrogenated to **73** and, on treatment with methanolic hydrochloric acid, underwent simultaneous ketal hydrolysis and ring closure to dl-3-methoxy-1,3,5,9(11)-estratetraen-17-one (**72**). (The analogous reaction sequence was also performed with p-methoxybenzyl chloride as one of the reactants; however, the final cyclization to provide the 2-methoxy analog of **72** did not readily occur [41]). Stereoselective hydrogenation then gave estrone methyl ether and its 18-methyl homolog; these two racemic estrogens were realized in yields of 32 and 5% overall from the trans hydrindane **68**. Further elaboration of this scheme should provide a greater variety of products.

7.4 CD Enamine Intermediates

The use of enamines and dienamines in total synthesis, pioneered by the Amsterdam group of Pandit, Huisman et al., has been reviewed in the context of heterocyclic steroid synthesis [42] (cf. Chapter 4, Section 4.2). The crucial intermediate dienamine **74** (Scheme 7-9) was readily obtained from the corresponding α,β-unsaturated ketone and pyrrolidine; however, **74** (n=1, R=O) was quite unstable and, in contrast to the D-homo analog, was difficultly purifiable. Therefore, the 17-ketone was reduced and the resulting alcohol acetylated to give the more easily handled **74** (n=1, R=β-OAc, α-H). This material, when treated with m-methoxybenzenediazonium fluoroborate (**75**), yielded a variety of products de-

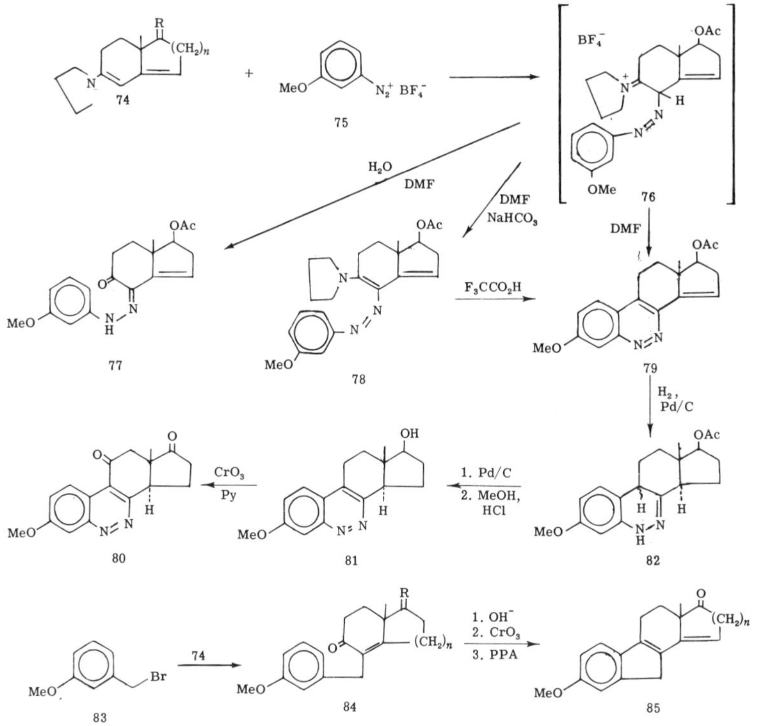

7.4 CD Enamine Intermediates

Scheme 7-9

pending upon the reaction conditions and work-up procedures employed [43–45]. Thus, **74** ($n=1$, R = β-OAc, α-H) and **75** were allowed to react in anhydrous methylene chloride at $-78°$. Careful hydrolysis with aqueous dimethylformamide gave two hydrazones, **77** (37%, felt to exist mainly in the enolic form) and a similar product resulting from electrophilic attack at C-15 instead of C-8 (40%, this material was actually isolated as the unhydrolyzed iminium salt). Alternatively, if anhydrous dimethylformamide was added and the resulting mixture warmed, the diazasteroid **79** was obtained directly in 17% yield. Furthermore, if the above dichloromethane–dimethylformamide reaction mixture was washed with sodium bicarbonate before warming, the isolated product (12%) was the azo material **78** rather than the steroid (it could, however, be converted into **79** in excellent yield by warming with trifluoroacetic acid). These interesting results, as well as comparable ones in the D-homo series, were felt to derive from the common intermediate **76**.

Thus, aqueous treatment hydrolyzed **76** to the corresponding ketone; whereas bicarbonate treatment removed the C-8 proton to give **78**;

warming in dimethylformamide allowed intramolecular nucleophilic attack by the aryl group to give, after loss of pyrrolidine, the diazasteroid. That the hydrazone **77** was not a precursor of **79** was indicated by the failure of the former to undergo acid-catalyzed cyclization to the latter; in fact, on acid treatment a unique rearrangement occurred to give benzoquinoline derivatives [44].

Further transformations in this series led to 6,7-diaza-11-oxoequilenin methyl ether (**80**). Catalytic hydrogenation of **79** reduced the Δ^{14}-bond and, additionally, gave partial reduction of the B ring. The product **82** was obtained in nearly quantitative yield and was dehydrogenated and converted to the alcohol **81** in approximately 50% yield. Chromium trioxide in pyridine provided a moderate yield (34%) of the diketone **80**, none of the 17-monoketone being isolated.

A further synthetic effort utilizing **74** as a starting material has also been reported [46]. In this instance the alkylating agent was m-methoxybenzyl bromide (**83**), and the product, **84** (n=1, R=β-OAc,α-H), was obtained in 85% yield. However, various attempts at cyclization were unsuccessful and, consequently, the 17-keto derivative was prepared. This material underwent ring closure with phosphoric acid–phosphorus pentoxide to provide the B-norsteroid **85** ($n=1,2$) in moderate yield.

7.5 Thiasteroids

One of the more novel approaches to heterocyclic steroids is outlined in Scheme 7-10, but has not been extensively developed [47–49]. The starting epoxide **86** was formed from the corresponding α,β-unsaturated ketone by hydrogen peroxide oxidation, and was condensed with the appropriate thiol **87**. At a pH of 7.5 the two combined to form crystalline **88** ($n=0,1,2$) in nearly quantitative yield. Dehydration to **89** ($n=0,1,2$,) followed immediately upon raising the pH of the solution above 8. An appreciable difference in reactivity of this vinyl thioether was noted depending on the value of n. Cyclodehydration of **89** ($n=0$) [or **88** ($n=0$)] with aluminum chloride in methylene chloride gave **92** in 48% overall yield from **86** [47]. From this it was possible to prepare the isomeric thiasteroids **94** and **95**, the latter (68%) by catalytic hydrogenation of **92** over palladium-on-calcium carbonate. To form **94**, it was necessary to first reduce the 17-keto group of **92** to the 17β-ol (sodium borohydride) and then hydrogenate. Reoxidation with dimethyl sulfoxide–dicyclohexylcarbodiimide afforded the CD-*trans* **94**. Both **94** and its preceding 17β-ol were demethylated (pyridine hydrochloride and boron tribromide-methyl-

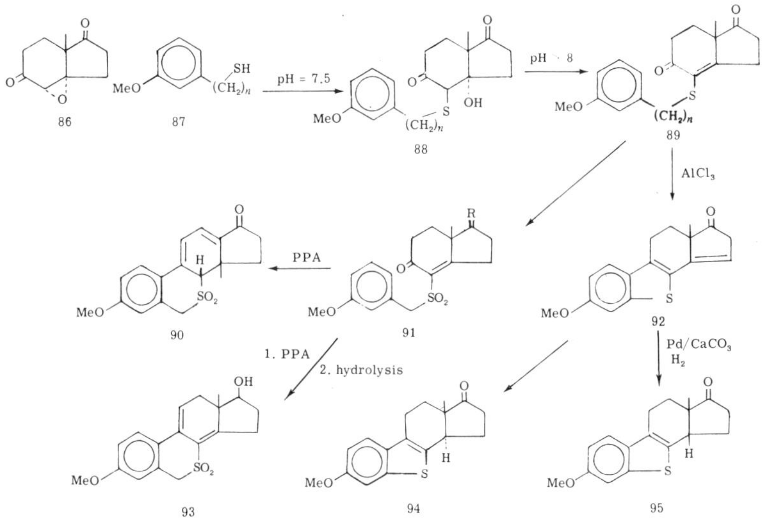

7.5 Thiasteroids

Scheme 7-10

ene chloride, respectively) to the corresponding 3-hydroxy compound. Thus, dl-B-nor-6-thiaequilenin (demethyl 94) was prepared from 86 in 10% overall yield [47]; similar compounds have been prepared starting from benzothiophene (cf. Chapter 3, Section 3.6.2).

Whereas the above cyclodehydration was readily effected in good (60%) yield, all attempts to perform a similar conversion on 89 ($n=1$) were negative. The tentative explanation advanced for this phenomenon invoked delocalization of the incipient cation by the vinyl thioether system to such an extent that nucleophilic attack by the aromatic ring was unfavorable. With the B-nor compound, formation of the aromatic benzothiophene system was sufficient driving force to overcome this detraction. Interestingly, it was possible to prepare the predicted cyclization product by first reducing 89 ($n=1$) to the 9ξ,17β-diol followed by treatment with refluxing acetic acid. The product, the 3-methyl ether of 7-thia-9ξ-estra-1,3,5(10),8(14)-tetraenediol 17β-acetate, was obtained in good yield. To decrease the above potential for charge delocalization, oxidation to the sulfone 91 (R=O) was carried out (hydrogen peroxide–acetic acid). Subsequent treatment with polyphosphoric acid indeed gave a cyclized product; but rather than the expected one, the rearranged 90 was isolated (10%) [49]. If, however, the 17-ketone was reduced and the resulting alcohol acetylated to give 91 (R=β-OAc,αH), polyphosphoric acid cyclization gave a good yield, after hydrolysis, of 93 in which methyl migration had not occurred.

Quite obviously, several details of this approach remain to be elucidated; but the merit has been proved and, in fact, extension to other heterocyclics would seem possible.

7.6 Miscellaneous Routes

The Diels-Alder reaction in total synthesis has been discussed at some length in Chapter 5 and will be only briefly mentioned here, as work in this area has been minimal. 4-Vinylindane (**96**) was cycloadded to several dienophiles, one example of which is illustrated in Scheme 7-11 [50,51]. These reactions were carried out in sealed tubes and the yields were only moderate (32% when benzoquinone was used in place of **97**). By using an excess of the quinone, the initial cycloadduct was partially dehydrogenated *in situ* to give **98** in which the A-ring substituent was only tentatively assigned the 3-position. Unfortunately, 4-vinylindane did not react with 5-methoxy-2-methylbenzoquinone; this product would presumably have possessed the C-10 angular methyl group [52]. In an effort to form more highly saturated steroid derivatives, the Spanish group employed **99** as the diene [53,54]. Thus, **100** was prepared in roughly 20% yield and, following borohydride reduction, dehydration, and hydrogenation, the 17-deoxyestrone methyl ether **101** was isolated. The CD ring fusion was felt to be *cis*, but the configuration at C-8 was uncertain.

Another fascinating preparation, based on the von Pechmann coumarin synthesis [55], is exemplified by the production of **104** [56]. Condensa-

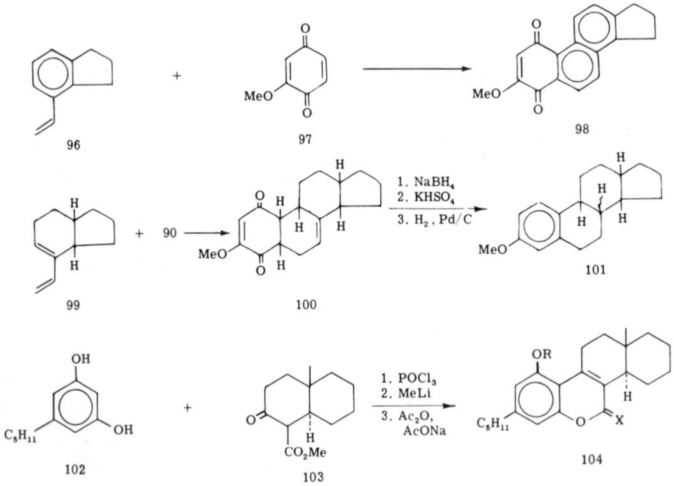

Scheme 7-11

7.6 Miscellaneous Routes

tion of olivetol (**102**) with **103** [57] in the presence of phosphorus oxychloride led to a 51% yield of the pyrone **104** (R=H, X=O). Treatment with methyllithium, followed by acetic anhydride–sodium acetate gave **104** (R=Ac, X=Me$_2$) in 34% yield. This material, a steroidal analog of tetrahydrocannabinol, was not prepared in the context of steroid total synthesis; however, it would seem feasible, by appropriate modifications of the A and CD fragments, to utilize the reaction as a convenient route to 6-oxaestranes.

Scheme 7-12 portrays a unique reaction sequence leading to a 5,10-diazasteroid [58]. The starting material **105** (R=Me, R'=Ac) was prepared via methyl bromoacetate alkylation of the dienamine **74** (R=β-OAc,α-H; Scheme 7-9). Hydrolysis to the keto acid (**105**; R=R'=H) and lactonization with acetic anhydride produced a 30:70 mixture of the nonseparable lactones **106** and **107** in excellent yield. This mixture was then refluxed in benzene with an equivalent amount of the tetrahydropyridazine **110**. This latter material resulted from a Diels–Alder reaction between 2,3-dimethoxy-1,3-butadiene and diethyl azodiformate and subsequent basic hydrolysis–decarboxylation. The diazasteroid **109** was isolated directly as a crystalline solid in 20% yield. The secosteroid **108** was the other major product; and, interestingly, it

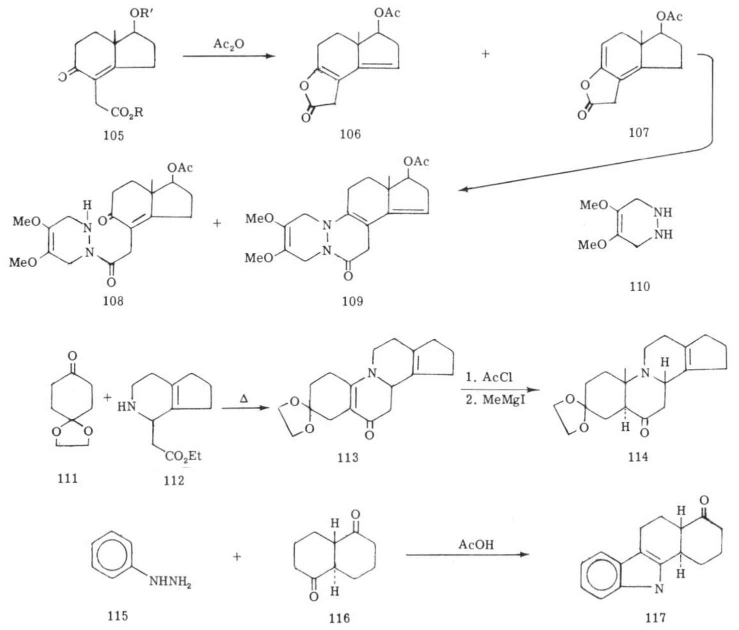

Scheme 7-12

could not be cyclized to **109** under a variety of conditions. These results were rationalized in the following manner. The pyridazine reacted with lactone **106** to give an intermediate β,γ-unsaturated ketone, whereas reaction with lactone **107** provided an α,β-unsaturated ketone (i.e., **108**). The lowered reactivity of the conjugated carbonyl group in this latter instance (as well as some possible unfavorable steric considerations) was enough to preclude cyclization. On the other hand, unconjugated ketones similar to the Δ^{14}-isomer of **108** are known to undergo facile cyclodehydration [59]. Presumably, then, by starting with pure **106**, the yield of diazasteroid could be increased considerably. The synthetic scheme appears promising.

Brief note should be made of the preparation of 9-azasteroids realized by Meyers and his co-workers [60,61]; as the method is essentially only a slight modification of Meyers' more extensive effort for preparing 8-azasteroids (cf. Scheme 4-22, Chapter 4), it receives only cursory treatment here. Combination of **111** and **112** (prepared from cyclopent-1-enylethylamine and diethyl malonate) in refluxing toluene provided **113**. Introduction of the C-10 methyl group was accomplished by first forming the O-acetyl derivative with acetyl chloride; the resulting $\Delta^{5,9}$-diene immonium salt was reacted with methyl Grignard affording an oil from which **114** could be separated. Also prepared were the analogous D-homo derivatives as well as those with an aromatic D ring and those lacking 3-substituents.

Finally, a reported synthesis of the 6-aza-B-nor-D-homosteroid **117** is presented [62]. Simple heating of *trans*-decalin-1,5-dione with phenylhydrazine provided **117** in unspecified yield together with several other components. Tilak and his collaborators have also been responsible for several similar heterocyclic B-norsteroids as discussed in Chapter 3 (Scheme 3-19; see also Scheme 4-27, Chapter 4). Other brief reports delineating pathways to cyclopentanophenanthrenes by the CD + A route have appeared [63,64], but they have little applicability to actual steroid total synthesis.

References

1. W. S. Johnson, D. K. Banerjee, W. P. Schneider, and C. D. Gutsche, *J. Amer. Chem. Soc.* **72**, 1426 (1950).
2. W. S. Johnson and L. J. Chinn, *J. Amer. Chem. Soc.* **73**, 4987 (1951).
3. W. S. Johnson, D. K. Banerjee, W. P. Schneider, C. D. Gutsche, W. E. Shelberg, and L. J. Chinn, *J. Amer. Chem. Soc.* **74**, 2832 (1952).
4. T. M. Dawson, J. Dixon, P. S. Littlewood, B. Lythgoe, and A. K. Sakena, *J. Chem. Soc., London* p. 2960 (1971), and other articles in this series.

5. W. S. Johnson, I. A. David, H. C. Dehm, R. J. Highet, E. W. Warnhoff, W. D. Wood, and E. T. Jones, *J. Amer. Chem. Soc.* **80**, 661 (1958).
6. W. L. Meyer, D. D. Cameron, and W. S. Johnson, *J. Org. Chem.* **27**, 1130 (1962).
7. K. H. Loke, G. F. Marrion, W. S. Johnson, W. L. Meyer, and D. D. Cameron, *Biochim. Biophys. Acta* **28**, 214 (1958).
8. W. S. Johnson, H. C. Dehm, and L. J. Chinn, *J. Org. Chem.* **19**, 670 (1954).
9. P. Bagchi and D. K. Banerjee, *J. Indian Chem. Soc.* **23**, 397 (1946).
10. K. Dimroth, *Angew. Chem.* **59**, 215 (1947).
11. W. S. Johnson, J. Szuszkovicz, and M. Miller, *J. Amer. Chem. Soc.* **72**, 3726 (1950).
12. W. S. Rapson and R. Robinson, *J. Chem. Soc., London* p. 1285 (1935).
13. J. F. Collins and H. Smith, *J. Chem. Soc., London* p. 4308 (1956).
14. A. J. Birch and H. Smith, *J. Chem. Soc., London* p. 4909 (1956).
15. I. N. Nazarov and S. I. Zav'yalov, *Izv. Akad. Nauk SSSR, Otd. Khim. Nauk* p. 1233 (1958); *Chem. Abstr.* **53**, 4242 (1959).
16. A. J. Birch, G. A. Hughes, and H. Smith, *J. Chem. Soc., London* p. 4774 (1958).
17. A. J. Birch, M. Kocor, and D. C. C. Smith, *J. Chem. Soc., London* p. 782 (1962).
18. D. J. Crispin and J. S. Whitehurst, *Proc. Chem. Soc., London* p. 356 (1962).
19. D. J. Crispin and J. S. Whitehurst, *Proc. Chem. Soc., London* p. 22 (1963).
20. D. J. Crispin, A. E. Vanstone, and J. S. Whitehurst, *J. Chem. Soc., London* p. 10 (1970).
21. H. Smith, G. A. Hughes, and B. J. McLoughlin, *Experientia* **19**, 178 (1963).
22. G. H. Douglas, J. M. H. Graves, D. Hartley, G. A. Hughes, B. J. McLoughlin, J. Siddall, and H. Smith, *J. Chem. Soc., London* p. 5072 (1963).
23. C. B. C. Boyce and J. S. Whitehurst, *J. Chem. Soc., London* p. 2022 (1959).
24. C. B. C. Boyce and J. S. Whitehurst, *J. Chem. Soc., London* p. 4547 (1960).
25. Y. Kitahara, A. Yoshikoshi, and S. Oida, *Tetrahedron Lett.* p. 1763 (1964).
26. D. J. France, J. J. Hand, and M. Los, *Tetrahedron* **25**, 4011 (1969).
27. D. J. France, J. J. Hand, and M. Los, *J. Org. Chem.* **35**, 468 (1970).
28. J. J. Hand and M. Los, *Chem. Commun.* p. 673 (1969).
29. K. Sakai and S. Amemiya, *Chem. Pharm. Bull.* **18**, 641 (1970).
30. A. J. Birch, *J. Chem. Soc., London* p. 1270 (1947).
31. S. Danishefsky and R. Cavanaugh, *J. Amer. Chem. Soc.* **90**, 520 (1968).
32. S. Danishefsky, A. Nagal, and D. Peterson, *Chem. Commun.* p. 374 (1972).
33. S. Danishefsky and A. Nagal, *Chem. Commun.* p. 373 (1972).
34. A. Nagal, Ph.D. Thesis, University of Pittsburgh, Pittsburgh, Pennsylvania, 1971.
35. H. H. Inhoffen, G. Quinkert, and S. Schütz, *Chem. Ber.* **90**, 1283 (1957).
36. H. H. Inhoffen and K. Irmscher, *Chem. Ber.* **89**, 1833 (1956).
37. R. S. Davidson, P. S. Littlewood, T. Medcalfe, S. M. Waddington-Feather, D. H. Williams, and B. Lythgoe, *Tetrahedron Lett.* p. 1413 (1963).
38. R. S. Davidson, S. M. Waddington-Feather, D. H. Williams, and B. Lythgoe, *J. Chem. Soc., London* p. 2534 (1967).
39. G. S. Grinenko, E. V. Popova, and V. I. Maksimov, *Zh. Org. Khim.* **5**, 1329 (1969).
40. G. S. Grinenko, E. V. Popova, and V. I. Maksimov, *Zh. Org. Khim.* **7**, 935 (1971).
41. G. S. Grinenko, S. D. Podobrazhnykh, E. V. Popova, A. N. Akalaev, and O. S. Anisimova, *Zh. Org. Khim.* **7**, 1893 (1971).

42. H. O. Huisman, *Angew. Chem., Int. Ed. Engl.* **10**, 450 (1971).
43. U. K. Pandit, M. J. M. Pollman, and H. O. Huisman, *Chem. Commun.* p. 527 (1969).
44. M. J. M. Pollmann, U. K. Pandit, and H. O. Huismann, *Rec. Trav. Chim. Pays-Bas* **89**, 941 (1970).
45. M. J. M. Pollmann, H. R. Reus, U. K. Pandit, and H. O. Huisman, *Rec. Trav. Chim. Pays-Bas* **89**, 929 (1970).
46. U. K. Pandit, K. de Jonge, K. Erhart, and H. O. Huisman, *Tetrahedron Lett.* p. 1207 (1969).
47. R. R. Crenshaw and G. M. Luke, *Tetrahedron Lett.* p. 4495 (1969).
48. R. R. Crenshaw, G. M. Luke, T. A. Jenks, G. Bialy, and M. E. Bierwagen, *J. Med. Chem.* **15**, 1162 (1972).
49. H. O. Huisman, *Proc. Int. Congr. Horm. Steroids, 3rd., 1970* p. 110 (1971).
50. M. Lora-Tamayo and J. Marin, *An. Real. Soc. Espan. Fis. Quim., Ser. B* **48**, 693 (1952).
51. M. Lora-Tamayo and C. Corrall, *An. Real. Soc. Espan. Fis. Quim., Ser. B* **53**, 45 (1957).
52. C. Corrall, *Rev. Real. Acad. Cienc. Exactas, Fis. Natur. Madrid* **51**, 103 (1957).
53. M. Soto Martinez, *Rev. Real. Acad. Cienc. Exactas, Fis. Natur. Madrid* **56**, 383 (1962).
54. A. Alberola, M. Lora-Tamayo, A. del Key, J. L. Soto, and M. Soto, *An. Real Soc. Espan. Fis. Quim., Ser. B* **59**, 151 (1963).
55. H. von Pechmann and C. Duisberg, *Ber. Deut. Chem. Ges.* **16**, 2119 (1883).
56. R. K. Razdan, H. G. Pars, F. E. Granchelli, and L. S. Harris, *J. Med. Chem.* **11**, 377 (1968).
57. G. Stork, P. Rosen, N. Goldman, R. V. Combs, and J. Tsuji, *J. Amer. Chem. Soc.* **87**, 275 (1965).
58. H. Evers, U. K. Pandit, and H. O. Huisman, *Syn. Commun.* **1**, 89 (1971).
59. E. R. de Waard, R. Neeter, U. K. Pandit, and H. O. Huisman, *Rec. Trav. Chim. Pays-Bas* **87**, 572 (1968).
60. A. I. Meyers, G. G. Munoz, W. Sobotka, and K. Baburao, *Tetrahedron Lett.* p. 254 (1965).
61. A. I. Meyers and W. N. Beverung, *Chem. Commun.* p. 877 (1968).
62. G. V. Bhide, N. R. Pai, N. L. Tikotkar, and B. D. Tilak, *Tetrahedron* **4**, 420 (1958).
63. R. A. Barnes and L. Gordon, *J. Amer. Chem. Soc.* **71**, 2644 (1949).
64. O. Süs, K. Möller, *Justus Liebigs Ann. Chem.* **593**, 91 (1955).

8
A + D → AD → ABCD

Although utilized by relatively few investigators, this method of total synthesis has given rise to a plethora of steroid derivatives, due primarily to the work of Smith and his collaborators which is presented in Section 8.1. The second section discusses various methods used for preparing 8-azasteroids, and Section 8.3 recounts the brief investigations of three groups leading to 8,13-diazasteroids. The final division, with one exception, is primarily of historical interest only and concerns some early attempts at total synthesis by a Monsanto research group.

8.1 The Smith-Hughes Approach

In a series of reports spanning the decade of the 1960's [1–23], H. Smith and G. A. Hughes and their fellow investigators revealed a novel approach to total synthesis, the heart of which is illustrated in Scheme 8-1.* One of the features lending attractiveness to this route is the wide variation possible in the starting materials and, therefore, the

---

* In fact, more than one approach was utilized including AB+D and A+CD; some of the references given above have little to do with actual synthesis and more with derivatization, but because they constitute a series of papers on total synthesis by the Wyeth research group they are included.

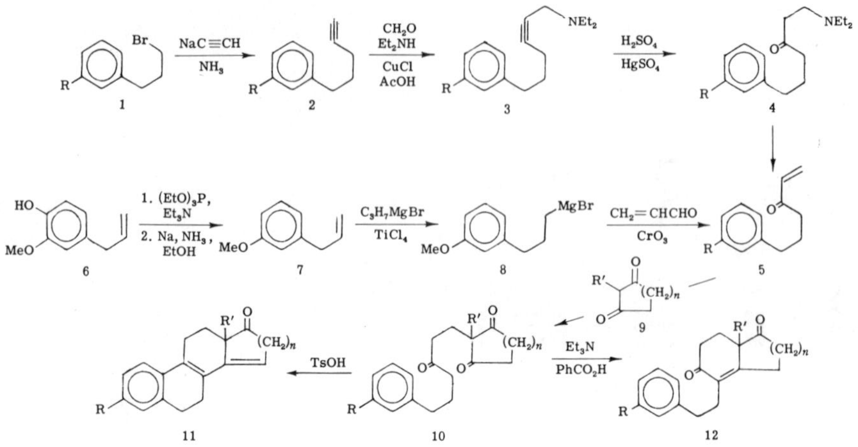

Scheme 8-1

products. The alkyne **2**, possessing various C-3 substituents, was prepared from the appropriate bromide **1**. Thus, **2** (R=OMe) was isolated in 90% yield from the reaction of 3-(m-methoxyphenyl)propyl bromide with sodium acetylide in liquid ammonia. In a similar manner, **1** (R=OH, or R=H) gave rise to the corresponding **2** (R=OH, or R=H) [4]. Still further experimentation led to the α-methyl homolog of **2** (R=OMe) [9]. Mannich condensation of **2** with formalin and diethylamine in the presence of acetic acid and cuprous chloride gave very good yields of **3** which was hydrated to the ketone **4** using sulfuric acid and mercuric sulfate. It was found that attempted distillation of **4** led to partial decomposition such that the distillate collected was a mixture of **4** and **5**, the composition of which depended upon the rate of collection. While pure **5** could be obtained by prolonged heating of **4** followed by acid extraction, the procedure was unnecessary, and in most instances a mixture of **4** and **5** was used for further reaction.

A French group has devised an interesting preparation of **5** uncontaminated with **4**, starting with the natural product eugenol (**6**) [24,25]. The hydroxyl group of **6** was removed in 75% yield by first forming the phosphate ester and then reducing with sodium–ammonia–ethanol. Treatment with isopropylmagnesium bromide in the presence of titanium tetrachloride gave the Grignard **8** (this conversion was also effected via hydroboration of **7** to the alcohol followed by bromination and Grignard formation) which was reacted with acrolein. The resulting allylic alcohol was readily oxidized to **5**.

The AD intermedate **10** was prepared by refluxing a mixture of **4**

8.1 The Smith-Hughes Approach

and **5** with the appropriate 2-alkylcycloalkane-1,3-dione (**9**) [5] in methanol containing a minimal amount of base. The 1,3-diones employed were both cyclopentyl and cyclohexyl derivatives with representative R′ groups being methyl, ethyl, isobutyl, and n-hexadecyl. Cyclization of **10** could be effected under a variety of acidic conditions. Refluxing in xylene with triethylammonium benzoate afforded **12** (80–90%). Alternatively, *p*-toluenesulfonic acid, polyphosphoric acid, or alcoholic hydrochloric acid treatment of **10** all led to the pentaene **11**. The preferred procedure was to combine 2.5 equivalents of the anhydrous sulfonic acid with the trione **10** in benzene at or below room temperature. When the cyclodehydration was complete, the precipitated acid hydrate was removed leaving a solution of **11** (35–50%, based on starting cycloalkanedione). Although the secosteroid **12** could be converted into **11** [26] (cf. Chapter 7, Section 7.2), it was not considered to be an intermediate in this sulfonic acid cyclodehydration since it could be recovered unchanged under conditions whereby **10** was readily transformed into **11** [4]. Rather, it was speculated that **10** was converted first into an ABD intermediate, thence into **11**. A considerable effort was expended developing the chemistry of **12**, but because this same material has been prepared via the A+CD route, such work is more appropriately discussed in Chapter 7.

Two other reports closely related to this synthesis deserve mention at this point. A Japanese group [27] arrived at the same pentaene (**11**) by a slightly modified route employing as the precursor the alkyne **2** (R=OMe). From this was formed the acetylenic Grignard which was reacted with formaldehyde. The resulting primary alcohol was brominated to give an analog of **3** (R=OMe; NEt$_2$ replaced by Br). Condensation with 2-methylcyclopentane-1,3-dione gave an AD intermediate [**10**, with a 9(11)-triple bond rather than the 9-ketone]. Polyphosphoric acid cyclization led directly to **11** (R=OMe; R′=Me; $n=1$) in over 20% yield from **2** (R=OMe).

A second interesting observation has been recorded by Eder *et al.* [28] who found that cyclodehydration of **10** (R=OMe; R′=Me; $n=1$) in the presence of L-phenylalanine and perchloric acid led to optically active **11**. This represents another example of asymmetric induction—a tool of rapidly increasing importance to synthetic chemists (cf. Chapter 12, Section 12.6).

Scheme 8-2 depicts a few of the basic reactions of the key pentaene. For simplicity of illustration one specific compound has been chosen; it should be remembered, however, that the conversions shown have, for the most part, been carried out on a multitude of derivatives of **13** (i.e., D-homo, 18-alkyl, and 3-hydroxy or deoxy compounds) [4,5]. Hydro-

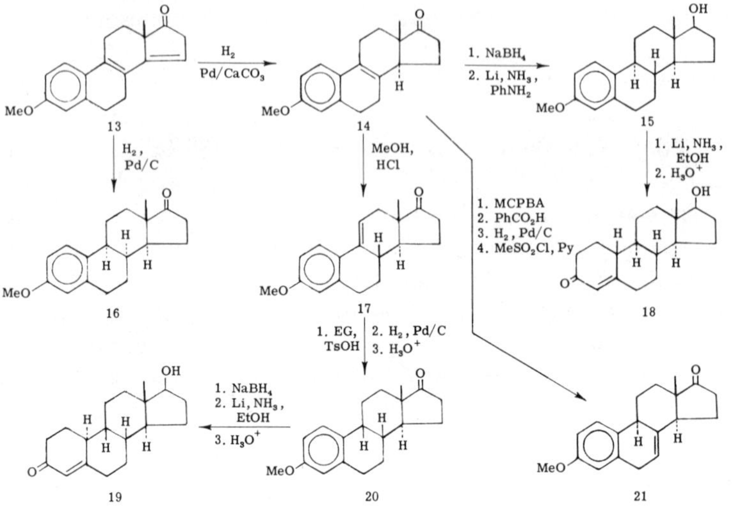

Scheme 8-2

genation of **13** to 8-dehydroestrone methyl ether could be accomplished with several catalysts including Raney nickel and 2% palladized calcium carbonate. The reduction was quite facile giving 80-90% yields of the CD-*trans* material **14**. To complete the synthesis of the 3-methyl ether of racemic estradiol (**15**), the ketone **14** was first reduced to the alcohol (sodium borohydride) and then treated with lithium-aniline-ammonia. More directly, but slightly less efficiently, **14** could be reduced to **15** by potassium in liquid ammonia. Chromic acid oxidation of **15** gave (±)-estrone 3-methyl ether, prepared in 18% overall yield from 3-*m*-methoxyphenylpropyl bromide [4].

Several of the isomers of estrone methyl ether were also prepared, two of which (**16** and **20**) are shown in Scheme 8-2. The 8α-product **16** was readily formed by catalytic hydrogenation of the pentaene **13** over 10% palladium-on-charcoal (55%) [2,4,10]. 9β-Estrone 3-methyl ether (**20**) was less directly obtained and required the intermediate **17**. This latter compound was obtained in excellent yield by refluxing **14** in methanolic hydrogen chloride. A similar isomerization in the D-homo series provided a mixture containing approximately equal amounts of the Δ^8- and $\Delta^{9(11)}$-products indicating subtle differences in olefin stability depending on the size of the D ring. The ethylene ketal of **17** was hydrogenated over palladium-on-charcoal in acetic acid and acetic anhydride to give, after hydrolysis, a mixture (ca. 50:50) of estrone methyl ether and the 9β-

8.1 The Smith-Hughes Approach

isomer **20**. Hydrogenation of **17** with 10% palladium-on-charcoal in ethanol gave estrone methyl ether as the only readily isolable product [5].

Mention must also be made of related work in which optically active **13** was produced [29]. The racemic acid **23** (X= -OCH$_2$CH$_2$O-; Scheme 8-3) was prepared from 2-methylcyclopentane-1,3-dione and ethyl acrylate; resolution was carried out with the aid of cinchonine. The optically active AD intermediate **10** (R=OMe, R'=Me, n=1; Scheme 8-1) was obtained as the 17-ethylene ketal. Cyclization and hydrolysis produced (−)-**13** in unspecified yield. Following the methodology established in the racemic series [4], this material was transformed into natural estrone 3-methyl ether.

The Wyeth group also reported the ready conversion of **14** (and derivatives thereof) into the methyl ethers of equilenin and equilin (**21**) [23]. Epoxidation of the Δ^8-unsaturation of **14** followed by acid-catalyzed rearrangement gave a good yield of 8α-hydroxy-**17**. Catalytic hydrogenation over 5% palladium-on-charcoal stereoselectively gave the 8α-hydroxy analog of **16**. The problem of selective dehydration of this material was solved with methanesulfonyl chloride in warm pyridine, and the result was racemic equilin methyl ether (**21**).

Scheme 8-2 also presents the preparation of 19-nortestosterone (**18**)

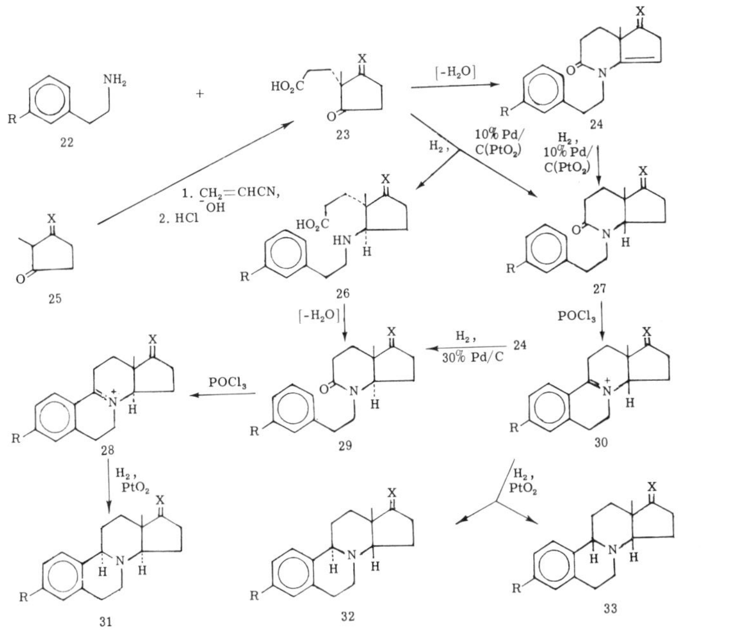

Scheme 8-3

and the isomeric 9β,10α-derivative **19**. Borohydride reduction of **20** gave the 17β-ol which was subjected to the action of lithium and ethanol in liquid ammonia. Subsequent acid hydrolysis produced **19** in moderate yield [7]. A similar reduction of **15** gave racemic 19-nortestosterone (**18**; 30%) [5]. It should, perhaps, be reemphasized that most of these reactions—and many additional ones—have been performed on a wide variety of steroid derivatives related to **13**. Too, although Schemes 8-1 and 8-2 concern racemic products, many have been resolved both chemically [16] and microbiologically [11,12]. Further experimentation has provided 18-alkyl-19-norpregnanes and corticoids.

8.2 8-Azasteroids

The total synthesis of 8-azasteroids has been accomplished by two groups by two different routes. Meyers' work has been considered in Chapter 4, while the present section is devoted to the work of the Warner-Lambert research group [30–36]. Scheme 8-3 illustrates the construction of three isomeric azaestrogens **31**, **32**, **33**. The D-ring component **23** was prepared from the appropriately substituted cyclopentane derivative **25** ($X=O$, H_2 or β-CO_2H) by Michael addition to acrylonitrile and subsequent hydrolysis to the acid. (Although Schemes 8-3 and 8-4 depict steroids derived from cyclopentane derivatives, a number of the corresponding D-homo materials have also been prepared [37]). Alternatively, **25** could be condensed with ethyl acrylate and hydrolyzed [38]. The A-ring portion **22** (R=OMe) was prepared from m-methoxyphenylacetic acid via the amide; additionally, some derivatives possessing 1,2-dimethoxy groups were prepared [31].

As shown in Scheme 8-3, the condensation could be effected in two ways. In the first, the amine and the acid were combined in refluxing xylene with azeotropic removal of water. The only isolated product was **24** (50–80%). This condensation failed when the ethyl ester of **23** was used in place of the acid. Reductive condensation of **22** and **23** ($X=H_2$ or O) in the presence of 10% palladium-on-charcoal yielded two new products, **26** ($X=H_2$, 25%; $X=O$, 15%) and **27** ($X=H_2$, 65%; $X=O$, 45%). This latter material was the exclusive product when **24** was subjected to catalytic reduction. Simple heating of **26** to its melting point effected cyclization to **29**. Thus, the CD-*cis* material **27** was readily prepared, whereas the more desirable *trans* product **29** was not. This problem was solved by the observation that reduction of **24** ($X=O$) to the 17β-ol followed by hydrogenation gave mixtures of **29** ($X=β$-OH,α-H)

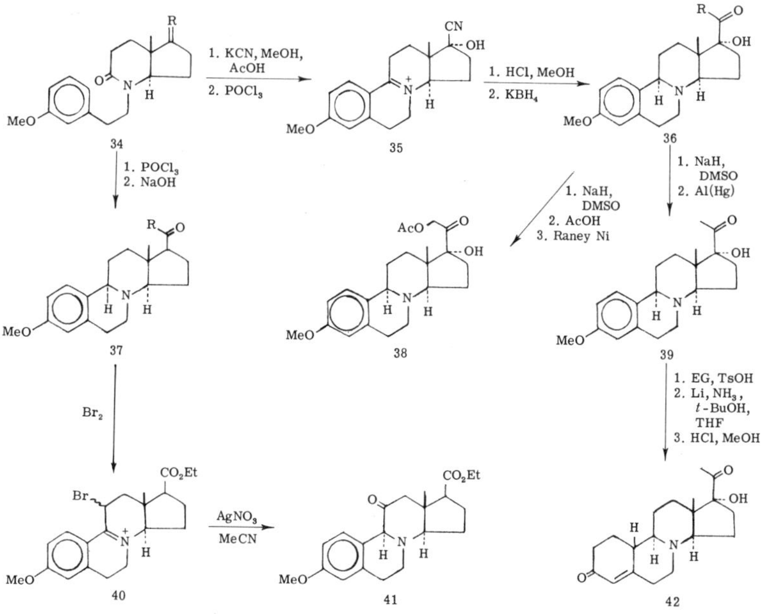

8.2 8-Azasteroids

Scheme 8-4

and **27** (X=β-OH,α-H). Conditions were ultimately found whereby the *trans* steroid precursor could be isolated in 75% yield [32]. Jones oxidation then gave ketone **29** (X=O) in overall yield of 27% based on **23** (X=O).

Interesting reversals of the above stereochemistry were observed when **23** (X=β-CO$_2$H,α-H) was employed in conjunction with platinum oxide [33]. The CD-*trans* amide **29** (X=β-CO$_2$H,α-H) was formed stereoselectively both from catalytic reduction (platinum oxide) of **24** (X=β-CO$_2$H,α-H) and from reductive condensation of **22** with **23** (X=β-CO$_2$H,α-H) over platinum oxide. Further differences in reactivity, depending on the C-17 substituent, were illustrated by the fact that unsaturated **24** underwent an unusual phosphorus oxychloride cyclization to give a poor yield of nonsteroidal product when X=O [32]; however, similar treatment of **24** (X=β-CO$_2$H,α-H) readily gave the desired 8-azasteroid skeleton [33]. The steroid nucleus of **28** (X=O,H$_2$) and **30** (X=O,H$_2$) was obtained, though, by phosphorus oxychloride cyclization of the saturated precursors **27** (X=O,H$_2$) and **29** (X=O,H$_2$). These quaternary salts, most easily isolated as the perchlorates, were reduced under a variety of conditions including platinum oxide, sodium–ammonia and zinc–acid. In all cases the CD *trans* **28** provided only one product, formulated as **31** (X=O,H$_2$), in excellent yield. Conversely, the CD-*cis* salt **30** gave two isomers **32** (X=O,H$_2$) and **33** (X=O,H$_2$).* As noted before, Meyers [39] has also prepared 8-azasteroids, some of which were identical to the above.

Further elaboration into 8-aza-19-norprogestogens and 11-substituted derivatives is presented in Scheme 8-4. Cyclization of **34** (R=β-CO$_2$H, α-H) was performed with phosphorus oxychloride in ethanol, and the intermediate salt was treated with dilute base to afford 9(11)-dehydro-**37** (R=OEt) (95%) [33]. This substrate was brominated at −80° to give a 1:1 mixture (95%) of the α- and β-epimers **40** [34] which was treated with silver nitrate in acetonitrile. The product, **41**, presumably formed via the intermediate nitrate ester, was isolated—as the perchlorate salt—in 54% yield. It was also possible to prepare the 11-acetoxy analogs of **40** by acetate displacement of the bromine, although in very poor yield (10%). Too, a method for introducing an 11-amino group was reported.

The pregnane side chain was elaborated by two reaction sequences [33]. In the first, 9(11)-dehydro-**37** (R=OEt) was reduced and hydrolyzed to the saturated free acid **37** (R=OH). Reaction with methyllithium gave **37** (R=Me) in impure form. This material was alternatively prepared from **37** (R=OEt) by reaction with dimsyl sodium, and cleavage of the resulting β-ketosulfoxide **37** [R=CH$_2$S(O)CH$_3$] with aluminum amalgam. For obtaining 17α-hydroxy derivatives, it was found expedient to treat **34** (R=β-CO$_2$H,α-H) with potassium cyanide and acetic acid in methanol. The cyanohydrin was cyclized by the usual phosphorus oxychloride treatment to **35** (61%, isolated as the perchlorate). Vigorous acid hydrolysis, esterification, and borohydride reduction led to **36** (R=OMe) in over 80% yield. Sulfoxide formation and reductive cleavage with aluminum amalgam as before gave **39** in 43% overall yield from **34** (R=O). The usual Birch reduction of the ketal, followed by hydrolysis gave 17α-hydroxy-8-aza-19-norprogesterone (**42**).

Finally, it was possible to introduce a 21-acetoxy group via a Pummerer rearrangement of the above formed β-ketosulfoxide [35]. Thus, heating the sulfoxide **36** [R=CH$_2$S(O)CH$_3$] in acetic acid led to the hemimercaptal acetate **36** [R=CH(SMe)OAc], which was desulfurized with Raney nickel to **38**. Yields greater than 50% were realized in this and similar conversions.

Other work concerned with total synthesis of 8-azasteroids by the

---

* Some confusion has existed concerning the assignment of configuration to these two isomers. The initial assignment [30] was later reversed, and has more recently been revised on the basis of X-ray analysis [36].

A+D route has been reported by Meyers [40–42] and his co-workers and by the Indian group led by Kessar [38,43]. Both groups reported initial efforts in this area that held some promise, but were abandoned in favor of alternate approaches that ultimately were successful.

8.3 8,13-Diazasteroids

Although three different groups have reported very similar work outlining the total synthesis of the title compounds, to date this work has not been elaborated to any degree, and the number of compounds prepared and derivatives thereof remains limited. The AD intermediate **48** (Scheme 8-5) was prepared in either of two ways, both employing **44** as the starting material. Condensation of **44** (R=R'=OMe) with β-bromopropionyl chloride provided the bromoamide **49** (R=R'=OMe) in 55% yield, and this in turn was reacted with potassium succinimide to give **48** (R=R'=OMe; 72%). Alternatively, the acid chloride **43**, prepared from succinimide and propiolactone as shown, could be condensed with **44** to give the necessary intermediate directly [44,45]. Under the usual conditions of the Bischler-Napieralski reaction, cyclization to the dihydroisoquinoline **50** (R=R'=OMe; R=OMe, R'=H; R=R'=H) readily occurred [46–48]. Initially, the synthesis was completed by acid hydrolysis and catalytic hydrogenation to the diamine **51** (R=R'=OMe; R=OMe, R'=H) which was condensed with ethyl 3-ethoxycarbonylpropionimidate hydrochloride [46]. The immediate product was 8(14)-dehydro-**52** (isolated as the perchlorate salt); however, this material

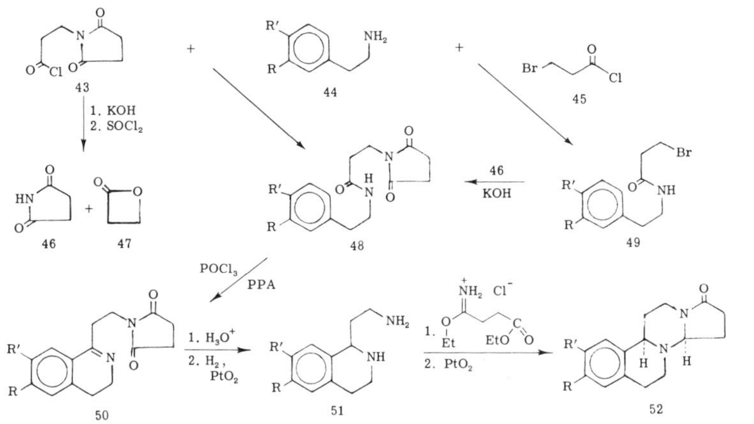

Scheme 8-5

could be reduced to **52** with either platinum oxide or sodium borohydride [47]. A more expeditious route to **52** was soon found; thus, reductive cyclization of **50** with platinum oxide [45] or sodium borohydride [47] afforded **52** directly in good yield (ca. 70%). That **52** (R=R'=OMe) possessed the 9α,14α configuration and thus an overall molecular structure quite similar to estrone was established by X-ray analysis [47]. By the steps outlined in Scheme 8-5, therefore, 2-methoxy-8,13-diaza-18-norestrone methyl ether was prepared in over 20% yield based on homoveratrylamine (**44**, R=R'=OMe). Further work established that the 17-ketone was readily removed by lithium aluminum hydride reduction.

8.4 Miscellaneous Routes

Scheme 8-6 illustrates two approaches which have provided the basic steroid skeleton, but for rather obvious reasons (primarily the lack of generality and poor yields) have not been extensively developed. As early as 1933, Sir Robert Robinson showed the feasibility of constructing the hexahydrochrysene derivative **56** [49,50], and this work was later clari-

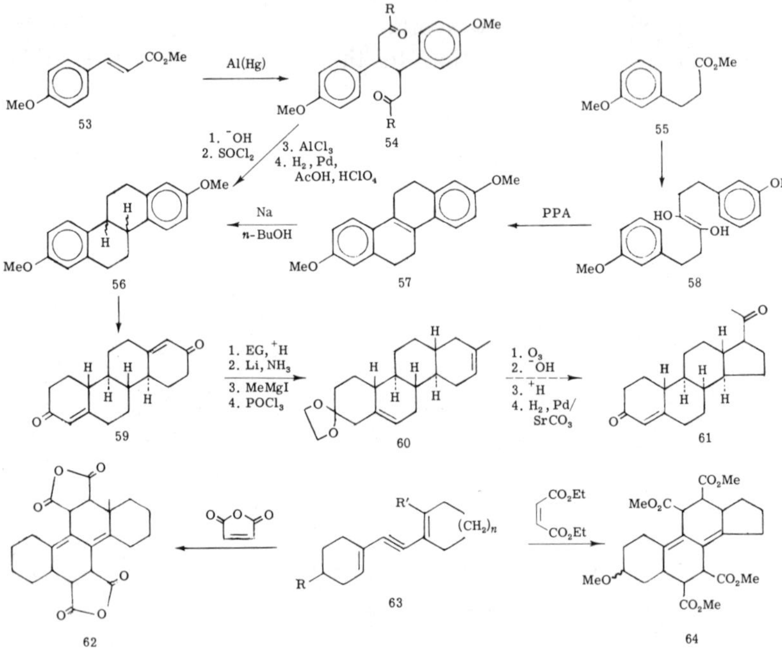

Scheme 8-6

8.4 Miscellaneous Routes

fied and expanded [51]. A mixture of the meso (24%) and racemic forms (18%) of the diester **54** (R=OMe) resulted from aluminum amalgam reduction of methyl *m*-methoxycinnamate (**53**). These forms were separated and individually hydrolyzed and converted to the diacid dichlorides **54** (R=Cl). Aluminum chloride catalyzed cyclization gave the 6,12-diketo derivatives of **56**; the meso material gave an 87% yield of the BC-*trans* product, the *dl*-form leading to the *cis* (86%). Hydrogenation with palladium in acetic acid–perchloric acid then gave excellent yields of the corresponding *cis* and *trans* **56**. An alternate approach to the same product started with methyl β-(*m*-methoxyphenyl)propionate (**55**) and, via the acyloin condensation product **58**, proceeded to the tetrahydrochrysene **57** [52]. This product, obtained from **55** in ca. 50% yield, could be hydrogenated to **56** by either sodium–ammonia or sodium–*n*–butanol; unfortunately, little stereoselectivity was observed in the process, with *trans* **56** being formed in slight preference to *cis* **56**. Birch reduction converted the *trans* tetracycle preferentially to the *anti-trans-anti* dione **59** which was monoketalized and reduced with lithium–ammonia to provide the CD-*trans* derivative [53]. The ketal **60** was then obtained by treatment with methyl Grignard and phosphorus oxychloride. The practicability of converting **60** into 18,19-bisnorprogesterone (**61**) was demonstrated with an aromatic A-ring analog of **60** by ozonolysis, cyclization, and hydrogenation; the actual conversion of **60** was, however, not reported.

Also outlined in Scheme 8-6 is a synthetic sequence remarkable for its simplicity [54–58]. The starting dienynes **63** were readily obtained from dehydration of the corresponding acetylenic glycols which were in turn prepared from acetylene and the appropriate cycloalkanones [59]. Condensation of **63** (R=H, R'=Me, $n=2$) with maleic anhydride at 130° gave a 1.9% yield of **62**, the actual stereochemistry of which was not determined [56]. A similar reaction of **63** (R=OMe, R'=H, $n=1$) with dimethyl fumarate produced **64** in moderate (45%) yield, but again the stereochemistry was unspecified [57]. It was determined that better yields were obtained when $n=2$ and when R'=H, thus dampening hopes of application to steroid total synthesis. Other dienophiles were employed, but only maleimide reacted appreciably (60%) [58]. Although some brief attempts were made to modify the tetracyclic products, no convenient decarboxylation procedure was found, and the overall method of synthesis was shortly abandoned.

The final synthetic approach in this section is shown in Scheme 8-7 [60]. 4-Methoxy-2-vinylpyridine (**65**) was prepared by a novel route starting with 2-ethylpyridine and proceeding through 4-methoxy-2-ethyl-

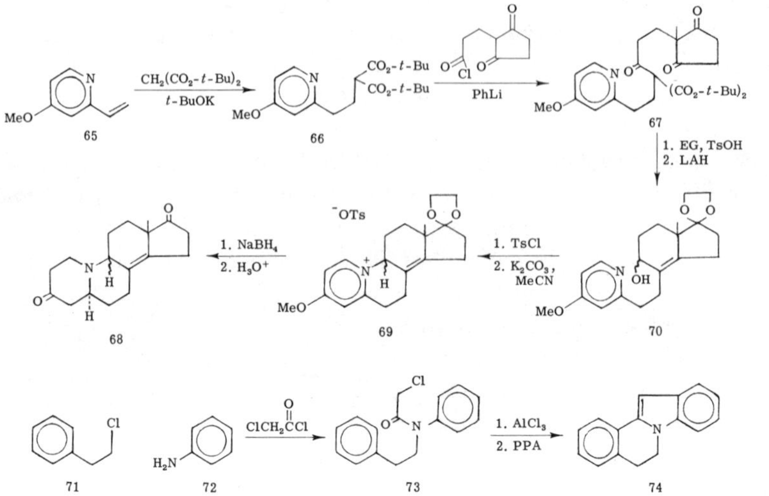

Scheme 8-7

pyridine-N-oxide, the rearrangement of which, in acetic anhydride, gave the α-acetoxy derivative. After hydrolysis and dehydration to **65**, Michael addition of di-t-butyl malonate gave **66**. Acylation was readily performed with the indicated acid chloride and phenyllithium to produce **67** (its similarity to **10**, Scheme 8-1 should be noted). Simultaneous decarboxylation and cyclization was effected by refluxing in benzene with p-toluensulfonic acid. The 9,10-secosteroid (9,17-dione of **70**) was thereby obtained in 57% yield from **65**; starting from 2-vinyl pyridine the analogous 3-deoxy derivative was also constructed. To effect ring closure it was first necessary to prepare the 17-ketal and then reduce with lithium aluminum hydride to **70**. This material was considered to be the 9β-alcohol to the near exclusion of the 9α. Tosylation followed by treatment with anhydrous potassium carbonate in acetonitrile afforded two crystalline salts, presumably the 9β- and 9α-isomers of **69**, with the former predominating (47:20%) Sodium borohydride reduction and acid hydrolysis then gave, stereospecifically, the corresponding 10-aza-19-nor-8(14)-androstene-3,17-diones (**68**; 9α-**69**→5α,9α-**68**; 9β-**69**→5β,9β-**68**). This overall sequence should be amendable to extension and modification; and the paucity of preparative methods leading to 10-azasteroids (see Chapter 12, Section 12.8 for the preparation of 19-nor-10-azatestosterone [61]) lends further attractiveness.

Also shown in Scheme 8-7 is an alluringly simple approach to the 8-aza-C-nor-D-homo steroid system [62]. Although this material does not closely resemble actual steroids, appropriate substitution and modification of the starting materials should lead to more desirable products. Thus, chloride displacement from 71 with aniline provided a 68% yield of N-phenyl-β-phenethylamine which was acylated with chloroacetyl chloride to give 73 (81%). Aluminum chloride catalyzed cyclization gave the intermediate 9,10-secosteroid (75%), and treatment with polyphosphoric acid led to 74. The same material was also prepared from 71 and m-(cyanomethyl)chlorobenzene.

References

1. G. A. Hughes and H. Smith, *Proc. Chem. Soc., London* p. 74 (1960).
2. G. A. Hughes and H. Smith, *Chem. Ind. (London)* p. 1022 (1960).
3. H. Smith, G. A. Hughes, G. H. Douglas, D. Hartley, B. J. McLoughlin, J. B. Siddall, G. R. Wendt, G. C. Buzby, Jr., D. R. Herbst, K. W. Ledig, J. McMenamin, T. W. Pattison, J. Suida, J. Tokolics, R. A. Edgren, A. B. A. Jansen, B. Gadsby, D. H. P. Watson, and P. C. Phillips, *Experientia* **19**, 394 (1963).
4. G. H. Douglas, J. M. H. Graves, D. Hartley, G. A. Hughes, B. J. McLoughlin, J. Siddall, and H. Smith, *J. Chem. Soc., London* p. 5072 (1963).
5. H. Smith, G. A. Hughes, G. H. Douglas, G. R. Wendt, G. C. Buzby, Jr., R. A. Edgren, J. Fisher, T. Foell, B. Gadsby, D. Hartley, D. Herbst, A. B. A. Jansen, K. Ledig, B. J. McLoughlin, J. McMenamin, T. W. Pattison, P. C. Phillips, R. Rees, J. Siddall, J. Suida, L. L. Smith, J. Tokolics, and D. H. P. Watson, *J. Chem. Soc., London* p. 4472 (1964).
6. D. Hartley and H. Smith, *J. Chem. Soc., London* p. 4492 (1964).
7. J. M. H. Graves, G. A. Hughes, T. Y. Jen, and H. Smith, *J. Chem. Soc., London* p. 5488 (1964).
8. G. H. Douglas, C. R. Walk, and H. Smith, *J. Med. Chem.* **9**, 27 (1966).
9. G. H. Douglas, G. C. Buzby, Jr., C. R. Walk, and H. Smith, *Tetrahedron* **22**, 1019 (1966).
10. G. C. Buzby, Jr., E. Capaldi, G. H. Douglas, D. Hartley, D. Herbst, G. A. Hughes, K. Ledig, J. McMenamin, T. Pattison, H. Smith, C. R. Walk, G. R. Wendt, J. Siddall, B. Gadsby, and A. B. A. Jansen, *J. Med. Chem.* **9**, 338 (1966).
11. L. L. Smith, G. Greenspan, R. Rees, and T. Foell, *J. Amer. Chem. Soc.* **88**, 3120 (1966).
12. G. Greenspan, L. L. Smith, R. Rees, T. Foell, and H. E. Albrum, *J. Org. Chem.* **31**, 2512 (1966).
13. G. C. Buzby, Jr., C. R. Walk, and H. Smith, *J. Med. Chem.* **9**, 782 (1966).
14. D. P. Strike, T. Y. Jen, G. A. Hughes, G. H. Douglas, and H. Smith, *Steroids*, **8**, 309 (1966).
15. G. A. Hughes and H. Smith, *Steroids* **8**, 547 (1966).
16. G. C. Buzby, Jr., D. Hartley, G. A. Hughes, H. Smith, B. W. Gadsby, and A. B. A. Jansen, *J. Med. Chem.* **10**, 199 (1967).
17. G. A. Hughes, T. Y. Jen, and H. Smith, *Steroids* **8**, 947 (1966).

18. G. C. Buzby, Jr., G. H. Douglas, C. Walk, and H. Smith, *Proc. Int. Congr. Horm. Steroids, 2nd, 1966* p. 28 (1967).
19. D. P. Strike, D. Herbst, and H. Smith, *J. Med. Chem.* 10, 446 (1967).
20. R. Rees, D. P. Strike, and H. Smith, *J. Med. Chem.* 10, 782 (1967).
21. D. R. Herbst and H. Smith, *Steroids* 11, 935 (1968).
22. B. Gadsby, M. R. G. Leeming, G. Greenspan, and H. Smith, *J. Chem. Soc., London* p. 2647 (1968).
23. R. P. Stein, G. C. Buzby, Jr., and H. Smith, *Tetrahedron* 26, 1917 (1970).
24. A. Horeau, L. Ménager, and H. Kagan, *C. R. Acad. Sci., Ser. C* 269, 602 (1969).
25. A. Horeau, L. Ménager, and H. Kagan, *Bull. Soc. Chim. Fr.* [5] p. 3571 (1971).
26. D. J. Crispin and J. S. Whitehurst, *Proc. Chem. Soc., London* p. 356 (1962).
27. T. Hiraoka and I. Iwai, *Chem. Pharm. Bull.* 14, 262 (1966).
28. U. Eder, G. Sauer, and R. Wiechert, *Angew. Chem.* 83, 492 (1971).
29. R. Bucourt, M. Vignau, and J. Weill-Raynal, *C. R. Acad. Sci., Ser. C* 265, 834 (1967).
30. R. I. Meltzer, D. M. Lustgarten, R. J. Stanaback, and R. E. Brown, *Tetrahedron Lett.* p. 1581 (1963).
31. R. E. Brown, D. M. Lustgarten, R. J. Stanaback, M. W. Osborne, and R. I. Meltzer, *J. Med. Chem.* 7, 232 (1964).
32. R. E. Brown, D. M. Lustgarten, R. J. Stanaback, and R. I. Meltzer, *J. Org. Chem.* 31, 1489 (1966).
33. R. E. Brown, D M. Lustgarten, R. J. Stanaback, and R. I. Meltzer, *J. Med. Chem.* 10, 451 (1967).
34. R. E. Brown, H. V. Hansen, D. M. Lustgarten, R. J. Stanaback, and R. I. Meltzer, *J. Org. Chem.* 33, 4180 (1968).
35. R. E. Brown, D. M. Lustgarten, and R. J. Stanaback, *J. Org. Chem.* 34, 3694 (1969).
36. R. E. Brown, A. I. Meyers, L. M. Trefonas, R. L. R. Towns, and J. N. Brown, *J. Heterocycl. Chem.* 8, 279 (1971).
37. Belgian Patent 642,060 (1964); *Chem. Abstr.* 63, 5714b (1965).
38. S. V. Kessar, A. L. Rampal, K. Kumar, and R. R. Jogi, *Indian J. Chem.* 2, 240 (1964).
39. A. H. Reine and A. I. Meyers, *J. Org. Chem.* 35, 554 (1970).
40. A. I. Meyers, J. Schneller, and N. K. Ralhan, *J. Org. Chem.* 28, 2944 (1963).
41. A. I. Meyers and N. K. Ralhan, *J. Org. Chem.* 28, 2950 (1963).
42. A. I. Meyers, B. J. Betrus, N. K. Ralhan, and K. B. Rao, *J. Heterocycl. Chem.* 1, 13 (1964).
43. S. V. Kessar, A. Kumar, and A. L. Rampal, *J. Indian Chem. Soc.* 40, 655 (1963).
44. G. Redeuilh and C. Viel, *Bull. Soc. Chim. Fr.* [6] p. 3115 (1969).
45. E. C. Taylor and K. Lenard, *Chem. Commun.* p. 97 (1967).
46. J. H. Burckhalter and H. N. Abramson, *Chem. Commun.* p. 805 (1966).
47. J. H. Burckhalter, H. N. Abramson, J. G. MacConnell, R. J. Thill, A. J. Olson, J. C. Hanson, and C. E. Nordman, *Chem. Commun.* p. 1274 (1968).
48. R. J. Thill, Ph.D. Thesis, University of Michigan, Ann Arbor, 1971.
49. G. R. Ramage and R. Robinson, *J. Chem. Soc., London* p. 607 (1933).
50. H. J. Lewis, G. R. Ramage, and R. Robinson, *J. Chem. Soc., London* p. 1412 (1935).
51. A. L. Wilds and R. E. Sutton, *J. Org. Chem.* 16, 1371 (1951).
52. A. J. Birch and H. Smith, *J. Chem. Soc., London* p. 1882 (1951).

References

53. G. Stork, H. N. Khastgir, and A. J. Solo, *J Amer. Chem. Soc.* **80**, 6457 (1958).
54. L. W. Butz, A. M. Gaddis, E. W. J. Butz, and R. E. Davis, *J. Amer. Chem. Soc.* **62**, 995 (1940).
55. L. W. Butz and L. M. Joshel, *J. Amer. Chem. Soc.* **63**, 3344 (1941).
56. L. M. Joshel, L. W. Butz, and J. Feldman, *J. Amer. Chem. Soc.* **63**, 3348 (1941).
57. L. W. Butz and L. M. Joshel, *J. Amer. Chem. Soc.* **64**, 1311 (1942).
58. F. E. Ray, E. Sawicki, and O. H. Brown, *J. Amer. Chem. Soc.* **74**, 1247 (1952).
59. P. S. Pinkney, G. A. Nesty, R. H. Wiley, and C. S. Marvel, *J. Amer. Chem. Soc.* **58**, 972 (1936).
60. H. J. Wille, U. K. Pandit, and H. O. Huisman, *Tetrahedron Lett.* p. 4429 (1970).
61. D. Bertin and J. Perronnet, *Bull. Soc. Chim. Fr.* [6] p. 117 (1969).
62. G. Van Binst and D. Tourwe, *Syn. Commun.* **1**, 295 (1971).

9
BC → ABC → ABCD

9.1 Synthesis of Epiandrosterone

The extensive and brilliant series of investigations on the total synthesis of nonaromatic steroids began as early as the late 1930's and were finally crowned with success in 1951, with the announcement of the synthesis of adrenocortical and sex hormones by the group of Robinson and Cornforth [1–3] and by Woodward [4,5]. The Woodward synthesis is described in Chapter 12. In the Robinson-Cornforth approach, the ring A was first added to the naphthalene derivatives containing B and C fragments of the steroidal skeleton. A reactive group was left in ring C, which was then utilized for the construction of ring D at at later stage.

The synthesis was carried out in several stages. The initial objective was the synthesis of the so-called Reich diketone **10** (Scheme 9-1), which had earlier been isolated by Reich as a degradation product of cholesterol and deoxycholic acid [3,6,7]. The next stage was the conversion of this diketone **10** into the Köster-Logemann ketone **15**, which in turn was available as a by-product during the large scale oxidation of cholesteryl acetate to dehydroepiandrosterone acetate [8]. In their actual synthesis, Robinson and Cornforth used the latter ketone obtained from the natural sources as the relay intermediate, for the construction of ring D.

9.1 Synthesis of Epiandrosterone

9.1.1 Reich Diketone

The starting material for the synthesis was 1,6-dihydroxy-naphthalene (**1**), which was converted into 5-methoxy-2-tetralone (**2**), by successive etherification, reduction, and hydrolysis (Scheme 9-1) [9,10]. After methylation with methyl iodide–sodium methoxide, the ring A was constructed using the Robinson ring annelation method. Ketone **3**, on treatment with 4-diethylamino-2-butanone methiodide and potassium ethoxide afforded the tricyclic ketone **4** in 71% yield. Cleavage of the ether linkage with hydrogen iodide in acetic acid gave the phenol **5**. In a later [11] modification, the tricyclic phenol **5** was synthesized by the same group of workers, using 5-acetamido-2-naphthol as the starting material.

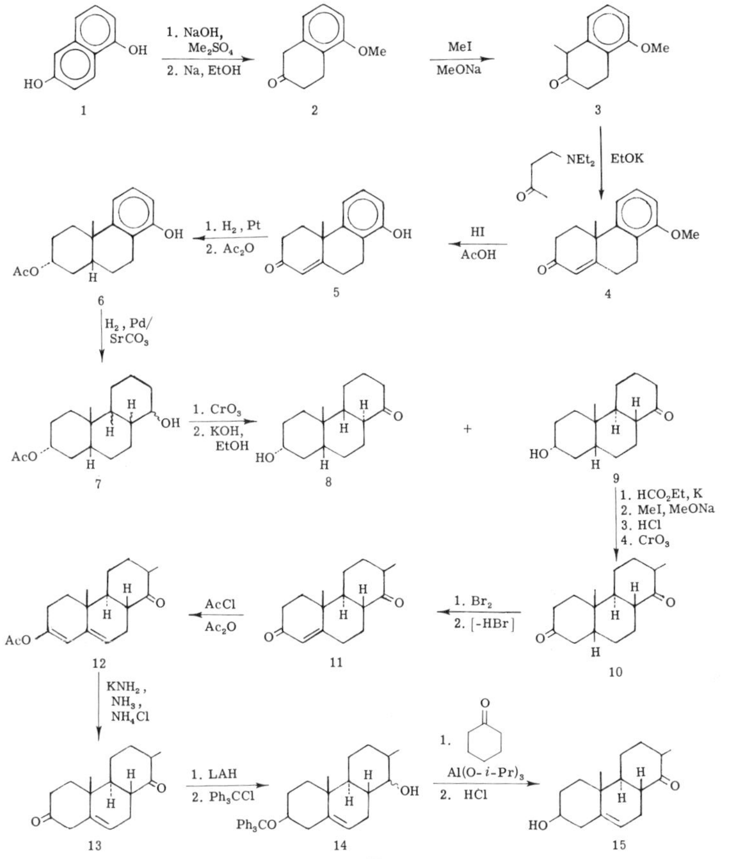

Scheme 9-1

Catalytic hydrogenation of the compound 5, followed by selective acetylation of the alcoholic hydroxyl group led to the AB-*cis* isomer **6**. Further hydrogenation using palladium-on-strontium carbonate as catalyst at 200°C reduced the aromatic ring to furnish a mixture of all four *dl*-hydroxy acetates **7**, two BC *cis* and two BC *trans*. Oxidation of this mixture with chromium trioxide was followed by alkaline hydrolysis to give a mixture of the isomeric hydroxy ketones **8** and **9**. The separation of the two isomers was effected by crystallization of the corresponding 3-hemisuccinates. Optical resolution of the two half-esters was effected via the brucine succinates of **9** and the *l*-menthoxy acetates of **8**. All four optically active hydroxy ketones were methylated at C-13 using a modified procedure of Sen and Mondal [12] and then oxidized at C-3 to give the diketones of the type **10**. One of these disastereomers **10**, derived from the (+)-form of the hydroxyketone **9**, was identical with the Reich diketone obtained as a degradation product of cholesterol and deoxycholic acid. Later these configurations were confirmed by the rotatory dispersion measurement studies by Klyne and Djerassi [13].

In their original studies Robinson and co-workers obtained a large amount of the undesired isomer of the hydroxyketone **8**. Further studies by Renfrow and Cornforth [14] revealed that by setting up a *trans* AB-ring junction initially, the reaction sequence could be controlled to give the desired *cis-anti-trans* isomer **9** as the major product.

9.1.2 Köster-Logemann Ketone

The Reich diketone **10** was the first relay compound to be used for the further synthesis of epiandrosterone. It was next converted to the "Köster-Logemann ketone" **15**, which was to serve as the second relay. Following a method developed by Billeter and Miescher [15], the diketone **10** was converted into the unsaturated diketone **11**. Treatment of the latter with a mixture of acetyl chloride and acetic anhydride afforded the enol acetate **12**. Ammonolysis of this compound with potassium amide in liquid ammonia followed by treatment with ammonium chloride [16,17] yielded the β,γ-unsaturated ketone **13**. Reduction with lithium aluminum hydride led to the diol, which was converted into the monotrityl derivative **14**. Oppenauer oxidation of the ring-C hydroxyl group followed by removal of the protecting group afforded the Köster-Logemann ketone **15**.

The next objective in the synthetic scheme was the construction of ring D onto ketone **15**, quantities of which were obtained from natural materials (Scheme 9-2). The hydroxyl group at C-3 was first protected by benzoylation, and a methoxy carbonyl group was then introduced by

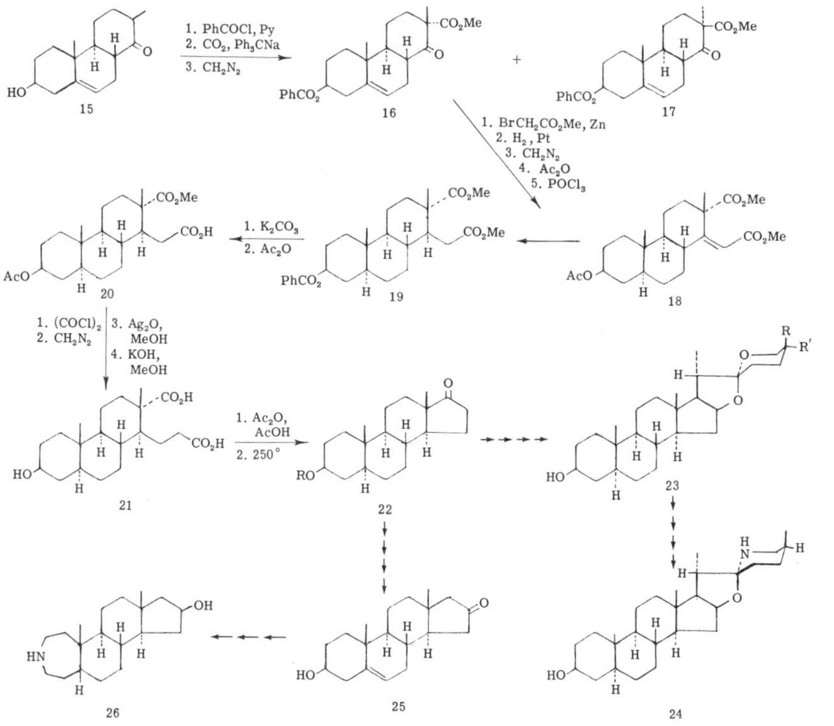

9.1 Synthesis of Epiandrosterone

Scheme 9-2

treatment of the benzoate with triphenylmethylsodium, followed by addition of carbon dioxide, acidification, and esterification. A mixture of two isomeric keto esters **16** and **17** was obtained in a ratio of 1:2.

After separation, the isomer **16** was converted into epiandrosterone acetate (**22**, R = Ac), following the method developed by Bachmann for the synthesis of equilenin (cf. Chapter 3, Scheme 3-2). Reformatsky reaction of **16**, followed by hydrolysis yielded a mixture of two epimeric dihydroxy half-esters. One of these two isomers was hydrogenated, esterified, acetylated, and dehydrated to give the diester **18**. After further hydrogenation, the resulting aetio-*allo*-bilianic ester was isolated as the benzoate **19**, identical with the material obtained from cholesterol.

Alkaline hydrolysis of the diester **19** followed by acetylation yielded the acetoxy half-ester **20**. This acid was now subjected to Arndt-Eistert reaction. Successive treatment of **20** with oxalyl chloride, diazomethane, ammoniacal silver nitrate in aqueous ethanol, and methanolic potassium hydroxide led to the homoacid **21**. The latter on acetylation, followed by pyrolysis and hydrolysis afforded epiandrosterone (**22**, R = H).

The synthesis of epiandrosterone by the Robinson-Cornforth approach was a classical example of the difficulties encountered during the early attempts of the total synthesis of nonaromatic steroids. The overall yield of the final product was less than 0.0004%. Nonetheless, this long series of studies explored many new avenues of synthetic possibilities, which later served as valuable guidelines to other workers in the field.

Also, since epiandrosterone can be transformed into testosterone, cholesterol, cortisone, androsterone, and many other steroidal derivatives, the above synthesis could be regarded as a formal synthesis of all these nonaromatic steroid hormones. Furthermore, epiandrosterone was the starting material for the synthesis of tigogenin (**23**, R=H, R'=Me) and neotigogenin (**23**, R=Me, R'=H), typical members of the large and important family of steroidal sapogenins [18]. The synthesis of these two compounds, in turn, constituted the total synthesis of a number of sapogenins including diosgenin, gitogenin, smilagenin, and hecogenin, because of their known interconversions. Also, in view of the interrelation between sapogenins and certain steroidal alkaloids, the synthesis of the former could mean the formal total synthesis of the corresponding alkaloids. This is illustrated by the conversion of neotigogenin to tomatidine (**24**) [19]. More recently, the Japanese workers [20] have completed the synthesis of samanine (**26**) from 3β-hydroxyandrost-5-en-16-one (**25**), the latter being obtained from epiandrosterone.

9.2 Other Methods

In later years, several modifications of the Robinson-Cornforth approach were made by different workers (Scheme 9-3). Thus Banerjee and co-workers [21,22], synthesized the intermediate diketone **11** from ethyl cyclohexanone carboxylate **27**. The latter was converted in several steps

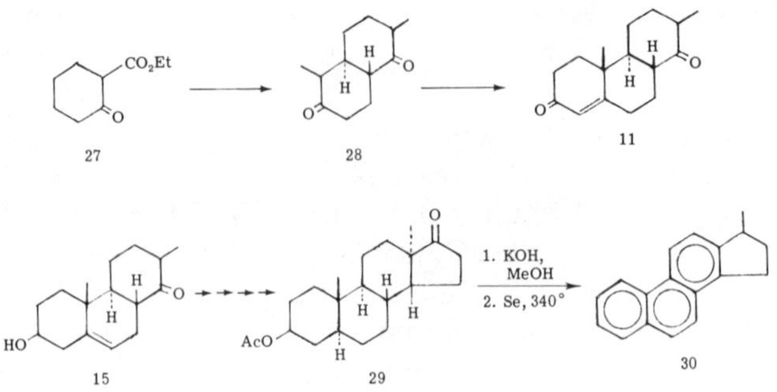

Scheme 9-3

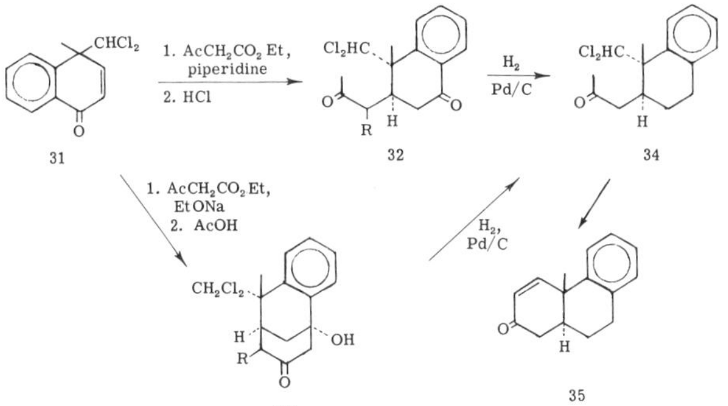

9.3 Synthesis According to Wilds

Scheme 9-4

to the diketone 28 which was transformed into the tricyclic ketone 11 in about 2.5% yield.

The Köster-Logemann ketone 15, was also utilized by Billeter and Miescher [23] for the synthesis of the tetracyclic 13α-compound 29. Hydrolysis of the latter followed by dehydrogenation yielded the Diels hydrocarbon, 3′-methyl-1,2-cyclopentenophenanthrene (30).

A totally different, stereospecific synthesis of the hydrophenanthrone system was developed by Wenkert and Stevens [24] (Scheme 9-4). The base-catalyzed Michael condensation of the naphthalenone 31 with ethyl acetoacetate afforded the keto esters 32 (R = COOEt) or 33 (R = COOEt) (depending on reaction conditions). Acid hydrolysis of either ester led to a mixture of the ketones 32 (R = H) and 33 (R = H), both of which, upon hydrogenation gave the same ketone 34. Treatment of the latter with sodium triphenylmethide led to the tricyclic ketone 35.

9.3 Synthesis According to Wilds

Another short route to the tricyclic intermediates for the synthesis of nonaromatic steroids was described by Wilds and co-workers [25,26] (Scheme 9-5). Alkylation of dihydroresorcinol (36, not shown) with 1-(N,N-diethylamino)-3-pentanone methiodide was followed by cyclization and hydrolysis to yield the diketone 37. Ring A was constructed by treatment of the latter compound with methyl vinyl ketone, followed by alkaline cyclization. The resulting tricyclic ketone 38, on catalytic hydrogenation in the presence of alkali yielded the diketone 39.

The dione 39 was converted into the 3-ethylenedioxy-13-hydroxymethylene derivative 40, which was then alkylated successively with methyl

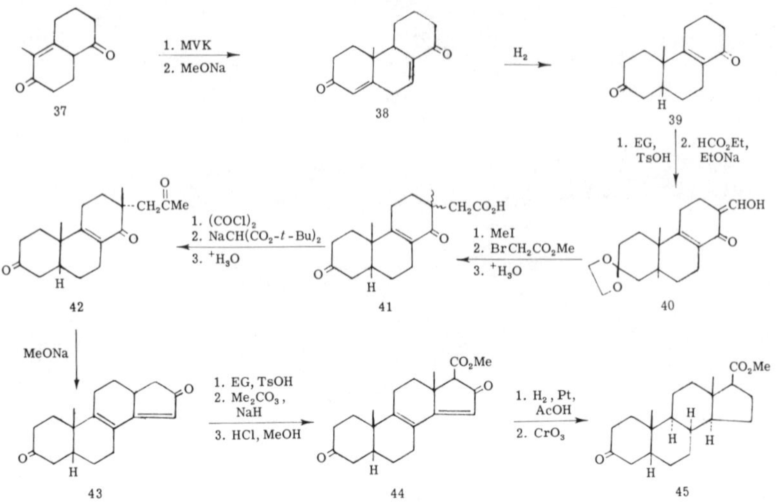

9 BC → ABC → ABCD

Scheme 9-5

iodide and methyl bromoacetate. The resulting diketo acid **41** was isolated in two isomeric forms. The acid chloride of one of them (13β) was converted into the methyl ketone **42**, which on cyclization with sodium methoxide, yielded the tetracyclic diketone **43**. Selective ketal formation at C-3, followed by successive reaction with methyl carbonate in the presence of sodium hydride and finally acid hydrolysis led to the keto ester **44**. The latter compound, on hydrogenation and oxidation yielded the keto ester **45**, which was identified as the methyl ester of dl-3-oxo-5β-etianoic acid. This acid had been previously converted into deoxycorticosterone and progesterone.

9.4 Synthesis of Cortisone

The elegance in the total synthesis of cortisone (**70**, R = H) by Sarett and co-workers [27–39] lies in the fact that a high degree of stereospecificity was maintained in each step. Besides this, no relay compound was involved in their reaction scheme. The Diels-Alder reaction of 3-ethoxy-1,3-pentadiene with benzoquinone (**46**) was followed by selective hydrogenation to give the diketone **47**. (Scheme 9-6). Successive reduction with lithium aluminum hydride and hydrolysis gave the diol **48**. Following Robinson's annelation method, ring A was constructed by the Triton B-catalyzed addition of methyl vinyl ketone to **48**. The unsaturated ketone system of the resulting product **49** was next protected by con-

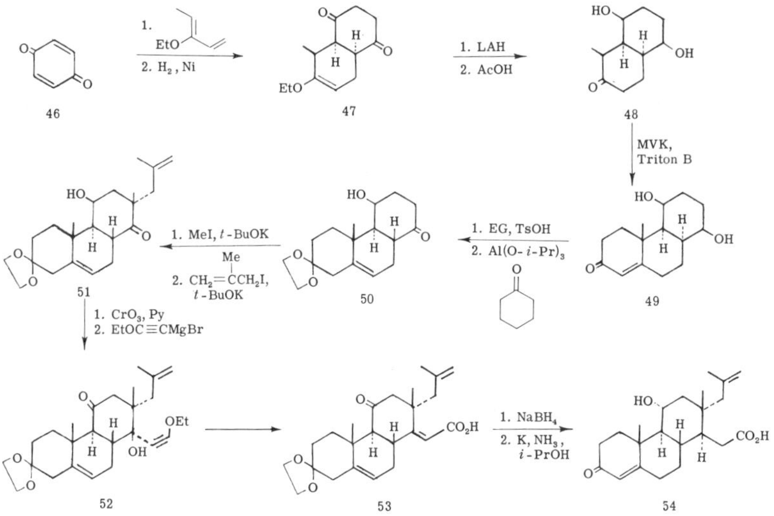

9.4 Synthesis of Cortisone

Scheme 9-6

version into the 3-ethylene ketal. Oppenauer oxidation of this latter compound was accompanied by epimerization at C-8 to give the so-called Sarett's ketone **50**. The synthesis of this last compound was a major achievement in the total synthesis of cortical hormones. Besides its utilization in the synthesis of cortisone, this compound served as the starting material for the synthesis of aldosterone by the Ciba group under the leadership of Wettstein, by the Dutch group under the leadership of Szpilfogel, and by the group of Reichstein (see Section 9.5).

The second stage of the synthesis involved the construction of ring D, and was achieved with remarkable stereoselectivity at almost every step. In addition, the acid sensitive protective 3-ketal group was retained until the final step, which necessitated the exclusion of strongly acidic media.

Successive alkylation of the Sarett's ketone **50** with methyl iodide and methallyl iodide in the presence of potassium *t*-butoxide took place in the desired stereospecific direction to give the keto alcohol **51**. The nearly quantitative oxidation of the C-11 hydroxyl group of **51** was carried out with chromic acid–pyridine complex to avoid cleavage of the protective ketal group at C-3. Condensation of the resulting product with ethoxyethynylmagnesium bromide yielded the carbinol **52**. Hydrolysis and dehydration of the latter yielded the acid **53**. Consecutive reduction with sodium borohydride and potassium–ammonia–isopropyl alcohol afforded the C-14β-acetic acid **54**, which contained the required stereochemical arrangement of the cortisone ring system.

No fewer than three different methods were developed by Sarett and co-workers for the construction of ring D from the hydroxy acid **54** (Scheme 9-7). In the first of them [33,35], reduction of the free acid **54** or its methyl ester with lithium aluminum hydride produced the diol **55** in good yield. Selective tosylation and subsequent oxidation with chromium trioxide–pyridine complex afforded the ketotosylate **57**. Treatment of the latter compound with one equivalent of osmium tetroxide yielded a mixture of the isomeric glycols, which on periodic acid oxidation yielded the diketotosylate **60**. The latter, on cyclization under the action of sodium methoxide, yielded initially the 17α-epimer of the diketone **63**. However, isomerization to the desired 11-keto derivative **64** could be readily effected by treatment of **63** with refluxing potassium carbonate in aqueous methanol.

A second method [38] for the conversion of the tricyclic diol **55** into the 11-ketoprogesterone derivative **64**, involved the initial oxidation of

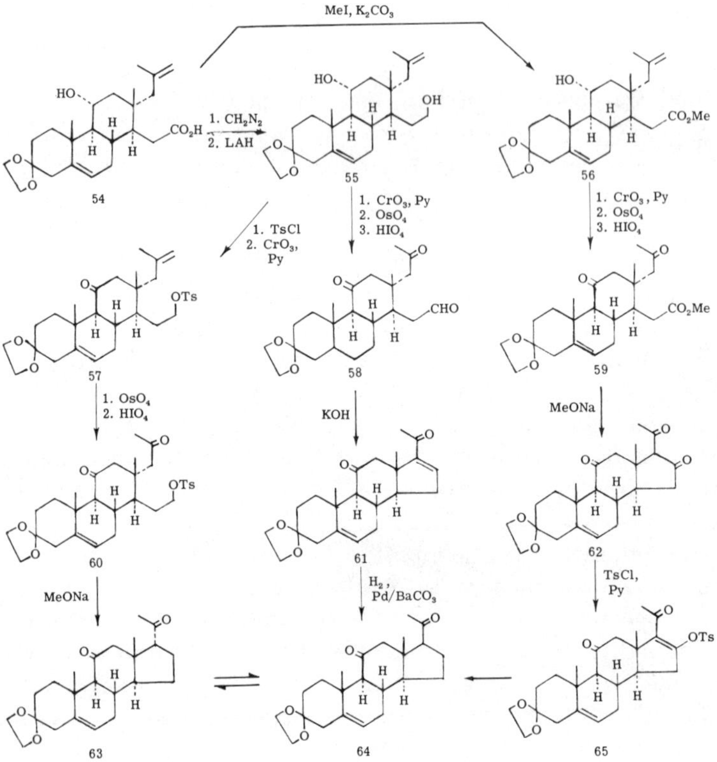

Scheme 9-7

9.4 Synthesis of Cortisone

the diol with chromic anhydride in pyridine. Selective hydroxylation of the methallyl double bond in the resulting keto aldehyde with osmium tetroxide, followed by cleavage of the glycol with periodic acid, afforded the diketo aldehyde 58. An intramolecular aldol condensation was effected by treatment of this aldehyde with potassium hydroxide. The resulting 3-ethylene ketal (61) of 16,17-dehydro-11-ketoprogesterone was hydrogenated in presence of palladium-on-barium carbonate to give 64.

In yet another method [39] for the closure of the ring D, the acid 54 was first esterified with methyl iodide and potassium carbonate to give the hydroxyester 56. It was followed by successive oxidation with chromium trioxide–pyridine complex, hydroxylation with osmium tetroxide, and periodic acid cleavage, which led the diketo ester 59 in excellent yield. Cyclization of this ester with sodium methoxide proceeded almost quantitatively to give the triketone 62. Treatment of this latter compound with p-toluenesulfonyl chloride in pyridine afforded a difficultly separable mixture of the enol tosylates, from which a small amount of the compound 65 could be isolated by chromatography. A one-step hydrogenolysis–hydrogenation with palladium-on-barium carbonate catalyst was used to convert both pure tosylate 65 and the mixture of tosylates to the 11-keto derivative 64. The overall yield of the diketone 64, from the acid 54 was about 50%, by this method.

The final stage in the synthesis of cortisone (70, R = H) involved the introduction of the 17α- and 21-hydroxyl groups into the diketone 64 (Scheme 9-8) [37,40–42]. Following the method of Ruschig [43,44], the diketone was condensed with methyl oxalate, and the resulting glyoxalate was hydrolyzed to the acid 66. The latter was resolved as the strychnine salts and the (+)-acid was treated with iodine and potassium acetate to produce the ketol acetate 67. Hydrolysis of the corresponding racemic mixture of 67 yielded racemic dehydrocorticosterone (68).

Introduction of the 17-hydroxy group was carried out by the cyanohydrin method developed by Sarett [45,46]. Treatment of the ketol acetate 67 with hydrogen cyanide and triethylamine afforded the 20-cyanohydrin, which on dehydration with phosphorous oxychloride led to the unsaturated nitrile 69. Selective hydroxylation of the latter with potassium permanganate was followed by hydrolysis to give dl-cortisone acetate (70, R = Ac). Similarly, application of the above mentioned 21-acetoxylation sequence to the 3-ketal (71) of dl-11β-hydroxyprogesterone (obtained from 54) provided dl-corticosterone acetate (72) [35,37].

Sarett and his group also studied the possibilities of optical resolution at various intermediate stages in their synthetic schemes (Scheme 9-8). Thus, the reduction of the ketone 50 afforded the diol 73, which was

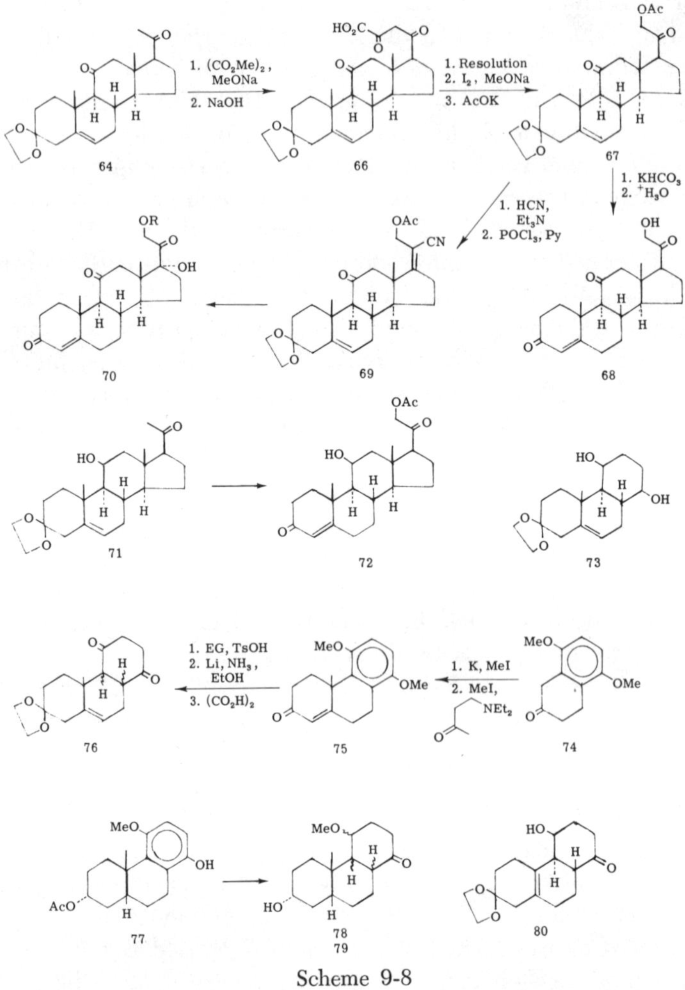

Scheme 9-8

resolved into its optical antipodes by crystallization of the brucine salts of the 14-hemisuccinates [47] or the strychnine salts of the 14-hemiphthalates [48]. Also, the oxidation of the hydroxyketone **50** with chromium trioxide–pyridine complex yielded the *anti-trans* diketone **76** (8β,9α). Two other isomers, the *anti-cis* diketone **76** (8α,9α) and *syn-trans* diketone **76** (8α9β) were also prepared by the Merck group [33,34]. However, attempts to synthesize the fourth isomer, *syn-cis* diketone were not successful [49].

The tricyclic diketone **76** (8β,9α) was also synthesized by Grob and co-workers [50–54] following a different approach. Methylation of the ketone 74 was followed by construction of ring A by Robinson's method to give the tricyclic ketone 75 in about 60% yield. The 3-keto group was protected as the ethylene ketal. Birch reduction followed by mild acid hydrolysis afforded the diketone 76 (8β,9α) in 54% yield.

In still another approach for the synthesis of tricyclic ABC intermediates, the ring C of the 3α,5β-derivative 77 was reduced by hydrogenation in the presence of a ruthenium catalyst [55,56]. Oxidation of the resulting product, followed by alkaline hydrolysis afforded a mixture of the two epimeric methoxyketones 78 (8α,9β,11α) and 79 (8β,9α,11β) in a ratio of 14:1. The formation of the unnatural 8α,9β,11α-isomer 78 could be explained by the preferential hydrogenation of the AB-*cis*-isomers from the β-direction.

The synthesis of the 19-nor analog of the Sarett's ketone was also studied by Pivnitskii and Torgov [57,58], who synthesized the ketol 80, an intermediate with potential application in the synthesis of 19-nor corticoids. However, no attempt toward the construction of the Ring D was made. In another series of studies, Hogg and co-workers [58a] achieved the synthesis of cortisone acetate from 11-ketoprogesterone. The latter compound was obtained by oxidation of 11α-hydroxyprogesterone, which itself was obtained by a novel microbiological oxidation of progresterone [58b]. Also, an elegant method for the elaboration of the pregnane and corticoid chains at position 17 in epiandrosterone acetate (22, R = Ac) has been described by the Syntex group [58c].

9.5 Synthesis of Aldosterone

The intense interest in the field of steroidal hormones was remarkably exemplified by the synthesis of aldosterone (96), which was first isolated in a pure state in 1953 [59,60]. Two years later, the Swiss group at Ciba achieved the total synthesis of this powerful adrenal cortical hormone [61]. By 1958, at least four more syntheses were announced by different groups [62-65].

9.5.1 Wettstein's First Synthesis of Aldosterone

The starting material in the total synthesis of aldosterone [61,66-68] by Schmidlin, Anner, Billeter, and Wettstein was the Sarett's ketone 50 (Scheme 9-9). Condensation of this compound with ethyl carbonate afforded the keto ester 81, which was alkylated with methallyl iodide to give the substituted keto ester 82. Subsequent reduction with sodium

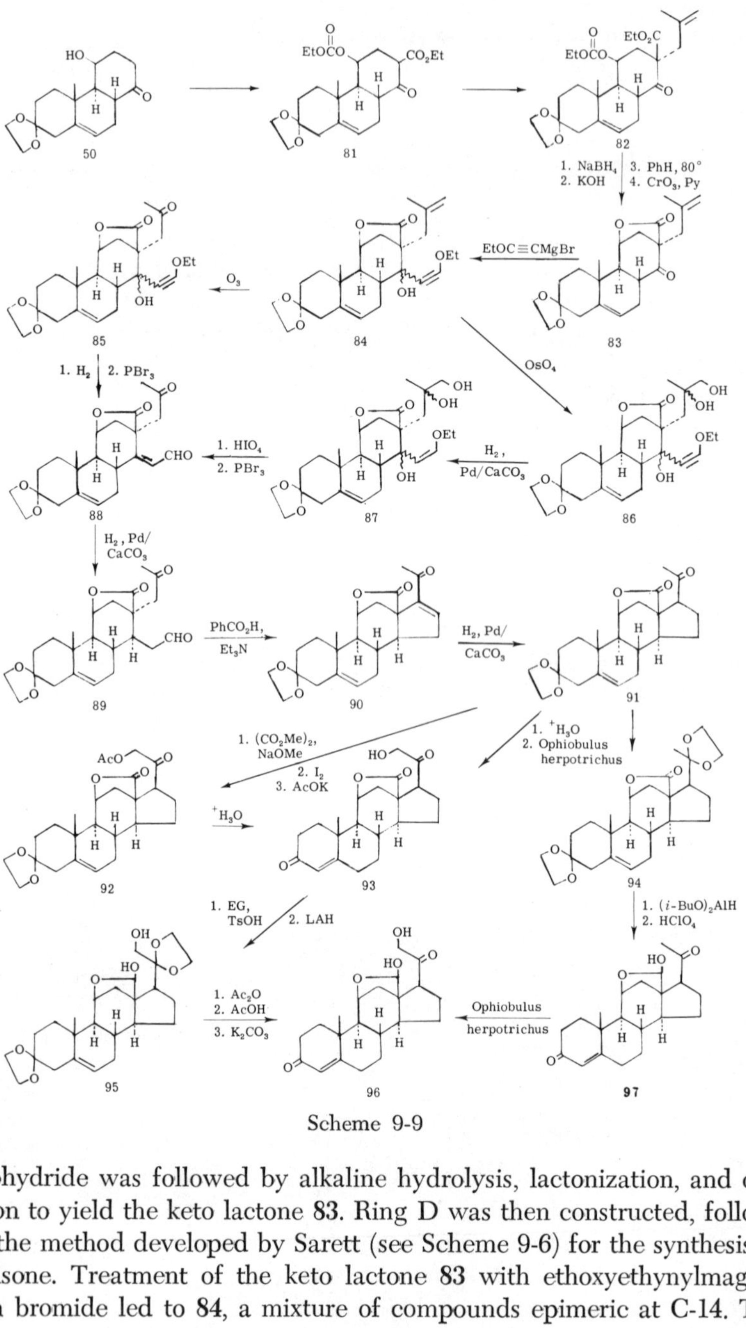

Scheme 9-9

borohydride was followed by alkaline hydrolysis, lactonization, and oxidation to yield the keto lactone **83**. Ring D was then constructed, following the method developed by Sarett (see Scheme 9-6) for the synthesis of cortisone. Treatment of the keto lactone **83** with ethoxyethynylmagnesium bromide led to **84**, a mixture of compounds epimeric at C-14. The

9.5 Synthesis of Aldosterone

14β-configuration of the hydroxy group was assigned to the predominating epimer. Oxidation of this product **84** with osmium tetroxide led to a mixture of triols **86**, which upon selective hydrogenation, afforded the vinylcarbinols **87**. The latter were converted into a mixture of *cis-trans*-isomeric aldehyde **88**, by successive treatment with periodic acid and phosphorous tribromide [61,67]. In an alternative method [67], the compound **84** was ozonized to give the hydroxy ketone **85**. Selective hydrogenation, followed by hydrolysis led to the keto aldehyde **88**.

Catalytic hydrogenation of **88** proceeded stereospecifically to give the aldehyde **89** as the major product. Cyclization with benzoic acid and triethylamine led to the tetracyclic ketone **90**, which was hydrogenated again to the keto lactone **91**.

The next stage in the synthesis was the introduction of the 21-hydroxy group. This was accomplished by following the method developed by Ruschig [43,44]. Condensation of the ketone **91** with dimethyl oxalate yielded the oxalyl derivative, which on subsequent iodination and treatment with potassium acetate led to the keto acetate **92**. Acid hydrolysis of the latter compound gave the racemic diketone **93**.

Later the Swiss group developed a microbiological hydroxylation method [69] for the introduction of the 21-hydroxy group, which was vastly more simple. The methyl ketone **91**, after deketalization, was treated with the fungus *Ophiobolus herpotrichus*, whereupon half of the material was converted to the desired *d*-enantiomer of the ketol **93**. The material not attacked by the fungus was isolated and identified as the *l*-enantiomer of **91**. In order to complete the synthesis of aldosterone, it was necessary to reduce the lactone group of the ketol **93**. This was accomplished by first protecting the keto groups as the diketal. Subsequent reduction with lithium aluminum hydride, acetylation, hydrolysis, and deacetylation afforded *d*-aldosterone (**96**). In another modification [70–72], the keto lactone **91** was transformed into the diketal **94**. Reduction of the lactone group with diisobutylaluminum hydride, followed by deketalization, yielded the diketone **97**. The latter, on microbiological hydroxylation with the fungus *Ophiobolus herpotrichus* gave the desired *d*-aldosterone (**96**).

9.5.2 Synthesis of Aldosterone by Reichstein

A highly ingenious method for the construction of ring D was developed by Reichstein and his group during their synthesis of aldosterone [62,73–76] (Scheme 9-10). Once again the starting material was the Sarett's ketone **50**, which upon successive condensation with ethyl carbo-

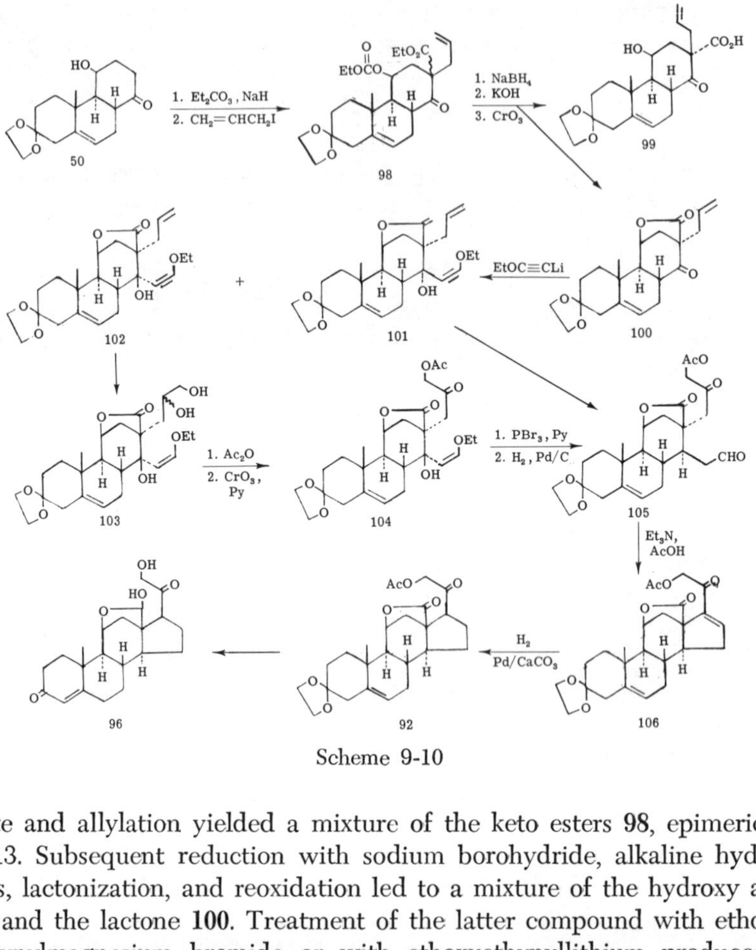

Scheme 9-10

nate and allylation yielded a mixture of the keto esters **98**, epimeric at C-13. Subsequent reduction with sodium borohydride, alkaline hydrolysis, lactonization, and reoxidation led to a mixture of the hydroxy acid **99** and the lactone **100**. Treatment of the latter compound with ethoxyethynylmagnesium bromide or with ethoxyethynyllithium produced a mixture of the isomeric carbinols **101** and **102**. Subsequently, both of these isomers were separately converted into the unsaturated aldehyde **105** by following the same sequence of reactions.

Oxidation of the lactone **102** with osmium tetroxide was followed by selective hydrogenation to give the triol **103**. The latter, upon selective acetylation and oxidation led to the ketol **104**. Treatment with phosphorus tribromide and catalytic hydrogenation yielded the aldehyde **105**. Cyclization afforded the tetracyclic product **106**, which upon hydrogenation, led to the previously known acetoxy ketone **92**. The conversion of the latter to aldosterone **96** has already been described in Scheme 9-9.

9.5.3 Szpilfogel's Synthesis of Aldosterone

The Organon group in Oss, Holland, working in close cooperation with the Swiss workers, announced a third synthesis of aldosterone in 1958 [63,77,78] (Scheme 9-11). The starting material was the Sarett's ketone **50**, which on formylation, followed by Michael condensation with acrolein yielded the dialdehyde **107**. Treatment of the latter with methanolic hydrochloric acid, followed by chromium trioxide oxidation of the resulting product yielded the keto lactone **108**. Introduction of the side chain at C-14 was accomplished by treatment of this lactone with ethoxyethynyllithium, affording a mixture of the two isomeric carbinols **109**. The two epimers were separated by chromatography, and the 14α-hydroxy isomer was used in the further synthesis. Selective hydrogenation followed by treatment with thionyl chloride and pyridine afforded the unsaturated aldehyde **110**. Catalytic hydrogenation of this latter compound, followed by mild hydrolysis gave the unstable dialdehyde **111**. Without isolation, the dialdehyde was cyclized by treatment with a catalytic amount of piperidine acetate in boiling benzene. The resulting aldehyde **112**, which could not be isolated in crystalline form, was

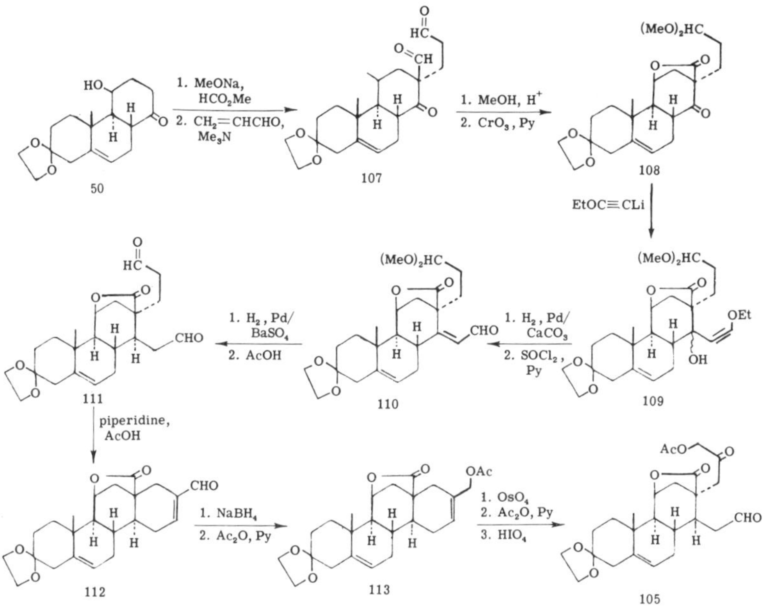

Scheme 9-11

reduced and acetylated to yield the acetate **113**. The latter was converted into the known tricyclic aldehyde **105** by oxidation, acetylation, and treatment with periodic acid. The conversion of the aldehyde **105** to aldosterone has already been described in Scheme 9-10. The overall yield of the aldehyde **105** from the Sarett's ketone **50** by the Organon method was 13% as compared to the yield of 2% by Reichstein's method. However, in spite of various modifications, cyclization of the keto aldehyde **105** proceeded with poor yield. The overall yield of aldosterone in the entire synthesis, based on Sarett's ketone **50**, was 0.0017%.

9.5.4 Wettstein's Second Synthesis of Aldosterone

The intensive search by the Ciba group for a better method of the synthesis of aldosterone [66–72,79–93] finally culminated in the announcement of their second synthesis [64,65] of this hormone in 1957–1958 (Scheme 9-12). Alkylation of the Sarett's ketone **50** with methallyl iodide [91] led to a dialkyl derivative **114**. Ozonolysis of the latter yielded a diketone, which, on treatment with pyridine cyclized to the dihydropyran **115**. Ring closure in the presence of strong base afforded the α,β-unsaturated steroidal ketone **116** in excellent yield. Since attempted hydrogenation of this compound gave a product with the unnatural 14β-configuration, the 14,15-double bond was temporarily protected by oxide formation.

Treatment of the resulting epoxide **117** with osmium tetroxide and subsequent periodic acid cleavage led to the acetoxy keto aldehyde **118**. Saponification of the 11-acetoxy group was accompanied by spontaneous cyclization to the hemiacetal **119**. After protecting the hydroxyl group as the tetrahydropyranyl derivative **120**, the epoxide ring was now opened up by catalytic hydrogenation. This was followed by dehydration to give the unsaturated ketone **122**.

The next phase in the synthesis involved the introduction of the side chain at C-17. Condensation with dimethyloxalate, followed by treatment with acetic anhydride afforded the enol acetate **124**. Attempted hydrogenation of the intermediate condensation product **123** or the enol acetate **124** led to mixture of various products [83]. However, better results were obtained by conversion of **124** to the morpholide **125**, which, on subsequent hydrogenation over palladium and reduction with sodium borohydride yielded the hydroxy derivative **126**, with the desired 14α-configuration.

It is interesting to note that the steric course of hydrogenation of the Δ^{14}-16-ketones such as **116** and **125** depended upon the presence or

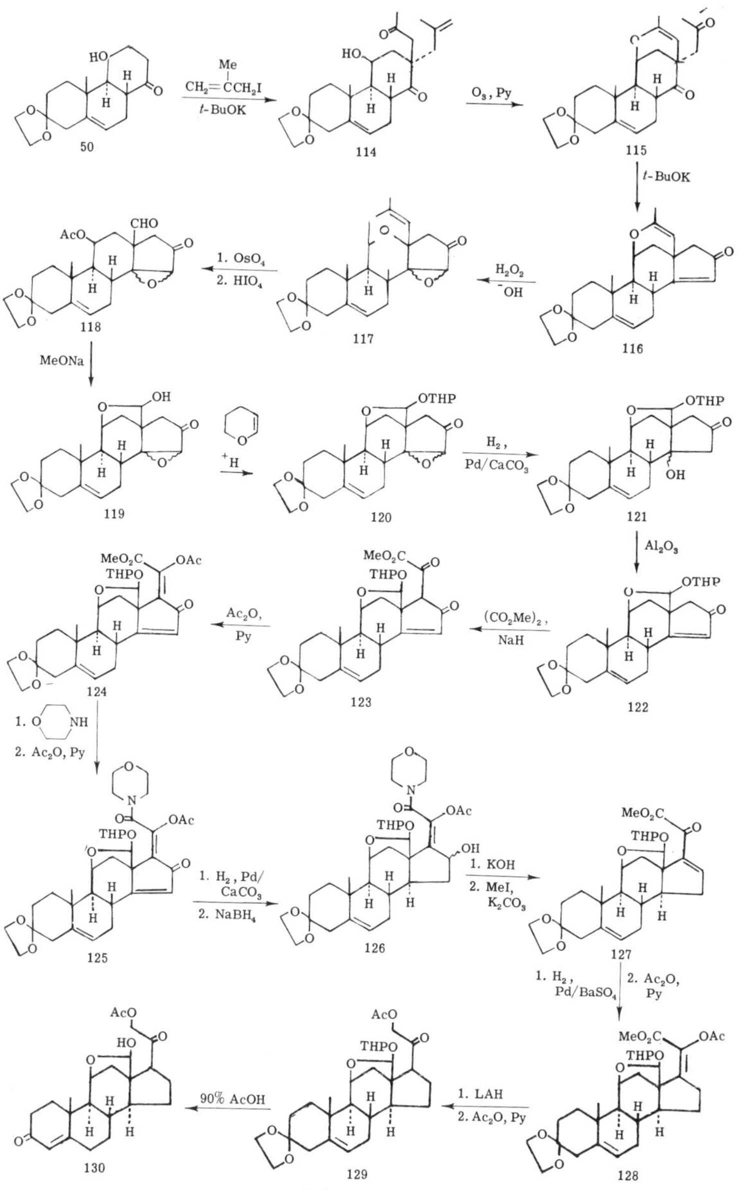

9.5 Synthesis of Aldosterone

Scheme 9-12

absence of a direct bond between the 11β-oxygen atom and carbon atom 18. In the absence of such direct bonds, the unnatural 14β-derivative was the main product, whereas the presence of such bonds facilitated the formation of the natural CD-*trans* products.

Both the morpholide and acetyl groups in the amide 126 were now removed by alkaline hydrolysis under conditions leading to concurrent dehydration. Subsequent methylation gave the unsaturated keto ester 127 which was hydrogenated and acetylated to the enol acetate 128. Reduction of this latter compound with lithium aluminum hydride and acetylation led to the keto acetate 129. This procedure was followed by hydrolysis of the two protective groups, whereupon the desired aldosterone acetate (130) was obtained in excellent yield. In another modification the unsaturated keto ester 127 could be converted to the final product 130, by successive reduction, enol acetylation and hydrolysis.

Many partial syntheses of aldosterone [94-102] have been described in the literature, the detailed discussion of which is beyond the scope of this chapter. However, two of these approaches deserve special mention for their ingenuity and wide applicability. The first of them was based on the discovery of Jeger and his collaborators of the 1,4 free radical transfer reaction [99,100,103-105]. The partial synthesis of aldosterone by Barton is another application of the 1,4 transfer reaction [101,102,106]. This brilliant synthesis involving the photochemical transformation of a nitrite ester to an oxime, is one of the most practical ways to obtain aldosterone acetate.

9.6 3β-Hydroxy-5α-pregnan-20-one, Latifoline, and Conessine

A straightforward, stereospecific total synthesis of racemic 3β-hydroxy-5α-pregnan-20-one (139) was developed by Nagata and his group [107, 108] in 1963 (Scheme 9-13). The starting material was 6-methoxy-1-tetralone (131), which was converted to 6-methoxy-1-methyl-2-tetralone (132) in three stages. The ring A was constructed following Robinson's ring extension method to give the tricyclic ketone 133 [109,110]. Modified Birch reduction, using lithium and alcohol in liquid ammonia was followed by acid hydrolysis and benzoylation to yield the unsaturated ketone 134 in an overall 66% yield. Alkylation of this ketone with 5-bromopentan-2-one ethylene ketal, and subsequent alkaline hydrolysis led to the diketone 135. The latter compound without purification was converted to the tetracyclic methyl ketone 136 by successive reduction, deketalization, and cyclization. The overall yield of 136 from the tricyclic conjugated ketone 134 was 52%. A hydrocyanation process, which had been so successfully developed by Nagata [111-116]

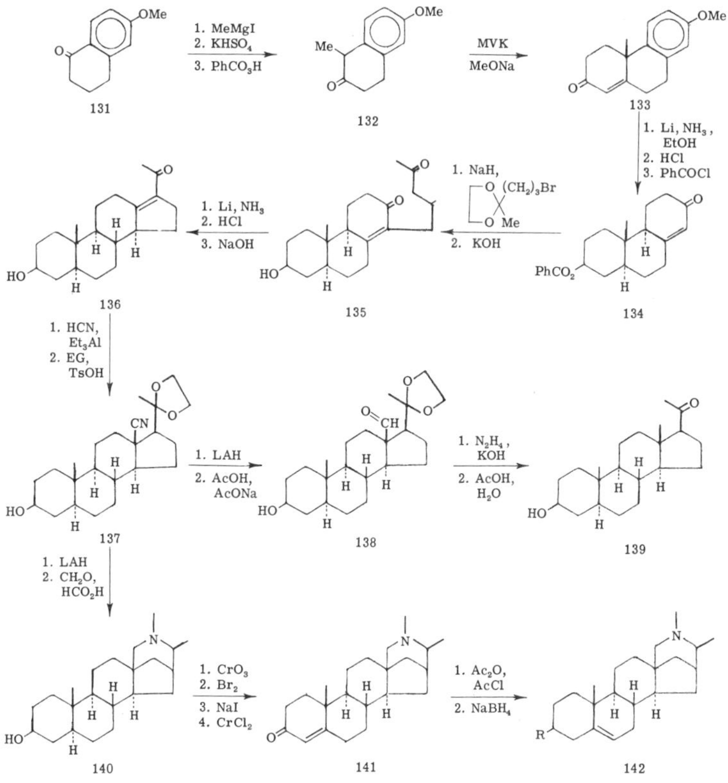

9.6 3β-Hydroxy-5α-pregnan-20-one, Latifoline, and Conessine

Scheme 9-13

was used for the introduction of the angular methyl group at C-13. Treatment of the ketone **136** with hydrogen cyanide and triethylaluminum yielded almost exclusively the 13β-cyano ketone, which was next converted to the ethylene ketal **137**. Reduction with lithium aluminum hydride and hydrolysis gave the 13β-formyl derivative **138**. This was followed by Huang-Minlon reduction and deketalization to produce the racemic 3β-hydroxy-5α-pregnan-20-one (**139**). The overall yield of the final product **139** from the tricyclic ketone **133** was 16.2%.

The racemic 13β-cyano-18-norpregnane derivative **137** also served as the starting material for the total synthesis of the alkaloids latifoline (**142**, R=OH) and conessine (**142**, R=NMe$_2$). Reduction with lithium aluminum hydride at 95–100°C afforded a pyrrolidine derivative, which on treatment with formaldehyde and formic acid yielded the racemic dihydrolatifoline **140**. Chromic acid oxidation led to the 3-ketone, which was converted into racemic conan-4-en-3-one (**141**) by bromination, reaction with sodium iodide in methyl ethyl ketone and reduction with

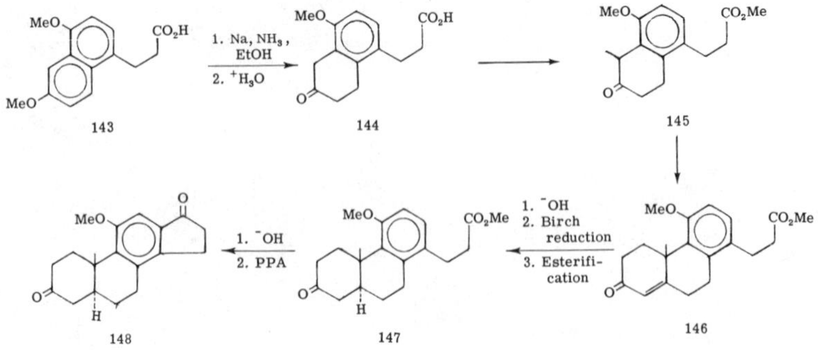

Scheme 9-14

chromous chloride [117]. The ketone **141**, on enolacetylation and subsequent sodium borohydride reduction afforded the racemic latifoline (**142**). Also, since the same ketone **141** had been previously converted to conessine (**142**, R=NMe$_2$) [118], this route constituted, in a formal sense, the total synthesis of this alkaloid.

9.7 Ring-C Aromatic Steroids

The total synthesis of a ring-C aromatic 18-nor steroid was accomplished by Chatterjee and Hazra [119] (Scheme 9-14). Reduction of the known acid **143** [120] with sodium and alcohol in liquid ammonia was followed by acid hydrolysis to afford the keto acid **144**. This was esterified, and the enamine of the ester was alkylated with methyl iodide to give the keto ester **145**. Treatment with the methiodide of 1-diethylaminobutan-3-one afforded the tricyclic ester **146**. The latter, on successive alkaline hydrolysis, Birch reduction, and esterification, furnished the keto ester **147**. Finally, alkaline hydrolysis and subsequent cyclization with polyphosphoric acid led to the C aromatic compound **148** in excellent yield.

References

1. H. M. E. Cardwell, J. W. Cornforth, S. R. Duff, H. Holtermann, and R. Robinson, *Chem. Ind. (London)* p. 389 (1951).
2. H. M. E. Cardwell, J. W. Cornforth, S. R. Duff, H. Holtermann, and R. Robinson, *J. Chem. Soc., London* p. 361 (1953).
3. J. W. Cornforth and R. Robinson, *Nature (London)* **160**, 737 (1947).
4. R. B. Woodward, F. Sondheimer, D. Taub, K. Heusler, and W. M. McLamore, *J. Amer. Chem. Soc.* **73**, 2403 (1951).

5. R. B. Woodward, F. Sondheimer, D. Taub, K. Heusler, and W. M. McLamore, *J. Amer. Chem. Soc.* **74**, 4223 (1952).
6. H. Reich, *Helv. Chim. Acta* **28**, 892 (1945).
7. J. W. Cornforth and R. Robinson, *J. Chem. Soc., London* p. 676 (1946).
8. H. Köster and W. Logemann, *Ber. Deut. Chem. Ges.* B **73**, 298 (1940).
9. J. W. Cornforth and R. Robinson, *J. Chem. Soc., London* p. 1855 (1949).
10. J. W. Cornforth, R. H. Cornforth, and R. Robinson. *J. Chem. Soc., London* p. 689 (1942).
11. J. W. Cornforth, O. Kauder, J. E. Pike, and R. Robinson, *J. Chem. Soc., London* p. 3348 (1955).
12. H. K. Sen and K. Mondal, *J. Indian Chem. Soc.* **5**, 609 (1928).
13. W. Klyne and C. Djerassi, *Chem. Ind. (London)* p. 988 (1956).
14. W. B. Renfrow and J. W. Cornforth, *J. Amer. Chem. Soc.* **75**, 1347 (1953).
15. F. Billeter and K. Miescher, *Helv. Chim. Acta* **33**, 388 (1950).
16. A. J. Birch, *J. Chem. Soc., London* p. 2325 (1950).
17. C. W. Shoppe and G. H. R. Summers, *J. Chem. Soc., London* p. 687 (1950).
18. Y. Mazur, N. Danieli, and F. Sondheimer, *J. Amer. Chem. Soc.* **82**, 5889 (1960).
19. F. C. Uhle and J. A. Moore, *J. Amer. Chem. Soc.* **76**, 6412 (1954).
20. K. Oka and S. Hara, *Tetrahedron Lett.* p. 1193 (1969).
21. A. Banerjee and B. K. Bhattacharyya, *J. Indian Chem. Soc.* **35**, 467 (1958).
22. A. Banerjee, B. B. Mukherjee, and B. K. Bhattacharyya, *J. Indian Chem. Soc.* **36**, 755 (1959).
23. J. R. Billeter and K. Miescher, *Helv. Chim. Acta* **34**, 2053 (1951).
24. E. Wenkert and T. E. Stevens, *J. Amer. Chem. Soc.* **78**, 5627 (1956).
25. A. L. Wilds, J. W. Ralls, W. C. Wildman, and K. E. McCaleb, *J. Amer. Chem. Soc.* **72**, 5794 (1950).
26. A. L. Wilds, J. W. Ralls, D. A. Tyner, R. Daniels, S. Kraychy, and M. Harnik, *J. Amer. Chem. Soc.* **75**, 4878 (1953).
27. L. H. Sarett, G. E. Arth, R. M. Lukes, R. E. Beyler, G. I. Poos, W. F. Johns, and J. M. Constantin, *J. Amer. Chem. Soc.* **74**, 4974 (1952).
28. L. H. Sarett, R. M. Lukes, G. I. Poos, J. M. Robinson, R. E. Beyler, J. M. Vandegrift, and G. E. Arth, *J. Amer. Chem. Soc.* **74**, 1393 (1952).
29. R. E. Beyler and L. H. Sarett, *J. Amer. Chem. Soc.* **74**, 1397 (1952).
30. R. M. Lukes, G. I. Poos, and L. H. Sarett, *J. Amer. Chem. Soc.* **74**, 1401 (1952).
31. R. E. Beyler and L. H. Sarett, *J. Amer. Chem. Soc.* **74**, 1406 (1952).
32. R. M. Lukes, G. I. Poos, R. E. Beyler, W. F. Johns, and L. H. Sarett, *J. Amer. Chem. Soc.* **75**, 1707 (1953).
33. L. H. Sarett, W. F. Johns, R. E. Beyler, R. M. Lukes, G. I. Poos, and G. E. Arth, *J. Amer. Chem. Soc.* **75**, 2112 (1953).
34. G. I. Poos, G. E. Arth, R. E. Beyler, and L. H. Sarett, *J. Amer. Chem. Soc.* **75**, 422 (1953).
35. W. F. Johns, R. M. Lukes, and L. H. Sarett, *J. Amer. Chem. Soc.* **76**, 5026 (1954).
36. G. E. Arth, G. I. Poos, R. M. Lukes, F. M. Robinson, W. F. Johns, M. Feurer, and L. H. Sarett, *J. Amer. Chem. Soc.* **76**, 1715 (1954).
37. G. I. Poos, R. M. Lukes, G. E. Arth, and L. H. Sarett, *J. Amer. Chem. Soc.* **76**, 5031 (1954).

38. G. I. Poos, W. F. Johns, and L. H. Sarett, *J. Amer. Chem. Soc.* **77**, 1026 (1955).
39. G. E. Arth, G. I. Poos, and L. H. Sarett, *J. Amer. Chem. Soc.* **77**, 3834 (1955).
40. Merck & Co., British Patent 785,682 (1957); *Chem. Abstr.* **52**, 14722 (1958).
41. Merck & Co., British Patent 785,685 (1957); *Chem. Abstr.* **52**, 14722 (1958).
42. Merck & Co., British Patent 785,686 (1957); *Chem. Abstr.* **52**, 14722 (1958).
43. H. Ruschig, *Angew. Chem.* **60**, 247 (1948).
44. H. Ruschig. *Chem. Ber.* **88**, 878 (1955).
45. L. H. Sarett, *J. Amer. Chem. Soc.* **70**, 1454 (1948).
46. L. H. Sarett, *J. Amer. Chem. Soc.* **71**, 2443 (1949).
47. R. M. Lukes and L. H. Sarett, *J. Amer. Chem. Soc.* **76**, 1178 (1954).
48. H. P. Uehlinger, C. Tamm, and T. Reichstein, *Helv. Chim. Acta* **40**, 2234 (1957).
49. P. A. Robins and J. Walker, *J. Chem. Soc., London* p. 1610 (1952).
50. C. A. Grob and W. Jundt, *Helv. Chim. Acta* **31**, 1691 (1948).
51. C. A. Grob and H. Wicki, *Helv. Chim. Acta* **31**, 1706 (1948).
52. C. A. Grob, W. Jundt, and H. Wicki, *Helv. Chim. Acta* **32**, 2427 (1949).
53. C. A. Grob and W. Jundt, *Helv. Chim. Acta* **35**, 2111 (1952).
54. C. A. Grob and O. Schindler, *Experientia* **10**, 367 (1954).
55. W. E. Newhall, S. A. Harris, F. W. Holly, E. L. Johnston, J. W. Richter, E. Walton, A. N. Wilson, and K. Folkers, *J. Amer. Chem. Soc.* **77**, 5646 (1955).
56. E. Walton, A. N. Wilson, A. C. Haven, C. H. Hoffman, E. L. Johnston, W. E. Newhall, F. M. Robinson, and F. W. Holly, *J. Amer. Chem. Soc.* **78**, 4760 (1956).
57. K. K. Pivnitskii and I. V. Torgov, *Izv. Akad. Nauk SSSR, Otd. Khim. Nauk* p. 1080 (1961); *Tetrahedron Lett.* p. 3671 (1964).
58. K. K. Pivnitskii and I. V. Torgov, *Zh. Obshch. Khim* **36**, 835 and 843 (1966).
58a. J. A. Hogg, P. F. Beal, A. H. Nathan, F. H. Lincoln, W. P. Schneider, B. J. Magerlein, A. R. Hanze, and R. W. Jackson, *J. Amer. Chem. Soc.* **77**, 4436 (1955).
58b. D. H. Peterson, H. C. Murray, S. H. Eppstein, L. M. Reineke, A. Weintraub, P. D. Meister, and H. M. Leigh, *J. Amer. Chem. Soc.* **74**, 5933 (1952).
58c. M. Biollaz, W. Haefliger, E. Velarde. P. Crabbé, and J. H. Fried, *Chem. Commun.* p. 1322 (1971).
59. S. A. Simpson, J. F. Tait, A. Wettstein, R. Neher, J. v. Euw, and T. Reichstein, *Experientia* **9**, 333 (1953).
60. S. A. Simpson, J. F. Tait, A. Wettstein, R. Neher, J. v. Euw, and T. Reichstein, *Helv. Chim. Acta* **37**, 1163 (1954).
61. J. Schmidlin, G. Anner, J. R. Billeter, and A. Wettstein, *Experientia* **11**, 365 (1955).
62. A. Lardon, O. Schindler, and T. Reichstein, *Helv. Chim. Acta* **40**, 666 (1957).
62a. W. S. Johnson, J. C. Collins, R. Pappo, and M. B. Rubin, *J. Amer. Chem. Soc.* **80**, 2585 (1958).
63. W. J. van der Burg, D. A. van Dorp, O. Schindler, C. M. Siegmann, and S. A. Szpilfogel, *Rec. Trav. Chim. Pays-Bas* **77**, 171 (1958).
64. A. Wettstein, P. Desaulles, K. Heusler, R. Neher, J. Schmidlin, H. Ueberwasser, and P. Wieland, *Angew. Chem.* **69**, 689 (1957).
65. K. Heusler, P. Wieland, H. Ueberwasser, and A. Wettstein, *Chimia* **12**, 121 (1958).

References

66. J. Schmidlin, G. Anner, J. R. Billeter, K. Heusler, H. Ueberwasser, P. Wieland, and A. Wettstein, *Helv. Chim. Acta* **40**, 1034 (1957).
67. J. Schmidlin, G. Anner, J. R. Billeter, K. Heusler, H. Ueberwasser, P. Wieland, and A. Wettstein, *Helv. Chim. Acta* **40**, 1438 (1957).
68. J. Schmidlin, G. Anner, J. R. Billeter, K. Heusler, H. Ueberwasser, P. Wieland, and A. Wettstein, *Helv. Chim. Acta* **40**, 2291 (1957).
69. E. Vischer, J. Schmidlin, and A. Wettstein, *Experientia* **12**, 50 (1956).
70. J. Schmidlin and A. Wettstein, *Helv. Chim. Acta* **45**, 331 (1962).
71. Ciba Ltd., French Patent 1,343,125 (1963); *Chem. Abstr.* **60**, 14573 (1964).
72. A. Wettstein, E. Vischer, and C. Meystre, U. S. Patent 2,844,513 (1958); *Chem. Abstr.* **53**, 1629 (1959).
73. J. v. Euw, R. Neher, and T. Reichstein, *Helv. Chim. Acta* **38**, 1423 (1955).
74. T. Reichstein, A. Wettstein, G. Anner, J. R. Billeter, K. Heusler, R. Neher, J. Schmidlin, H. Ueberwasser, and P. Wieland, Swiss Patent 341,495 (1959); *Chem. Abstr.* **55**, 624 (1961).
75. T. Reichstein, A. Wettstein, G. Anner, J. R. Billeter, K. Heusler, R. Neher, J. Schmidlin, H. Ueberwasser, and P. Wieland, Swiss Patent 341,496 (1959); *Chem. Abstr.* **55**, 625 (1961).
76. T. Reichstein, A. Wettstein, G. Anner, J. B. Billeter, K. Heusler, R. Neher, J. Schmidlin, H. Ueberwasser, and P. Wieland, West German Patent 1,088,486 (1960); *Chem. Abstr.* **55**, 27429 (1961).
77. S. A. Szpilfogel, W. J. van der Burg, C. M. Siegmann, and D. A. van Dorp, *Rec. Trav. Chim. Pays-Bas* **75**, 1043 (1956).
78. S. A. Szpilfogel, W. J. van der Burg, C. M. Siegmann, and D. A. van Dorp, *Rec. Trav. Chim. Pays-Bas* **77**, 157 (1958).
79. A. Wettstein, G. Anner, K. Heusler, H. Ueberwasser, P. Wieland, J. Schmidlin, and J. R. Billeter, West German Patent 1,076,685 (1960); *Chem. Abstr.* **56**, 6041 (1962).
80. A. Wettstein, G. Anner, K. Heusler, H. Ueberwasser, P. Wieland, J. Schmidlin, and J. R. Billeter, U. S. Patent 2,994,694 (1961); *Chem. Abstr.* **56**, 1509 (1962).
81. P. Wieland, K. Heusler, H. Ueberwasser, and A. Wettstein, *Helv. Chim. Acta.* **41**, 74 (1958).
82. P. Wieland, K. Heusler, H. Ueberwasser, and A. Wettstein, *Helv. Chim. Acta* **41**, 416 (1958).
83. K. Heusler, P. Wieland, and A. Wettstein, *Helv. Chim. Acta* **41**, 997 (1958).
84. P. Wieland, K. Heusler, and A. Wettstein, *Helv. Chim. Acta* **43**, 2066 (1960).
85. A. Wettstein, H. Ueberwasser, and P. Wieland, West German Patent 1,086,227 (1960); *Chem. Abstr.* **55**, 27432 (1961).
86. A. Wettstein, K. Hesuler, H. Ueberwasser, and P. Wieland, Ciba Ltd., West German Patent 1,086,228 (1960); *Chem. Abstr.* **55**, 27430 (1961).
87. A. Wettstein, K. Heusler, H. Ueberwasser, and P. Wieland, U. S. Patent 3,002,970 (1958); *Chem. Abstr.* **56**, 6052 (1962).
88. A. Wettstein, H. Heusler, H. Ueberwasser, and P. Wieland, U. S. Patent 3,002,971 (1958); *Chem. Abstr.* **56**, 6055 (1962).
89. A. Wettstein, K. Heusler, H. Ueberwasser, and P. Wieland, West German Patent 3,002,972 (1958); *Chem. Abstr.* **56**, 6036 (1962).
90. A. Wettstein, K. Heusler, and P. Wieland, West German Patent 1,137,011 (1962); *Chem. Abstr.* **58**, 4625 (1963).
91. K. Heusler, P. Wieland, and A. Wettstein, *Helv. Chim. Acta* **42**, 1586 (1959).

92. A. Wettstein, K. Heusler, H. Ueberwasser, and P. Wieland, *Helv. Chim. Acta* **40**, 323 (1957).
93. K. Heusler, H. Ueberwasser, P. Wieland, and A. Wettstein, *Helv. Chim. Acta* **40**, 787 (1957).
94. L. Velluz, G. Muller, R. Bardoneschi, and A. Poittevin, *C. R. Acad. Sci.* **250**, 725 (1960).
95. M. E. Wolff, J. F. Kerwin, F. F. Owings, B. B. Lewis, B. Blank, A. Magnani, and V. Georgian, *J. Amer. Chem. Soc.* **82**, 4117 (1960).
96. M. E. Wolff, J. F. Kerwin, F. F. Owings, B. B. Lewis, and B. Blank, *J. Org. Chem.* **28**, 2729 (1963).
97. L. Labler and F. Sorm. *Chem. Ind. (London)* p. 1114 (1961).
98. P. K. Bhattacharyya, S. P. Dhage, P. C. Parthasarathy, and B. R. Prema Madyastha, *Int. Symp. Chem. Natur. Prod., 8th, 1972* Abstract, p. 250 (1972).
99. K. Heusler, J. Kalvoda, C. Meystre, P. Wieland, G. Anner, A. Wettstein, G. Cainelli, D. Arigoni, and O. Jeger, *Experientia* **16**, 21 (1960).
100. K. Heusler, J. Kalvoda, C. Meystre, P. Wieland, G. Anner, A. Wettstein, G. Cainelli, D. Arigoni, and O. Jeger, *Helv. Chim. Acta* **44**, 502 (1961).
101. D. H. R. Barton and J. M. Beaton, *J. Amer. Chem. Soc.* **82**, 2641 (1960).
102. D. H. R. Barton and J. M. Beaton, *J. Amer. Chem. Soc.* **83**, 4083 (1961).
103. G. Cainelli, M. I. Mihailović, D. Arigoni, and O. Jeger, *Helv. Chim. Acta* **42**, 1124 (1959).
104. P. Buchschacher, M. Cereghetti, H. Wehrli, K. Schaffner, and O. Jeger, *Helv. Chim. Acta* **42**, 2122 (1959).
105. H. Wehrli, M. Cereghetti, K. Schaffner, and O. Jeger, *Helv. Chim. Acta* **43**, 367 (1960).
106. D. H. R. Barton, J. M. Beaton, L. E. Geller, and M. M. Pechet, *J. Amer. Chem. Soc.* **82**, 2640 (1960).
107. W. Nagata, T. Terasawa, and T. Aoki, *Tetrahedron Lett.* p. 865 (1963).
108. W. Nagata, T. Terasawa, and T. Aoki, *Tetrahedron Lett.* p. 869 (1963).
109. F. H. Howell and D. A. H. Taylor, *J. Chem. Soc., London* p. 1248 (1958).
110. F. H. Howell and D. A. H. Taylor, *J. Chem. Soc., London* p. 1607 (1959).
111. W. Nagata, T. Terasawa, S. Hirai, and K. Takeda, *Tetrahedron Lett.* p. 27 (1960).
112. W. Nagata, T. Terasawa, S. Hirai, and K. Takeda, *Tetrahedron* **13**, 295 (1961).
113. W. Nagata, I. Kikkawa, and K. Takeda, *Chem. Pharm. Bull.* **9**, 79 (1961).
114. W. Nagata, M. Yoshioka, and S. Hirai, *Tetrahedron Lett.* p. 461 (1962).
115. W. Nagata, *Nippon Kagaku Zasshi* **90**, 837 (1969).
116. W. Nagata, M. Yoshioka, and T. Terasawa, *J. Amer. Chem. Soc.* **94**, 4672 (1972).
117. G. Rosenkranz, O. Mancera, J. Gatica, and C. Djerassi, *J. Amer. Chem. Soc.* **72**, 4077 (1950).
118. J. A. Marshall and W. S. Johnson. *J. Amer. Chem. Soc.* **84**, 1485 (1962).
119. A. Chatterjee and B. G. Hazra, *Chem. Commun.* p. 618 (1970).
120. A. Chatterjee and B. G. Hazra, *Tetrahedron Lett.* p. 73 (1969).

10
BC → BCD → ABCD

10.1 Introduction

Ever since the pioneering works of Robinson and Bachmann in the late 1930's it has been realized that any success in the industrial production of steroids by totally synthetic routes would depend primarily on the stereospecificity of reactions in each step. The stereospecific synthesis of *trans*-benzohydrindane derivatives by Banerjee and co-workers could be regarded as an important advance in this direction [1]. Following the earlier method developed by Johnson [2], they synthesized the keto ester **3** (Scheme 10-1) which, in turn was converted into the tricyclic derivative **8**. It should be mentioned that both the intermediate isomeric ketones **6** (R=Me) and **9** had been synthesized earlier by Bachmann [3] and Robinson [4], following the classical procedure developed for the synthesis of equilenin (see Chapter 3, Scheme 3-2).

Later, Velluz and his group [5–11] at Roussel-UCLAF using the *trans*-benzohydrindanes **8** and **9** as the starting material, accomplished the first successful stereospecific total synthesis of steroids on an industrial scale. By 1960, they were able to complete the total synthesis of a host of optically active steroids including 19-nortestosterone (**21**), cortisone (**28**, R=H), and estradiol (**30**). One of the major contributing factors in their success stemmed from their ability to resolve the racemates in

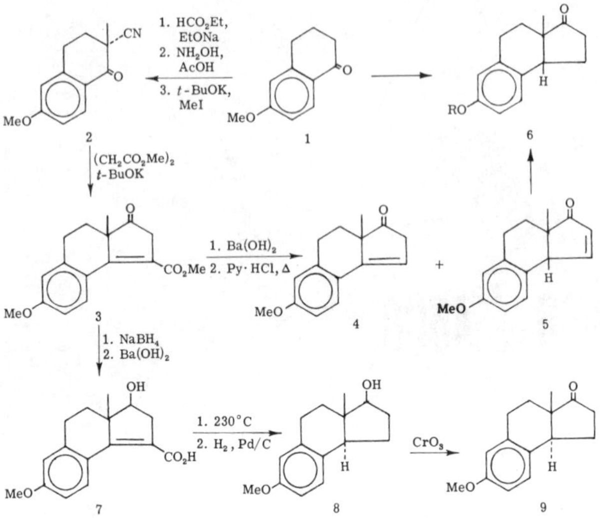

Scheme 10-1

the earliest possible stage in the synthetic scheme. Another significant modification was the extensive use of the concept of convergence of synthetic schemes [5,6]. In this approach two fragments of the future steroids containing the latent functional groups are formed initially. The combination of these two fragments is then carried out at a later stage. Thus the formation of expensive intermediates is delayed as long as possible and the overall yield in the synthesis is considerably increased. Some of the key intermediates obtained by BC→BCD route are also prepared by CD→BCD route. However, most of the syntheses involving the common intermediates will be described in the present chapter.

10.2 *Trans*-Benzohydrindane Derivatives

The starting material for the synthesis of the key intermediate **8** by Banerjee and co-workers [1] was 6-methoxytetralone (**1**, Scheme 10-1). Following Johnson's method for the synthesis of equilenin (see Chapter 3, Scheme 3-3), this ketone was converted into the cyanoketone **2**, by successive formylation, formation of an isoxazole and methylation. Stobbe condensation with dimethyl succinate in the presence of potassium *t*-butoxide yielded the tricyclic keto ester **3**. This was followed by reduction with sodium borohydride and saponification to give the hydroxy acid **7**. Decarboxylation by heating and catalytic reduction afforded the

10.3 19-Nortestosterone

trans-benzohydrindane derivative **8** in excellent yield. Oxidation of the latter led to the formation of the ketone **9**. On the other hand, the unsaturated keto ester **3**, upon saponification and subsequent decarboxylation yielded a mixture of Δ^{14}- and Δ^{15}-isomers **4** and **5**. Catalytic hydrogenation of **5** was followed by demethylation to give the phenolic ketone **6** (R=H) [1,5,9,12]. The above procedure was also extended to the synthesis of the 18-methyl homolog of **8** [1,13,14]. As mentioned earlier, the syntheses of the CD *cis*-isomer **6** (R=Me) and its CD *trans*-isomer **9**, as well as their higher homologs had been previously achieved by Bachmann [3] and Robinson [4] .These early syntheses were achieved by an adaptation of the sequence of reactions developed by Bachmann for the construction of the D ring in equilenin (see Chapter 3, Scheme 3-2).

10.3 19-Nortestosterone

In the first large scale synthesis [8,9] of 19-nortestosterone (**21**), resolution was accomplished at the stage of the carboxylic acid **7** via its salt with L-(+)-*threo*-2-amino-1-*p*-nitrophenyl-1,3-propanediol. Following the method of Banerjee, the resolved acid **7** was converted to the tricyclic alcohol **8** (Scheme 10-2). The benzoate **10**, obtained from **8** was now subjected to Birch reduction [8,12] to produce the methoxydiene **11**. The latter on treatment with methanolic hydrochloric acid gave the conjugated ketone **12** (R=PhCO). On the other hand, mild hydrolysis with oxalic acid led to the nonconjugated ketone **13**, which could be isomerized to **12** (R=PhCO) by the action of strong acid. Later studies showed that resolution of the racemates in this series could also be carried out by oxidation of the 17-hydroxy group in **12** (R=H) to the corresponding ketone by *Pseudomonas testosteroni* whereupon only the natural *d*-enantiomer was obtained [15]. The resolution could also be effected by hydrolysis of the 17-acetate **12** (R=Ac) by a pancreatic enzyme when only the *l*-enantiomer reacted [16].

Initial difficulties were encountered during the construction of ring A. All attempts to condense the unsaturated ketone **12** (R=H) or the corresponding benzoate **12** (R=PhCO) with methyl vinyl ketone led to the undesired phenalene derivatives **17** (R=H) and **17** (R=PhCO), respectively. Better results were obtained by condensation of the ketones **12** (R=PhCO) and **13** with 1,3-dichloro-2-butene, when the corresponding products **18** and **14** were obtained in good yield [5,8,9]. The compound **14** could be easily isomerized to **18**, which was also synthesized via the pyrrolidine enamine **15**. Acid hydrol-

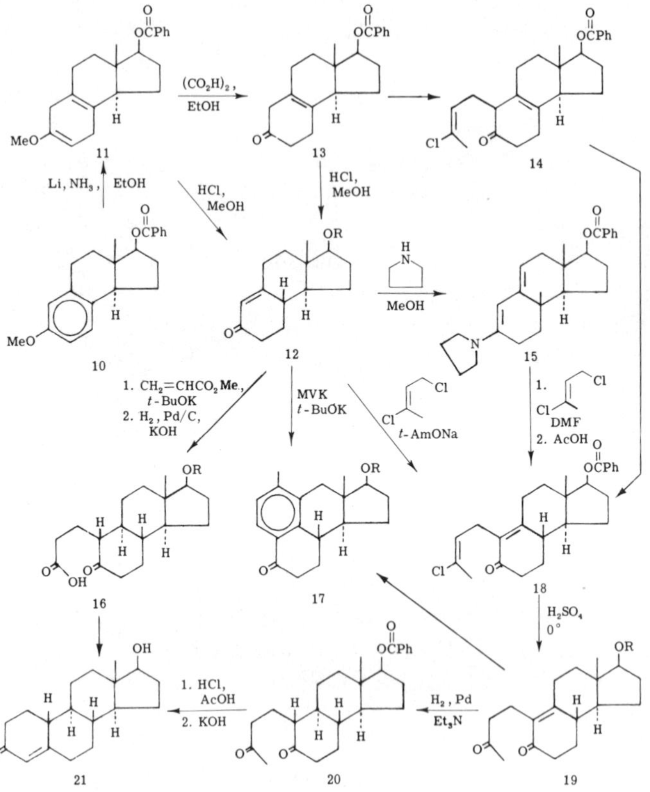

Scheme 10-2

ysis of the product **18** afforded the diketone **19** (R=PhCO) which later served as the starting material for the synthesis of a whole series of new steroidal compounds. Compounds like **19** were synthesized later by the D→CD→BCD→ABCD approach [5,17]. The stereochemistry of the catalytic hydrogenation of the double bond in the diketone **19** (R=PhCO) depended on the pH of the reaction medium. When the hydrogenation was carried out in acidic medium, the resulting mixture was found to contain the $9\beta,10\beta$-isomer **31** as the major product (see Scheme 10-3). However, in alkaline medium, the major product obtained had the $9\alpha 10\alpha$-structure, which was readily epimerized through the enolization of the 5-oxo group, into the $9\alpha,10\beta$-isomer **20**. Cyclization and hydrolysis of the latter afforded 19-nortestosterone (**21**) [8,9,17–22].

19-Nortestosterone (**21**) has also been obtained by various modifications [12,23–25] of the procedure described above. Chinn and Dryden [12]

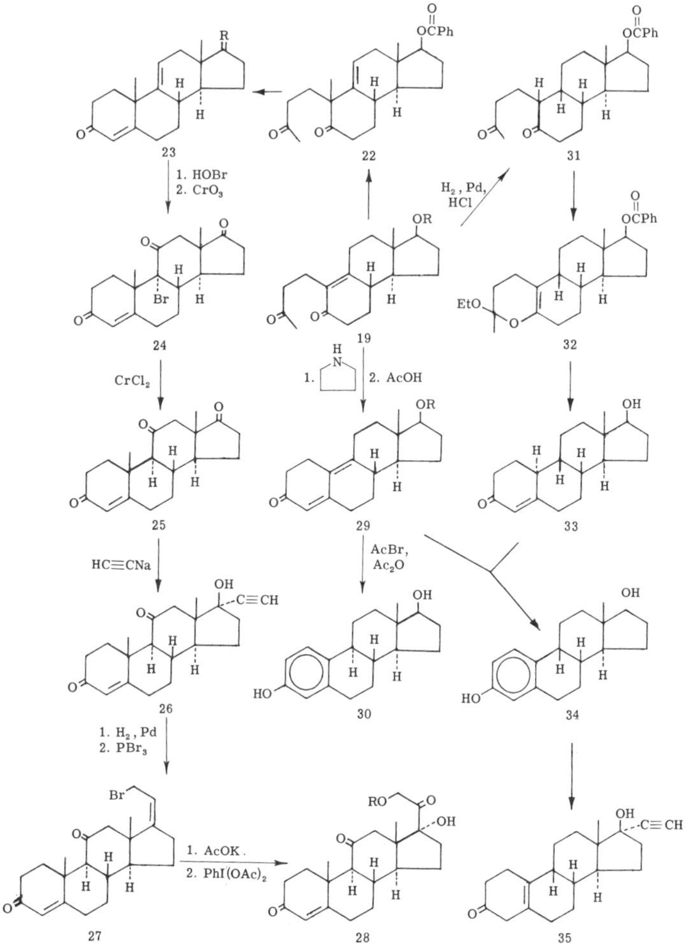

10.3 19-Nortestosterone

Scheme 10-3

described the condensation of the ketone **12** (R=H) with methyl acrylate in presence of potassium *t*-butoxide. The resulting keto acid was reduced by lithium and ammonia, or better by catalytic hydrogenation in presence of palladized carbon to give the *dl*-acid **16** (R=H). The resolution of the latter into its enantiomers could be effected via its salt with amphetamine. The final stage in the synthesis was the conversion of the acetate **16** (R=Ac) of the seco acid to 19-nortestosterone. This was achieved by following previously known methods [26–28]. On refluxing **16** (R=Ac) with acetic anhydride and sodium acetate in an N_2 atmosphere, an intermediate enol lactone was obtained, which on successive treatment with methylmagnesium iodide, acetic anhydride, hydrochloric acid, and methanolic sodium hydroxide afforded 19-nortestosterone (**21**).

10.4 Adrenocortical Steroids

The synthesis of the tricyclic intermediate **19** (R=PhCO) opened up the route to the industrial preparation of several adrenocortical hormones [9,29]. Protection of the carbonyl group in the side chain as the ethylene ketal was followed by stereospecific angular methylation with methyl iodide in presence of sodium *t*-amylate and subsequent hydrolysis to give the diketone **22** (Scheme 10-3). Cyclization and saponification afforded the keto alcohol **23** (R=α-H,β-OH) which on oxidation yielded $\Delta^{4,9(11)}$-androstadiene-3,17-dione (**23**, R=O). The next stage, the conversion of this latter compound into adrenosterone (**25**) was achieved via the 9α,11β-bromohydrin and oxidation of the 11-hydroxy group to give the compound **24**. Subsequent debromination led to adrenosterone (**25**), which had also been obtained by other routes [30–33]. Following the method of Ruzicka [34] and Hogg [35], **25** was then converted into cortisone (**28**, R=H). Ethynylation of the former compound yielded **26**, which on selective hydrogenation and treatment with phosphorus tribromide afforded the bromide **27**. Subsequent acetylation and oxidative hydroxylation with iodosobenzene diacetate led to cortisone acetate **28** (R=Ac) in about 10% yield, based on the diketone **23** (R=O).

10.5 Estrogens

Some other points of interest in the synthetic procedures developed by the group of Velluz deserve discussion. These concern the direct cyclization of the intermediate **19** (R=PhCO), which was found to depend on the nature of the reagents [9]. Treatment of this compound with Triton B led to the phenalene **17** (R=PhCO) (Scheme 10-2). However, cyclization with sodium *t*-amylate afforded the $\Delta^{4,9(11)}$-dione **29** (R=PhCO) (Scheme 10-3). The latter could also be obtained by successive treatment of the diketone **19** (R=PhCO) with pyrrolidine and aqueous acetic acid. The next stage was aromatization of the ring A, which could be effected by heating with palladized carbon in ethanol or, better, by treatment with a mixture of acetyl bromide and acetic anhydride at room temperature. The resulting product, upon hydrolysis afforded estradiol (**30**), possessing physiological activities similar to that of the natural hormone [9,22,36–40].

10.6 10-Allyl Steroids

In another modification [41], catalytic hydrogenation of the diketone 19 (R=PhCO) in the presence of an acid led to a mixture containing the 9β,10β-isomer 31 as a major product. Separation of the products was effected by taking advantage of the preferential formation of the dihydropyran derivative 32, from the 9β-isomer 31, whereas the 9α-isomer remained intact. Saponification and subsequent treatment with hydrochloric acid afforded 19-nor-9β,10α-testosterone (33). The 10α-configuration was determined by X-ray diffraction studies. It is interesting to note that both the retro 9β,10α-compound 33 and the dienic structure 29 underwent microbiological aromatization by the action of the same microorganism *Arthrobacter simplex* to give 9-isoestradiol (34). Subsequent Birch reduction and ethynylation led to the carbon-9 epimer of norethynodrel (35) [6].

10.6 10-Allyl Steroids

The diketone 19 (R=PhCO) also served as the starting material for the synthesis of 10-allyl steroids [42–48] (Scheme 10-4). Direct allylation of the ketal 36, obtained from 19 (R=PhCO), was followed by hydrolysis and cyclizaton to yield the 10β-allyl derivative 37. Catalytic hydrogenation of the latter led to the corresponding 10β-propyl analog [42,43].

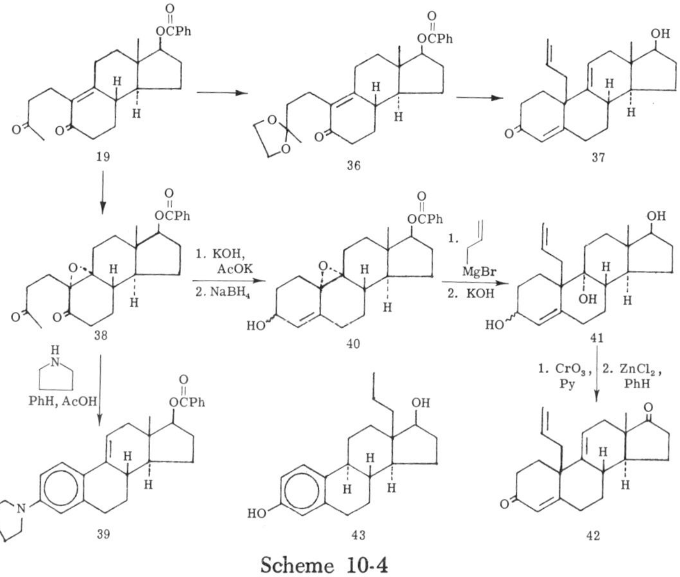

Scheme 10-4

In another modification [42,44–48], the diketone 19 (R = PhCO) was first converted to the epoxide 38, by the action of perphthalic acid. The 9α,10α-configuration of this epoxide was established by X-ray diffraction as well as by circular dichroism studies. An attempt to cyclize this compound, by heating with pyrrolidine led to the ring-A aromatic derivative 39. However, treatment of the epoxide 38 with potassium hydroxide and potassium acetate led to the cyclized product, which on reduction with sodium borohydride afforded 40. Reaction of the latter with allylmagnesium bromide resulted in nucleophilic attack on the β-face to give the allyltriol 41. Oxidation of the latter and subsequent dehydration yielded the $\Delta^{9(11)}$-diketone 42. The above general route was also used for the synthesis of 13-propyl-norestradiol (43) [49].

10.7   19-Norprogesterone and 11-Hydroxy Steroids

The historic observation by Allen and Ehrenstein [50,51] in 1944, of the high progesterone-like activity of a 19-norsteroid has been a stimulating factor in the development of a series of clinically useful steroidal hormones. Both partial synthesis and totally synthetic methods have been developed for the preparation of 19-norprogesterone (49). In the BC→BCD→ABCD approach [52–55], the starting material was the tricyclic ketone 9, ethynylation of which afforded the compound 44 (Scheme 10-5). Acetylation and hydration of the latter led to the ketoacetate 45, which on Birch reduction yielded the hydroxyketone 46. Subsequent oxidation and treatment with pyrrolidine yielded the enamine 47. Following the methods described in Scheme 10-2 (15→19, R = PhCO), 47 was next converted to the triketone 48. Unlike the compound in the 17-benzoate series (e.g., reduction of 19 (R = PhCO) to 20 and 31) the catalytic hydrogenation of 48 was found to be independent of the pH of the medium. The resulting saturated triketone, with the 9α-configuration, was cyclized to give 19-norprogesterone (49).

In another facile synthesis [56] of 19-norprogesterone (49) the starting material was estrone-3-methyl ether (50, R = Me) which was converted to the olefin 51 by Wittig reaction. Successive hydroboration, Birch reduction, and hydrolysis gave the keto alcohol 52, oxidation of which led to 19-norprogesterone in good yield. In another modification, estrone 50 (R = H) was first converted to the triene 54 by known methods [57,58]. Modified Birch reduction using lithium in ammonia, followed by successive treatment of the resulting product with methanolic hydrochloric acid and oxidation led to 49.

The keto alcohol 46 has also been used for the synthesis of the

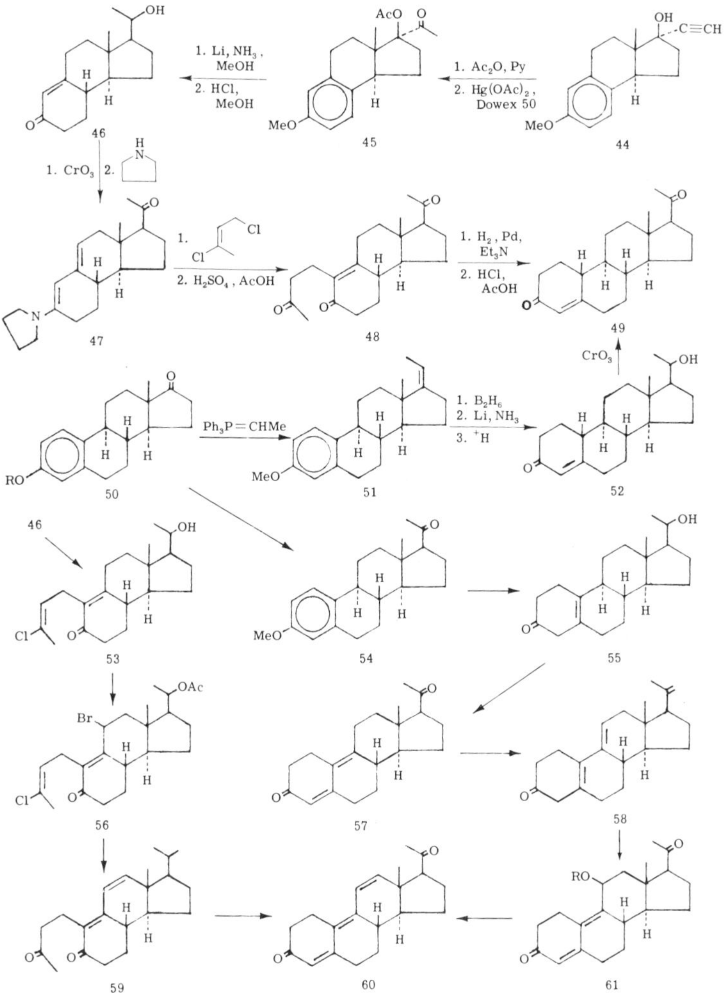

10.7 19-Norprogesterone and 11-Hydroxy Steroids

Scheme 10-5

9(10),11(12)-diene (**60**) of 19-norprogesterone [52,59]. Acetylation of **46** was followed by treatment with pyrrolidine, and the resulting enamine was condensed with 1,3-dichloro-2-butene to give the ketone **53**. Subsequent enolacetylation and bromination led to the ketone **56**, which was converted to the methyl ketone **59** by dehydrobromination and hydrolysis. Cyclization was effected by treatment with sodium *t*-amylate and the resulting triene, on saponification and oxidation afforded the diketone **60**. In still another modification [52] the hydroxyketone **55**, obtained during the Birch reduction of **54**, could be converted to the diene **57** by following the method of Perelman and co-workers [60]. A one-step bromination of the 5(10)-3-ketone **55** led to the $\Delta^{4,9}$-analog, treatment of which with chromium trioxide yielded the diketone **57**. Conversion of the latter compound into **58** was followed by treatment with molecular oxygen in ethanol with a small amount of triethylamine to form the 11β-hydroperoxide **61** (R=OH). Reduction with trimethyl phosphite led to the intermediate 11β-hydroxy derivative **61** (R=H). Dehydration with sulfuric acid afforded **60**.

The above general route developed by the French group was also utilized for the synthesis of 11-hydroxy steroids and steroidal lactones [61–65] (Scheme 10-6). The benzoate **62**, obtained from the tricyclic ketone **12** (R=PhCO), was converted to the epoxide **63** by the action of perphthalic acid. The latter compound, on treatment with formic acid, gave the diketone **66** in 62% yield. Alternatively the epoxide **63** could be initially converted to the intermediate **64** or to the hydroxyketone **65**. Both **64** and **65** furnished the diketone **66** on treatment with formic acid. The conversion of **66** to the hydroxy ketone **67** was achieved by selective ketalization, reduction of the 11-oxo group, and removal of the protective group. On formylation, the compound **67** gave the hydroxymethylene intermediate **70**. Subsequent heating with triethylamine, led to the 11,19-cyclohemiacetal, which on chromic acid oxidation afforded the lactone **69**. This compound could also be obtained directly from the hydroxyketone **67** by condensation with diethylcarbonate in presence of sodium ethoxide. For the construction of ring A, the lactone **69** was first condensed with methyl vinyl ketone to give the diketone **68**. Cyclization via the pyrrolidine enamine afforded the keto lactone **71** (R=PhCO).

The general reaction sequence described above has also been extended for the synthesis of cortisone derivatives [62]. Thus the lactone **72** (R=H), obtained from **71** (R=H) was converted to the 17-oxo compound **75** by oxidation with the chromium trioxide–pyridine complex. The side chain at C-17 was introduced by Wittig reaction and this was followed by oxidation with osmium tetroxide and treatment with hydrogen peroxide to give the hydroxyketone **74**. Introduction of the 21-acetoxyl group, and subsequent hydrolysis afforded the lactone **77**.

In another series of reactions [62], the hydroxylactone **72** (R=H) was converted to the trihydroxy compound **73**, by treatment with lithium aluminum hydride. Catalytic hydrogenation followed by treatment with N-bromosuccinimide led to the diketone **76**. Wolf-Kishner reduction, cleavage of the protective ketal group, and acetylation furnished **79**. The latter compound has also been obtained from ketone **78** by a different route [62,66].

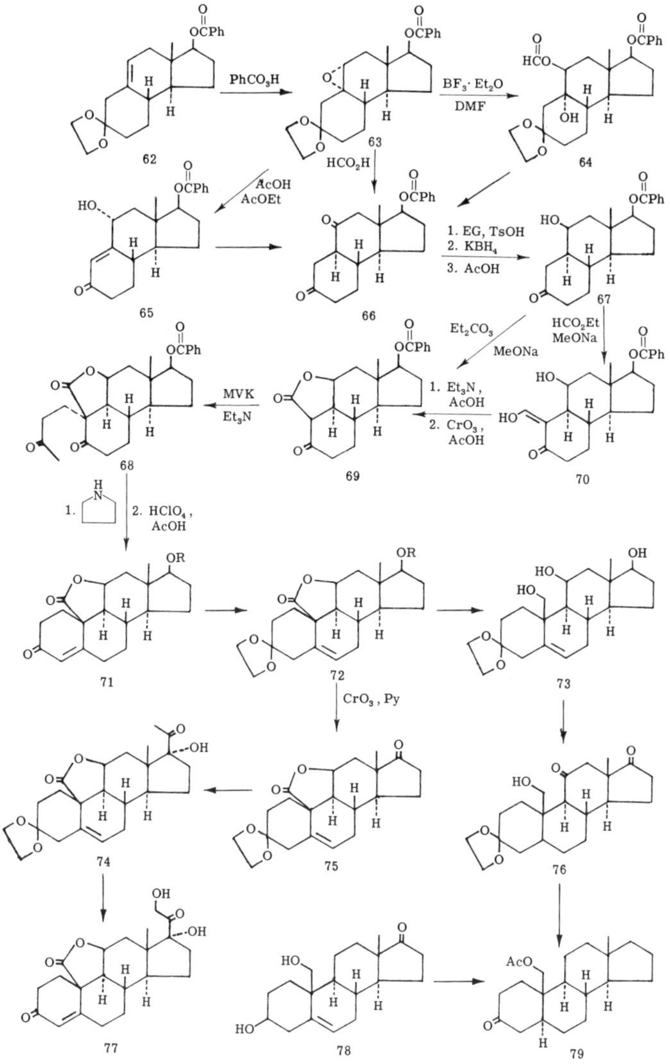

10.8 Trienes   211

Scheme 10-6

10.8 Trienes

A large series of highly unsaturated steroidal derivatives with considerable biological interest has been synthesized from the tricyclic ketone **12** (R = PhCO) (Scheme 10-7). Compound **18**, obtained from **12** (R = PhCO) was converted to the enolacetate **80** by treatment with acetic

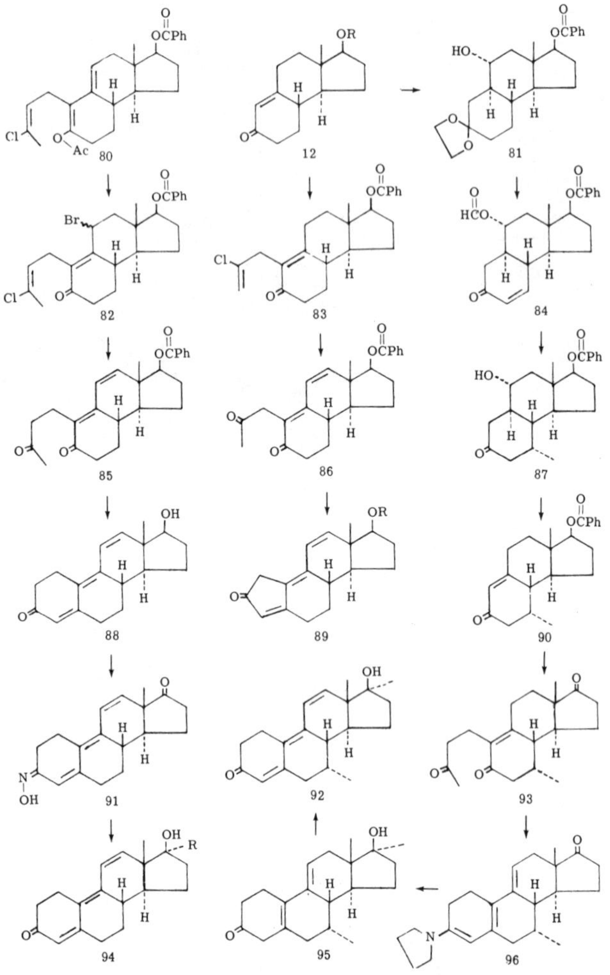

Scheme 10-7

anhydride and *p*-toluenesulfonic acid [67–69]. Bromination of the latter yielded **82**, which on dehydrobromination and hydrolysis gave the diketone **85**. The latter, on base-catalyzed cyclization, followed by alkaline hydrolysis yielded the trienone **88**. After protecting the 3-oxo group as the oxime, the 17-hydroxyl group was oxidized to give the ketone **91**. Grignard reaction, and subsequent removal of the oxime group afforded the 17α-methyl derivative **94** (R = Me), which was found to be fifty times more active as an androgen than methyltestosterone. In another modification [58], the ketone **91** was converted to **94** (R = C≡CH) by successive ethynylation and removal of the oximino group. As a contraceptive agent, the latter compound was found to be more active than norethynodrel (17α-ethynyl-17β-hydroxy-19-norandrost-5(10)-en-3-one).

The above method has also been extended for the synthesis of A-norsteroids [70]. Thus condensation of the pyrrolidine enamine of the ketone **12** (R = PhCO) with 1,2-dichloro-2-propene led to the dienone **83**. Subsequent enolacetylation, bromination, dehydrobromination, and hydrolysis afforded the diketone **86**. Ring closure with potassium t-amylate yielded the trienone **89** (R = PhCO). Saponification followed by acetylation led to the biologically active compound **89** (R = Ac).

10.9 7α-Methyl Steroids

A number of compounds with methyl groups at various positions of the steroidal skeleton have shown interesting biological properties [71,72]. The general method developed earlier was successfully applied for the synthesis of many methylated steroids. Thus the 7α-methyl triene derivative **92** was synthesized [52,70] from the key intermediate **12** (R = PhCO). The latter on ketalization and hydroboration yielded the 11α-hydroxy derivative **81**. Successive deketalization, treatment with formic acid, bromination, and dehydrobromination led to the conjugated ketone **84**. Subsequent treatment with methylmagnesium bromide led to the 7α-methyl derivative, which was next converted to the hydroxyketone **87**. Mesylation of **87** was followed by treatment with lithium bromide in dimethylformamide to yield the unsaturated ketone **90**. After saponification and oxidation, the side chain was built up by following the general route outlined earlier. The product **93** was cyclized with pyrrolidine to the ketone **96**, which was converted to ketone **95** in a few steps. Finally, dehydrogenation of this last compound with 2,3-dichloro-5,6-dicyanobenzoquinone led to the triene **92**. Later, the above general method was extended to the syntheses of various novel compounds methylated at carbons 2 [41], 6 [73], and 7 [6] of the steroid skeleton.

10.10 Miscellaneous Steroidal Derivatives

It was shown in the early 1960's by Perelman and co-workers [59] and by Fried and associates [74] that the introduction of a $\Delta^{4,9}$-bis-dehydro system in the 19-norsteroids resulted in the preparation of exceptionally potent oral antiestrogens. These compounds could be synthesized by a one-step bromination of $\Delta^{5(10)}$-3-ketosteroids in pyridine solution (as illustrated in the conversion of **55** to **57**, Scheme 10-5). The availability of optically active intermediate **19** (R=PhCO) on an industrial scale opened an attractive approach to this series of claudogenic products [6,52,75–77] (Scheme 10-8). Treatment of the diketone **19** (R=PhCO) with pyrrolidine led to the enamine **97b** in excellent yield. The latter compound was treated with a mixture of acetic acid and water for 15 minutes at room temperature to afford the $\Delta^{5(10),9(11)}$-diene **99b**. On the other hand, treatment of the enamine **97b** with acetic acid for 30 minutes, followed by addition with water yielded a mixture of the $\Delta^{5(10),9(11)}$-diene **99b** (8%) and $\Delta^{4,9(10)}$-diene **98b** (87%). The compound **98b** on treatment with pyrrolidine yielded the enamine **97b** [59]. Similarly, the compound **99d** possessing considerable progestational activity [78], was obtained from the dienone **98d**. Treatment of **98d** with pyrrolidine gave the enamine **97d**, which on cleavage with formic acid afforded **99d** [76,78].

The diene **99a** served as a key intermediate for the preparation of a series of interesting steroidal derivatives with substituted C ring [6,52,79].

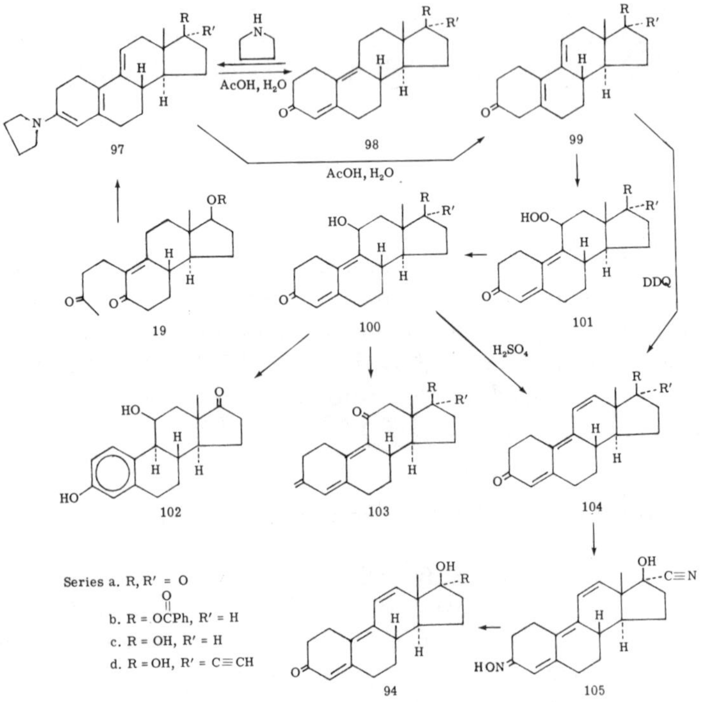

Series a. R, R' = O
b. R = OCPh, R' = H
c. R = OH, R' = H
d. R = OH, R' = C≡CH

Scheme 10-8

10.10 Miscellaneous Steroidal Derivatives

Treatment of 99a with molecular oxygen in ethanol in the presence of a small amount of triethylamine yielded the 11β-hydroperoxide 101a. This was followed by reduction with trimethyl phosphite to give the 11β-hydroxy derivative 100a [77,80,81]. Aromatization of the latter compound with palladium hydroxide led to the 11β-hydroxyestrone (102). In another series of reactions [77] the diene 99d was converted to the 11β-hydroxy derivative 100d, oxidation of which yielded the 3,11-diketone 103d. Conjugation of 98a and 98d has also been extended to the C-11 position, thus providing access to a series of potent progestional compounds. This was accomplished by the dehydration of 100a to give triene 104a. Alternatively, the intermediate 104a could also be prepared from the diene 99a, by reaction with 2,3-dichloro-5,6-dicyanobenzoquinone (DDQ) [52,82]. The triene 104a, on treatment with acetone cyanohydrin yielded the 17-cyanohydrin, which was next converted to the 3-oxime 105. Subsequent treatment with mild base, ethynylation, and removal of the oxime protective group afforded the triene 94 (R = C≡CH).

A novel method for angular alkylation at C-10 of the 5(10),9(11)-estradiene derivatives was developed by the Roussel-UCLAF group [83–85] (Scheme 10-9). Thus, the diene 112b, on treatment with m-chloroperbenzoic acid, afforded a mixture of 5α,10α-epoxide 113a (41%) and 5β,10β-epoxide 113b (12%). A small amount of the 9α,11α-isomer 116 was also produced. When the epoxide 113a was successively treated with methylmagnesium bromide and ammonium chloride solution, a *trans* diaxial opening of the oxirane occurred, which allowed the stereospecific introduction of the 10β-methyl group. The resulting product 110 was converted to 9(11)-dehydrotestosterone (111, R = Me; cf. formula 23, Scheme 10-3) by the action of methanolic hydrochloric acid. The synthesis of this compound has also been achieved by other routes [9,86,87] (see Schemes 10-3 and 10-10). Further extension of this method led to the synthesis of 111 (R = Et) and 111 (R = C≡CH).

Later, this approach was also used for the total synthesis of corticosteroids [83]. Thus treatment of 112a with excess potassium cyanide in acetic acid and methanol afforded the cyanohydrin 109 in excellent yield. The next step in the synthesis was the epoxidation with m-chloroperbenzoic acid. The presence of 17β-cyano 17α-hydroxy group facilitated the attack of peracid toward the α-side of the steroid molecule and preferably on the 5(10)-double bond. The resulting 5α,10α-epoxide 106 was obtained in about 65% yield. A versatile method was designed for the construction of the pregnane structure. The hydroxyl group in 106 was first protected as the trimethylsilyl ether. This was followed by

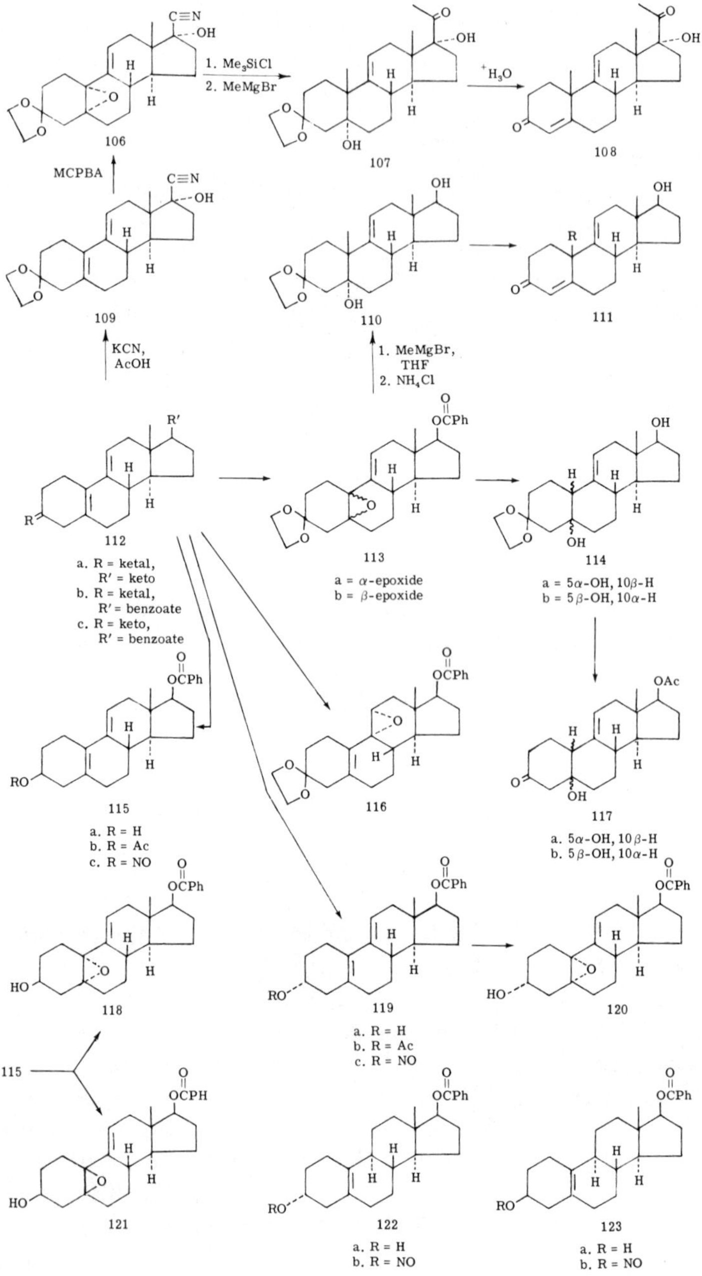
106
107
108
109
110
111
112
113
114
115
116
117
118
119
120
121
122
123
1. Me3SiCl
2. MeMgBr
+H3O
MCPBA
KCN,
AcOH
1. MeMgBr,
THF
2. NH4Cl
a. R = ketal,
R′ = keto
b. R = ketal,
R′ = benzoate
c. R = keto,
R′ = benzoate
a = α-epoxide
b = β-epoxide
a = 5α-OH, 10β-H
b = 5β-OH, 10α-H
a. R = H
b. R = Ac
c. R = NO
a. 5α-OH, 10β-H
b. 5β-OH, 10α-H
a. R = H
b. R = Ac
c. R = NO
a. R = H
b. R = NO
a. R = H
b. R = NO

Scheme 10-9

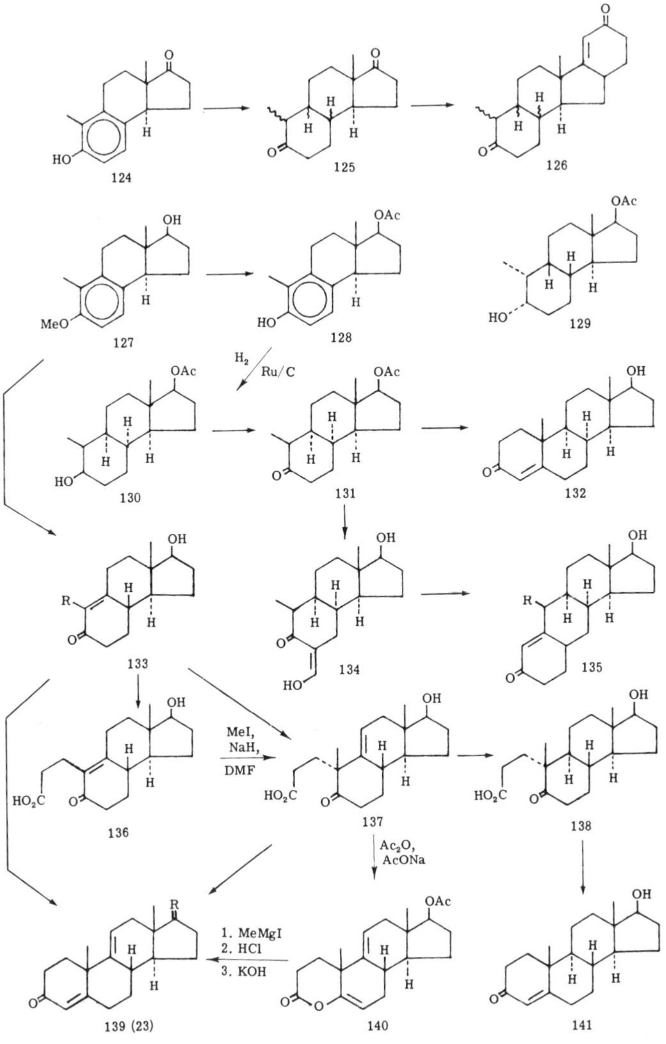

10.10 Miscellaneous Steroidal Derivatives

Scheme 10-10

treatment with methylmagnesium bromide in refluxing tetrahydrofuran, when two simultaneous reactions took place. The oxirane ring was opened with stereospecific introduction of the 19-methyl group. This was accompanied by condensation of the Grignard reagent with the nitrile, thus forming the pregnane side chain. Treatment of the resulting crude product with aqueous ammonium chloride led to the concomittant hydrolysis of the intermediate imine and of the protective trimethylsilyl ether. The overall yield of the product **107** was about 66%. Treatment with aqueous acid yielded the conjugated ketone **108**.

In a study directed toward establishing the configuration of the epoxides **113a** and **113b**, these two compounds were reduced with lithium aluminum hydride [84]. The resulting alcohols **114a** and **114b** were subsequently acetylated and hydrolyzed to give the hydroxyketones **117a** and **117b**, respectively. Finally, the configurations were established by circular dichroism studies. The French group also studied [84] the steric influence of the C-3 hydroxyl group during the epoxidation of 5(10)-double bonds in 5(10)9(11)-estradienes. Reduction of the ketone **112c** with sodium borohydride led to a mixture of the alcohols **115a** and **119a**, which were separated via the acetates **115b** and **119b**. The hydroxy compounds were converted to the nitrites **115c** and **119c**. Comparison of the circular dichroism data of **115c** and **119c** with those of the nitrites **122b** and **123b** established the configuration of the corresponding alcohols **115a** and **119a**. The nitrites **122b** and **123b** were obtained from the alcohols **122a** and **123a**, which in turn were obtained as a mixture by the sodium borohydride reduction of the corresponding 3-ketosteroid. As expected, the major product was the 3α-alcohol **122a**. It had been previously shown that hydride attack on an unhindered 6-membered ring ketone ordinarily leads to the equatorial alcohol [88–90]. Also, studies on the conformational preference effects in ring A of 5(10)-unsaturated steroids as well as of ring-B aromatic steroids have established that hydride reduction of the 3-ketones gives the 3α-alcohols as the major products [91-95]. Epoxidation of the diene **119a** with m-chloroperbenzoic acid yielded the $5\alpha,10\alpha$-epoxide **120** as the major product. However, similar epoxidation of **115** led to a mixture of the $5\alpha 10\alpha$-epoxide **118** and $5\beta,10\beta$-epoxide **121** in almost equal ratio.

10.11 Isosteroids

The versatile stereospecific synthesis of *dl-trans*-benzohydrindane derivatives developed by Banerjee has also been extended to the preparation of 8-isotestosterone (**132**) [14,96] (Scheme 10-10). This compound, obtained by Djerassi and co-workers by a multistep transformation from diosgenin, presents an interesting conformational feature in that the B ring exists in the boat form [97–99].

Demethylation of the compound **127**, was followed by preferential acetylation to give the phenol acetate **128**. Hydrogenation in the presence of 5% ruthenium-on-carbon catalyst under 300 atmospheres and at

10.11 Isosteroids

100–110°C led to a mixture of products, which were separated by chromatography. On the basis of Linstead's hypothesis of catalyst hindrance [100], the major product was assigned the stereochemistry shown in **130**. Since the axially oriented β-methyl group would hinder the β-approach of the catalyst, hydrogenation was likely to take place mainly from the α-side of the molecule. In addition, a small amount of the isomeric hydroxyacetate **129** was also isolated. Oxidation of the $8\alpha,9\alpha$-hydroxy acetate **130** furnished the keto acetate **131**. Since the latter compound remained unaffected after treatment with potassium t-butoxide in t-butanol, the 10-methyl group was assigned the equatorial β-configuration. Axial alkylation at the methine carbon atom in **131**, expected to be aided by hyperconjugation, should take place from the less hindered side of the *cis*-decalone system of rings B and C. The resulting product, could then be cyclized by forcing the B ring into a boat conformation. Treatment of **131** with methyl vinyl ketone in presence of Triton B proceeded in a completely stereospecific fashion and afforded the desired *dl*-8-isotestosterone (**132**) in about 21% yield. Additional evidence for the structural assignment of **132** was obtained by blocking the 6-methylene group in the keto acetate **131** as the methylanilinomethylene derivative, and thus directing the alkylation exclusively to the 10-methine carbon. Subsequent condensation with methyl vinyl ketone and removal of the blocking group yielded a material identical with the one obtained by direct alkylation. However, conversion of **131** to its hydroxymethylene derivative **134** and subsequent condensation of the latter with methyl vinyl ketone led to a different keto alcohol. A linear structure **135** (R = Me) has been assigned to this tetracyclic compound, which could be considered as an anthracene analog of 8-isotestosterone. The above method was later extended to the synthesis of *dl*-8-iso-19-noranthratestosterone (**135**, R = H) [101,102]. An unsuccessful attempt to convert 8-isotestosterone to testosterone also was made. The above method for the construction of the A ring represented a considerable improvement upon the old procedure of Martin and Robinson [4,103]. As early as 1943, these workers converted the hydroxyketone **124** to **125**. But attempts to attach the ring A by Robinson's method led to the formation of the undesired product **126**.

The tricyclic intermediate **127** also served as the starting material for a remarkably short synthesis of *dl*-9(11)-dehydrotestosterone (**23**, R = α-H, β-OH) and *dl*-testosterone (**141**) [87]. As described before, the former compound had been synthesized earlier by the Roussel-UCLAF group (see Schemes 10-3 and 10-9). Lithium ammonia–ethanol reduction of the tricyclic compound **127** led to the unsaturated keto alcohol **133** (R = Me).

The methylanilinomethylene derivative of **133** (R=Me), on condensation with acrylonitrile in presence of Triton B, followed by refluxing with potassium hydroxide yielded the keto acid **137** as the major product. Catalytic hydrogenation of **137** afforded the saturated acid **138**. The acid chloride of the latter compound was condensed with di-*t*-butylethoxy magnesiomalonate, and the resulting product was successively treated with acetic acid–monochloracetic acid and sodium hydroxide to give *dl*-testosterone (**141**) Similarly the acid chloride of the unsaturated keto acid **137** was converted to *dl*-9(11)-dehydrotestosterone **23** (R=α-H,β-OH). The latter compound could also be obtained from the condensation product of the unsaturated keto alcohol **133** (R=Me) with vinyl methyl ketone in presence of Triton B.

The keto acid **137** has also been synthesized by a different route [12]. Condensation of the keto alcohol **133** (R=H) with methyl acrylate in presence of potassium *t*-butoxide, followed by hydrolysis, led to the unsaturated acid **136**. Treatment of the latter with methyl iodide in presence of sodium hydride, followed by alkaline hydrolysis, yielded the product **137** [31,32]. In another modification, **137** was converted into the enol lactone **140** by refluxing with sodium acetate in acetic anhydride. Subsequent Grignard reaction, hydrolysis, and cyclization by known methods led to **23** (R=α-H,β-OH). Conversion of the latter compound to adrenosterone (**25**) has been described before (see Scheme 10-3) [9,31–33,104].

The syntheses of other 8-isosteroids have been studied by several groups [101,105–111]. Thus the conversion of the optically active key intermediate **8** to the ketone **142** (Scheme 10-11) was accomplished [105–108] by following the usual method (cf. transformation of **127** to **131**, Scheme 10-10). Michael condensation of the keto alcohol **142** with methyl vinyl ketone afforded the diol **143**. The latter, when treated with zinc chloride at 100°C gave a mixture of three compounds, which were assigned the structures **144, 145,** and **146**.

In an independent study, Buzby and associates [111] carried out experiments along similar lines previously developed by Torgov [109] and synthesized the ketone **145**. Reduction of the *dl*-estrapentaene **151** with sodium borohydride was followed by hydrogenation in presence of palladized charcoal to give the alcohol **148** (R=Me, R'=α-H,β-OH). Alternatively, this compound could be obtained by the sodium borohydride reduction of the corresponding 8α-estrone methyl ether **148** (R=Me, R'=O). Reduction with lithium and ethanol in liquid ammonia followed by treatment with oxalic acid in aqueous methanol led to the previously mentioned nonconjugated ketone **145**.

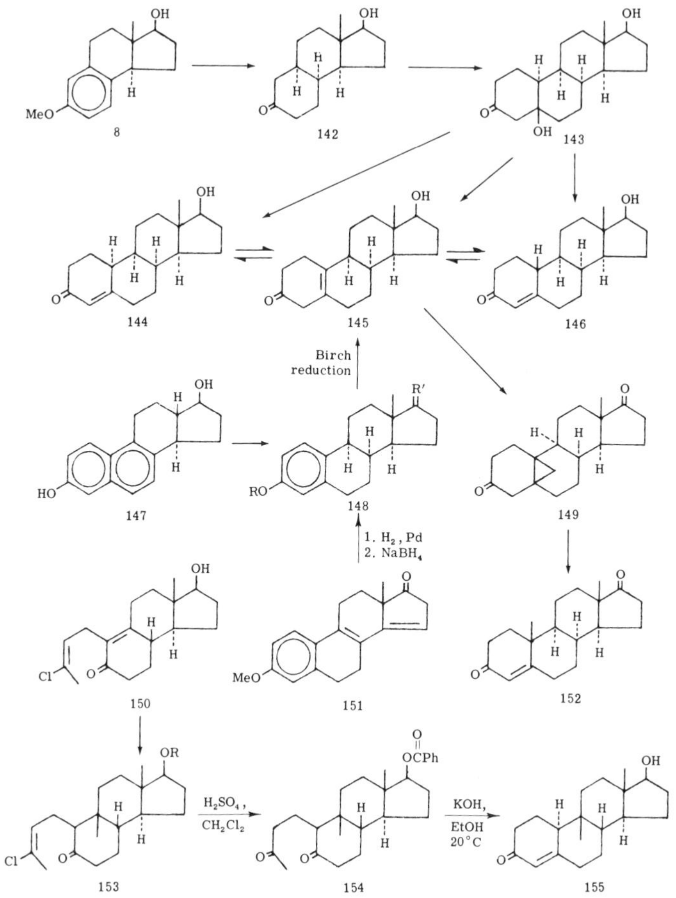

10.11 Isosteroids

Scheme 10-11

The syntheses of both **145** and *d*-8-iso-10-iso-19-nortestosterone (**144**) were achieved by Banerjee by following a different route [101,102]. Hydrogenation of *dl*-dihydroequilenin (**147**) with Raney nickel in an alkaline medium yielded *dl*-8-isoestradiol **148** (R=H, R'=α-H,β-OH). Birch reduction of the methyl ether **148** (R=CH₃ R'=α-H,β-OH), followed by hydrolysis with oxalic acid, led to *dl*-8-iso-17β-hydroxy-19-norandrost-5(10)-en-3-one (**145**). Treatment of **145** with methanolic hydrochloric acid gave a mixture of the starting material **145** and the α,β-isomer **144**, both of which were found to be identical with the materials obtained by the French group. However, it is interesting to note that the third isomer **146**, which was isolated by the latter group and was assigned the configuration 8-iso-19-nortestosterone, exhibited entirely different ORD curves with negative cotton effect from that of the positive ORD curve of d-8-isotestosterone (132) [101,102]. Later, this apparent discrepancy was clarified by the French group [107,108], who synthesized 144, 145, and 146 by following an unambiguous route. Following the method developed by Torgov [109], the ketone 151 was converted to 145 by three successive reductions. Treatment of the latter with aqueous methanolic hydrochloric acid afforded an equilibrium mixture of the starting material 145 and two isomeric ketones 144 and 146 in an approximate ratio 2·0:0·9:0·1. Stereochemical assignments were made by studying the ORD data of closely related compounds.

In an independent study, Birch and Subba Rao [112] also synthesized the nonconjugated ketone 145 and coverted it to dl-8α-androst-4-ene-3, 17-dione (152) via the intermediate 149.

A facile synthesis of 19(10→9β)abeo-10α-testosterone (155) was described by Bull and Tuinman [113,113a]. The seco compound 150, prepared from the tricyclic ketone 12 (R=H), was treated with dimethyl copper lithium in ether to give the 9β-methyl derivative 153 (R=H). The corresponding 17β-benzoate 153 (R=PhCO) on treatment with sulfuric acid in dichloromethane yielded the diketone 154. Annelation of the latter with potassium hydroxide in ethanol afforded a mixture from which the 9β,10α-ketone 155 could be isolated in 47% yield.

10.12   11-Oxygenated Steroids and Conessine

A remarkably short and flexible route to 11-oxygenated steroids was developed by Stork and co-workers [114] (Scheme 10-12). 6-Methoxy-1-tetralone (156, R=H) was first converted via the hydroxymethylene ketone to the 2-methyl derivative 157 (R=H). Condensation of this compound with 4-diethylamino-2-butanone methiodide in presence of potassium t-butoxide led to the tricyclic ketone 161 (R=H). Reduction of the latter with sodium borohydride yielded the corresponding alcohol, which was reduced catalytically over palladium-on-strontium carbonate to give a single saturated alcohol. Finally Birch reduction using lithium and liquid ammonia, followed by treatment with concentrated hydrochloric acid, yielded α,β-unsaturated ketone 160 as a single isomer.

The benzoate of 160 was next alkylated with 3-benzyloxybutyl bromide (or iodide) to give mainly the monoalkylated ketone 162, which was alkylated again with methyl iodide. Transformation to the tetracyclic ketone 166 (R=O) was next carried out in several steps without isolation of

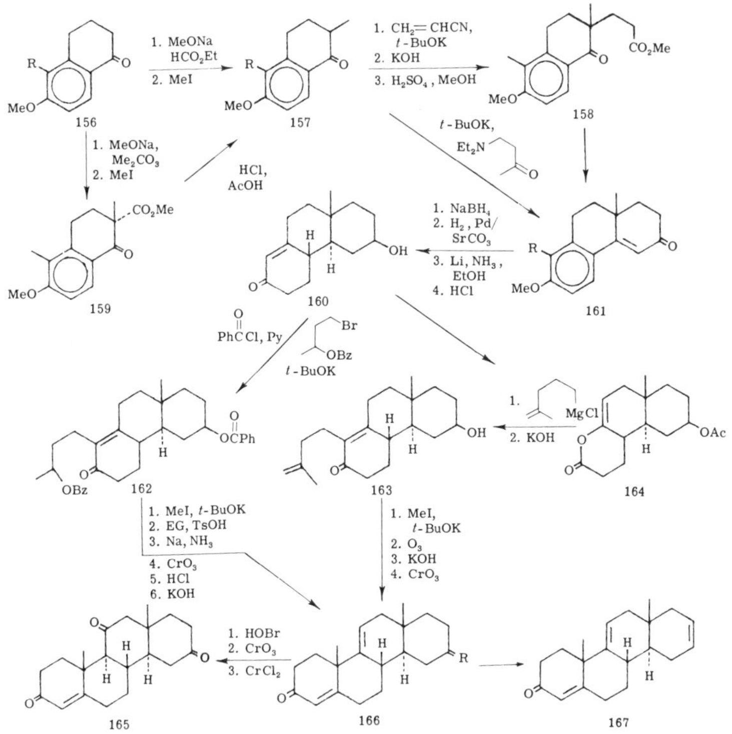

10.12 11-Oxygenated Steroids and Conessine

Scheme 10-12

intermediates. Successive ketalization, reductive elimination of the benzyl group, oxidation, deketalization, and base-catalyzed cyclization yielded a mixture containing almost equal amounts of **166** (R=O) and its 10α-epimer. Reduction of this diketone **166** (R=O) with sodium borohydride and reoxidation with manganese dioxide led to 16-hydroxy-D-homo-4,9(11)-androstadien-3-one (**166**, R=H,OH). Tosylation followed by refluxing with collidine yielded D-homo-4,9(11),16-androstatrien-3-one (**167**). The identity of this compound with an authentic sample [115] established the stereochemical assignments of the diketone **166** (R=O). The introduction of an 11-keto group was readily accomplished by transformation to 9,11-bromohydrin, oxidation to the 9-bromo-11-keto compound, and debromination to the desired D-homo-4-androstene-3,11,16-trione (**165**).

In another modification of the preparation of the diketone **166**, the tricyclic keto alcohol **160** was acetylated and ozonized. The enol lactone **164** of the resulting keto acid was treated with the Grignard derivative from 5-chloro-2-methyl-1-pentene. This was followed by cyclization of the resulting product with base to give **163**, isolated as its *p*-bromobenzoate. Alkylation of **163** with methyl iodide yielded a mixture of C_{10}-epimers in a ratio 2:1, and these were separated as *p*-nitrobenzoates. The predominant epimer had the natural 10β-configuration, as was shown by its subsequent conversion to the same ketone **166** (R = H,OH) described earlier.

In a closely similar approach to the tricyclic homolog **161** (R = Me), the tetralone **156** (R = Me) was used as the starting material [116]. The methoxycarbonylation and methylation of this compound gave the keto ester **159**. The latter compound was next converted to **157** (R = Me). Successive condensation with acrylonitrile, hydrolysis, and methylation yielded the keto ester **158**. Conversion of **158** to the corresponding methyl ketone and subsequent cyclization led to the tricyclic ketone **161** (R = Me). No attempt was made to construct the ring A.

The above approach for the construction of the BCD system was later modified and successfully applied to a stereospecific total synthesis of conessine [117] (Scheme 10-13). In this case, the control of the C-10 stereochemistry was achieved by the use of an 8-iso (*cis* BC) structure, which was eventually inverted to the natural 8β-configuraton. Reaction of 5-methyl-6-methoxy-α-tetralone with dimethyl carbonate in presence of sodium hydride in benzene gave the β-keto ester **168**. Successive addition of methyl isopropenyl ketone and ring closure led to the tricyclic ketone **169**. Catalytic hydrogenation to the dihydro alcohol, treatment with phosphorus oxychloride, and dehydrochlorination yielded the unsaturated ester **170**.

Transformation of the ring D was carried out by ozonolysis of **170** followed by cyclization of the resulting keto aldehyde to the methyl ketone **171**. The acetyl group was protected as the ethylene ketal. This was followed by reduction of the carbomethoxy group, tosylation, and catalytic hydrogenation to give the ketotosylate **172**. The latter compound, on heating with hydroxylamine hydrochloride in pyridine, yielded the nitrone **173**. Catalytic hydrogenation in presence of rhodium-on-charcoal, cleavage of the phenolic ether, and acetylation afforded phenol **174** (R = Ac), with the desired stereochemistry for the pyrrolidine ring. It may be mentioned that, in an independent study directed toward the synthesis of conessine, Mukharji and co-workers [118] achieved the synthesis of a closely related tetracyclic intermediate **174** (R = Me) by a different route. Further hydrogenation of **174** (R = Ac), in presence of ruthenium oxide, followed by chromic acid oxidation, furnished the ketone **175**. Treatment of **175** with acrylonitrile in presence of Triton B

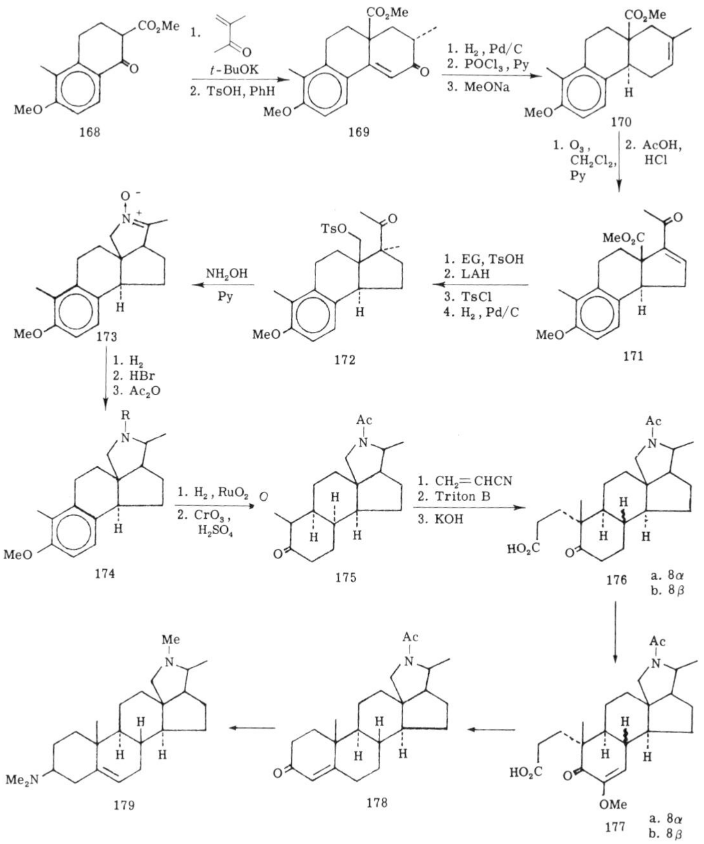

10.12 11-Oxygenated Steroids and Conessine 225

Scheme 10-13

gave the cyanoethyl derivative which was readily converted to the keto acid **176a**.

The next objective was the transformation of the 8α-compound **176a** to its natural 8β-epimer. Successive bromination, alkaline hydrolysis, and oxidation with Benedict solution afforded the 5,6-diketone, which was next converted to the enol ether **177a**. Refluxing with 10% aqueous sodium hydroxide furnished the 8β-epimer **177b**. Subsequent hydrogenation, treatment with calcium in liquid ammonia, reacetylation, and oxidation led to the desired 8β-epimer **176b**. Reaction of the latter with acetic anhydride and sodium acetate yielded the corresponding enol lactone, which was readily converted to the pentacyclic ketone **178** by treatment with methylmagnesium iodide. The compound **178** was treated with dimethylamine and *p*-toluenesulfonic acid to give the corresponding enamine. Subsequent reduction, deacetylation, and *N*-methylation furnished dl-conessine (**179**).

10.13 Miscellaneous Synthetic Routes

A novel, stereospecific approach to the construction of the BCD ring system, via intramolecular alkylation of enol chlorides was developed by Lansbury and co-workers [119] (Scheme 10-14). Selective reduction of the enone **180** with lithium in ammonia at −75° provided the *trans-anti* ketone **181** in 65% yield. Successive treatment with methyllithium and

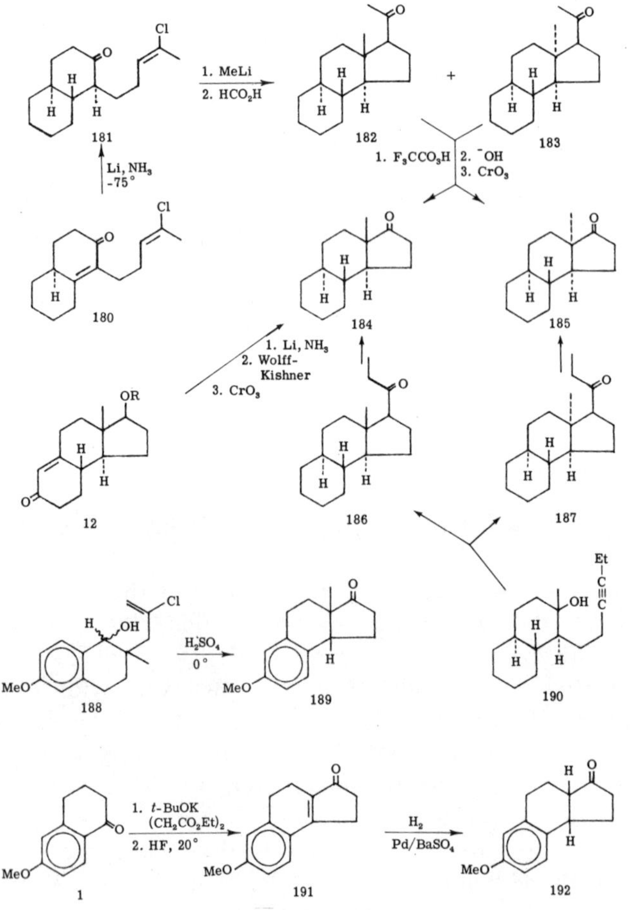

Scheme 10-14

10.13 Miscellaneous Synthetic Routes

formic acid afforded a 2:1 mixture of tricyclic ketones **182** and **183** in about 90% yield. Degradation of this mixture led to the *trans*-fused ketone **184** and its *cis* epimer **185** in a ratio of about 65:35. Additional proof of the structure of **184** was obtained by its synthesis from the known hydroxyketone **12** (R=H). In another modification [120] of the above approach, the acetylenic derivative **190** was subjected to solvolysis in both formic and trifluoroacetic acids using temperatures ranging from $-15°$ to reflux and reaction times of 0.5 to 12 hr. This was followed by hydrolysis to give a mixture of the propionylcyclopentanes **186** (major) and **187**. Further proof of the structures of **186** and **187** was furnished by their degradation to *trans*- and *cis*-hydrindanones **184** and **185**, respectively, as was done with **182** and **183**.

Another variation of the above approach utilizing intramolecular cyclization is exemplified by the synthesis of the *cis* ketone **189**, which was achieved by treatment of the alcohol **188** with sulfuric acid at 0°C [121]. As described in Chapter 3 (Scheme 3-6), this method was utilized for the stereospecific synthesis of *cis*-16-equilenone methyl ether. The synthesis of the tricyclic 18-nor compound **192** has been described by several groups [122–124]. The starting material used was 6-methoxy-1-tetralone (**1**), which was converted to the unsaturated ketone **191** in two steps. In another variation [124], Sarett's method for the construction of ring D (developed during the synthesis of cortisone; see Chapter 9, Schemes 9-6 and 9-7) was utilized for the synthesis of **192** from **1**. Hydrogenation of **191** led to the tricyclic ketone **192**, which could also be obtained in 94% yield by reduction of **191** with sodium in liquid ammonia [124]. However, attempts to introduce an angular methyl group in **192** were not successful [122,123].

The versatile intermediate **12** (R=H) was used as the starting material for the synthesis of the A-furano steroids of the type **197** [125] (Scheme 10-15). Successive reduction of **12** with lithium tri-*t*-butoxyaluminum hydride, acetylation, and epoxidation gave the oxide **193**. Reaction of this compound with vinylmagnesium bromide afforded **194** in excellent yield. Treatment of **194** with trifluoroperacetic acid, followed by oxidation with Jones' reagent led to the diketone **195**. Rearrangement of the latter compound to the furan **196** was effected by treatment with boron trifluoride etherate. Finally, the ketone **196** was converted into 17β-alcohol **197** (R=H) and 17-ethynyl compound **197** (R= C≡CH) by treatment with lithium tri-*t*-butoxyaluminum hydride and the acetylene–ethylenediamine complex, respectively.

In another study directed toward the preparation of oxacyclic steroid analogs, the synthesis of the dipyran **200** was achieved starting from the allyl ether (**198**) of 2,6-dihydroxynaphthalene [126]. Conversion of **198**

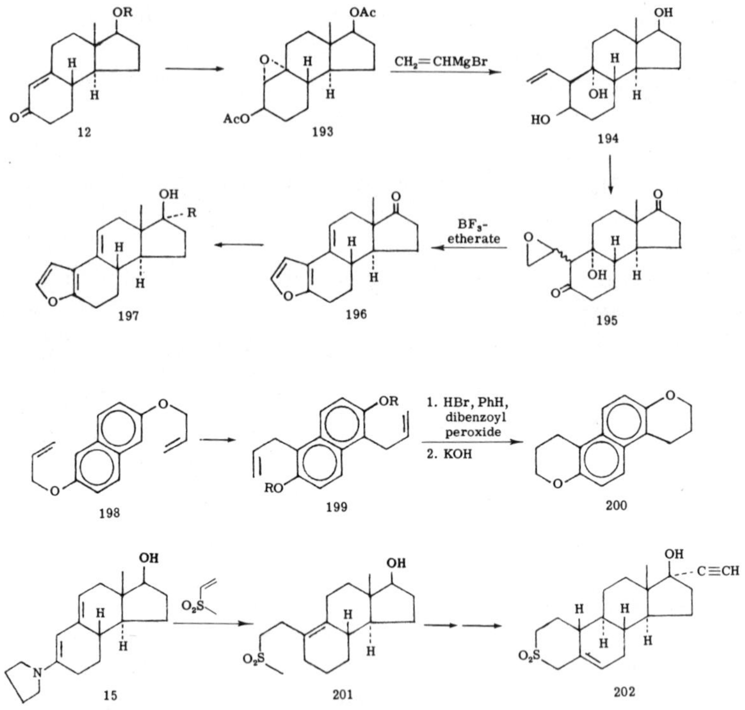

Scheme 10-15

into 1,5-diallyl-2,6-dihydroxynaphthalene (**199**, R=H) was effected by refluxing with N,N-dimethylaniline; this was next acetylated to **199** (R=Ac). Addition of hydrogen bromide under anti Markownikoff conditions, followed by treatment with methanolic potassium hydroxide afforded the dipyran **200**.

The general route developed by the Roussel-UCLAF group was also extended to the synthesis of thiasteroids [127]. Thus the enamine **15** (obtained from **12**, R=H) was converted to the interesting compound of **202** via the intermediate **201**.

References

1. D. K. Banerjee, S. Chatterjee, C. N. Pillai, and M. V. Bhatt, *J. Amer. Chem. Soc.* **78**, 3769 (1956).
2. W. S. Johnson, J. W. Peterson, and C. D. Gutsche, *J. Amer. Chem. Soc.* **69**, 2942 (1947).
3. W. E. Bachmann and D. G. Thomas, *J. Amer. Chem. Soc.* **64**, 94 (1942).
4. R. H. Martin and R. Robinson, *J. Chem. Soc., London* p. 491 (1943).

5. L. Velluz, J. Valls, and G. Nominé, *Angew. Chem., Int. Ed. Engl.* 4, 181 (1965).
6. L. Velluz, J. Mathieu, and G. Nominé, *Tetrahedron, Suppl.* 8, 495 (1966).
7. L. Velluz, *C. R. Acad. Sci.* 253, 1643 (1961).
8. L. Velluz, G. Nominé, J. Mathieu, E. Toromanoff, D. Bertin, J. Tessier, and A. Pierdet, *C. R. Acad. Sci.* 250, 1084 (1960).
9. L. Velluz, G. Nominé, and J. Mathieu, *Angew. Chem.* 72, 725 (1960).
10. R. Bucourt and D. Hainaut, *Bull. Soc. Chim. Fr.* p. 1366 (1965).
11. Roussel-UCLAF, Belgian Patent 626,561 (1963); *Chem. Abstr.* 60, 10621 (1964).
12. L. J. Chinn and H. L. Dryden, Jr., *J. Org. Chem.* 26, 3904 (1961).
13. D. K. Banerjee and S. K. Balasubramanian, *J. Org. Chem.* 23, 105 (1958).
14. D. K. Banerjee, V. Paul, S. K. Balasubramanian, and P. S. N. Murthy, *Tetrahedron* 20, 2487 (1964).
15. C. J. Sih and K. C. Wang, *J. Amer. Chem. Soc.* 85, 2135 (1963).
16. Y. Kurosawa, H. Shimojima, and Y. Osawa, *Steroids, Suppl.* 1, 185 (1965).
17. G. Nominé and D. Bertin, French Patent A 76,518 (1961); *Chem. Abstr.* 58, 5763 (1963).
18. G. Nominé, R. Bucourt, and A. Pierdet, Belgian Patent 610,383 (1962); *Chem. Abstr.* 57, 13646 (1962).
19. L. Velluz, G. Nominé, G. Amiard, V. Torelli, and J. Cérêde, *C. R. Acad. Sci.* 257, 3086 (1963).
20. G. Nominé, J. Tessier, and A. Pierdet, French Patent 1,360,155 (1964); *Chem. Abstr.* 61, 13383 (1964).
21. G. Nominé, A. Pierdet, R. Bucourt, and J. Tessier, U. S. Patent 3,138,617 (1964); *Chem. Abstr.* 61, 12061 (1964).
22. Roussel-UCLAF, Belgian Patent 629,251 (1963); *Chem. Abstr.* 60, 13297 (1964).
23. Y. Ozawa, H. Shimoshima, and K. Chuma, Japanese Patent 21,080/63 (1963); *Chem. Abstr.* 60, 2870 (1964).
24. K. Nakuma, Y. Ozawa, and H. Shimojima, Japanese Patent 11,988/63 (1963); *Chem. Abstr.* 60, 594 (1964).
25. L. J. Chinn and H. L. Dryden, *Abstr. 134th Nat. Amer. Chem. Soc. Meet.*, (1958).
26. M. Uskokovic and M. Gut, *J. Org. Chem.* 22, 996 (1957).
27. J. A. Hartman, A. J. Tomasewski, and A. S. Dreiding, *J. Amer. Chem. Soc.* 78, 5662 (1956).
28. S. Kushinsky, *J. Biol. Chem.* 230, 31 (1958).
29. L. Velluz, G. Nominé, J. Mathieu, E. Toromanoff, D. Bertin, R. Bucourt, and J. Tessier, *C. R. Acad. Sci.* 250, 1293 (1960).
30. G. Stork, A. Brizzolara, H. Landesman, J. Szmuszkovicz, and R. Terrell, *J. Amer. Chem. Soc.* 85, 207 (1963).
31. J. A. Vida and M. Gut, *J. Org. Chem.* 30, 1244 (1965).
32. J. A. Vida and M. Gut, *Abstr. 148th Nat. Amer. Chem. Soc. Meet.*, p. 40S (1964).
33. J. Fried and E. F. Sabo, *J. Amer. Chem. Soc.* 75, 2273 (1953).
34. L. Ruzicka and P. Müller, *Helv. Chim. Acta* 22, 416 (1939).

35. J. A. Hogg, P. F. Beal, A. H. Nathan, F. H. Lincoln, W. P. Schneider, B. J. Magerlein, A. R. Hanze, and R. W. Jackson, *J. Amer. Chem. Soc.* **77**, 4436 (1955).
36. L. Velluz, G. Nominé, J. Mathieu, E. Toromanoff, D. Bertin, M. Vignau, and J. Tessier, *C. R. Acad. Sci.* **250**, 1510 (1960).
37. G. Nominé, R. Bucourt, and M. Vignau, U. S. Patent 3,085,098 (1963); *Chem. Abstr.* **60**, 592 (1964).
38. Roussel-UCLAF, Belgian Patent 634,308 (1963); *Chem. Abstr.* **61**, 1916 (1964).
39. C. Snozzi and B. Goffinet, French Patent 1,290,876 (1962); *Chem. Abstr.* **58**, 1519 (1963).
40. Roussel-UCLAF, British Patent 914,737 (1963); *Chem. Abstr.* **58**, 14068 (1963).
41. L. Velluz, G. Nominé, R. Bucourt, A. Pierdet, and J. Tessier, *C. R. Acad. Sci.* **252**, 3903 (1961).
42. G. Nominé, R. Bucourt, and A. Pierdet, *C. R. Acad. Sci.* **254**, 1823 (1962).
43. G. Nominé, R. Bucourt, and A. Pierdet, U. S. Patent 3,141,025 (1964); *Chem. Abstr.* **61**, 10744 (1964).
44. G. Nominé, R. Bucourt, and A. Pierdet, French Patent 1,282,638 (1962); *Chem. Abstr.* **57**, 16702 (1962).
45. G. Nominé, D. Bertin, R. Bucourt, and A. Pierdet, U.S. Patent 3,055,885 (1962); *Chem. Abstr.* **58**, 3483 (1963).
46. Laboratoires Français de Chimiothérapie, French Patent 1,269,641 (1960); *Chem. Abstr.* **56**, 14372 (1962).
47. Roussel-UCLAF, Belgian Patent 628,790 (1963); *Chem. Abstr.* **60**, 10753 (1964).
48. Roussel-UCLAF, British Patent 936,894 (1963); *Chem. Abstr.* **61**, 5725 (1964).
49. L. Velluz, G. Nominé, R. Bucourt, A. Pierdet, and P. Dufay, *Tetrahedron Lett.* p. 127 (1961).
50. W. M. Allen and M. Ehrenstein, *Science* **100**, 251 (1944).
51. M. Ehrenstein, *J. Org. Chem.* **9**, 435 (1944).
52. J. Mathieu, *Proc. Int. Symp. Drug Res.* 1967 p. 134 (1900).
53. R. Bucourt, J. Tessier, and G. Nominé, *Bull. Soc. Chim. Fr.* p. 1757 (1962).
54. R. Bucourt, J. Tessier, and G. Nominé, *Bull. Soc. Chim. Fr.* p. 1923 (1963).
55. G. Nominé, R. Bucourt, and J. Tessier, U.S. Patent 3,102,145 (1963); *Chem. Abstr.* **61**, 7082 (1964).
56. A. M. Krubiner and E. P. Oliveto, *J. Org. Chem.* **31**, 24 (1966).
57. C. Djerassi, L. Miramontes, and G. Rosenkranz, *J. Amer. Chem. Soc.* **73**, 3540 (1951).
58. C. Djerassi, L. Miramontes, and G. Rosenkranz, *J. Amer. Chem. Soc.* **75**, 4440 (1953).
59. G. Nominé, R. Bucourt, J. Tessier, A. Pierdet, G. Costerousse, and J. Mathieu, *C. R. Acad. Sci.* **260**, 4545 (1965).
60. M. Perelman, E. Farkas, E. J. Fornefeld, R. J. Kraay, and R. T. Rapala, *J. Amer. Chem. Soc.* **82**, 2402 (1960).
61. L. Velluz, R. Bucourt, M. Vignau, E. Toromanoff, and G. Nominé, *Justus Liebigs Ann. Chem.* **669**, 153 (1963).
62. R. Bucourt and G. Nominé, *Bull. Soc. Chim. Fr.* p. 1537 (1966).
63. G. Nominé and R. Bucourt, French Patent 1,334,950 (1963); *Chem. Abs.* **60**, 3040 (1964).

64. Roussel-UCLAF, Belgian Patent 612,588 (1962); *Chem. Abstr.* **57**, 14968 (1962).
65. G. Nominé and R. Bucourt, French Patent 1,304,009 (1962); *Chem. Abstr.* **58**, 10269 (1963).
66. J. Tadanier, *J. Org. Chem.* **28**, 1744 (1963).
67. Roussel-UCLAF, French Patent M 1,958 (1963); *Chem. Abstr.* **60**, 3040 (1964).
68. Roussel-UCLAF, Dutch Patent 6,411,300 (1965); *Chem. Abstr.* **63**, 10031 (1965).
69. L. Velluz, G. Nominé, R. Bucourt, and J. Mathieu, *C. R. Acad. Sci.* **257**, 569 (1963).
70. L. Velluz, G. Nominé, J. Mathieu, R. Bucourt, L. Nédélec, M. Vignau, and J. C. Gasc, *C. R. Acad. Sci.* **264**, 1396 (1967).
71. P. D. Klimstra and F. B. Colton, in "Contraception: The Chemical Control of Fertility" (D. Lednicer, ed.), p. 69. Marcel Dekker, New York, 1969.
72. J. A. Campbell, S. C. Lyster, G. W. Duncan, and J. C. Babcock, *Steroids* **1**, 317 (1963).
73. L. Velluz, R. Bucourt, M. Vignau, E. Toromanoff, and G. Nominé, *Justus Liebigs Ann. Chem.* **669**, 153 (1963).
74. J. H. Fried, T. S. Bry, A. E. Oberster, R. E. Beyler, T. B. Windholz, J. Hannah, L. H. Sarett, and S. L. Steelman, *J. Amer. Chem. Soc.* **83**, 4663 (1961).
75. G. Nominé, R. Bucourt, and M. Vignau, U.S. Patent 3,052,672 (1962); *Chem. Abstr.* **57**, 16702 (1962).
76. G. Nominé and R. Bucourt, U.S. Patent 3,033,856 (1962); *Chem. Abstr.* **57**, 11270 (1962).
77. R. Joly, J. Warnant, J. Jolly, and J. Mathieu, *C. R. Acad. Sci.* **258**, 5669 (1964).
78. Roussel-UCLAF, French Patent M 1,830 (1963); *Chem. Abstr.* **60**, 593 (1964).
79. D. Bertin and A. Pierdet, *C. R. Acad. Sci.* **264**, 1002 (1967).
80. J. J. Brown and S. Bernstein, *Steroids* **8**, 87 (1966).
81. J. J. Brown and S. Bernstein, *Steroids* **1**, 113 (1963).
82. M. Heller, R. H. Lenhard, and S. Bernstein, *Steroids* **10**, 211 (1967).
83. J. C. Gasc and L. Nédélec, *Tetrahedron Lett.* p. 2005 (1971).
84. L. Nédélec, *Bull. Soc. Chim. Fr.* p. 2548 (1970).
85. L. Nédélec and J. C. Gasc, *Bull. Soc. Chim. Fr.* p. 2556 (1970).
86. F. W. Heyl and M. E. Herr, *J. Amer. Chem. Soc.* **77**, 488 (1955).
87. D. K. Banerjee, K. M. Damodaran, P. S. N. Murthy, and V. Paul, *Abstr., Int. Symp. Chem. Natur. Prod., 8th, 1972,* p. 223 (1972).
88. O. R. Vail and D. M. S. Wheeler, *J. Org. Chem.* **27**, 3803 (1962).
89. T. L. Eggerichs, A. C. Ghosh, R. C. Matejka, and D. M. S. Wheeler, *J. Chem. Soc., London* p. 1632 (1969).
90. A. C. Ghosh, K. Mori, A. C. Rieke, S. K. Roy, and D. M. S. Wheeler, *J. Org. Chem.* **32**, 722 (1967).
91. S. G. Levine and A. C. Ghosh, *Tetrahedron Lett.* p. 39 (1969).
92. S. G. Levine, N. H. Eudy, and C. F. Leffler, *J. Org. Chem.* **31**, 3995 (1966).
93. A. D. Cross, E. Denot, R. Acevedo, R. Urquiza, and A. Bowers, *J. Org. Chem.* **29**, 2195 (1964).
94. W. F. Johns, *J. Org. Chem.* **29**, 1490 (1964).

95. K. H. Palmer, C. E. Cook, F. T. Ross, J. Dolar, M. E. Twine, and M. E. Wall, *Steroids* **14**, 55 (1969).
96. D. K. Banerjee, V. Paul, and S. K. Balasubramanian, *Tetrahedron Lett.* p. 23 (1960).
97. C. Djerassi, A. J. Manson, and H. Bendas, *Tetrahedron* **1**, 22 (1957).
98. C. Djerassi, R. Riniker, and B. Riniker, *J. Amer .Chem. Soc.* **78**, 6362 (1956).
99. C. Djerassi, R. Riniker, and B. Riniker, *J. Amer. Chem. Soc.* **78**, 6377 (1956).
100. R. P. Linstead, W. E. Doering, S. B. Davis, P. Levine, and R. R. Whetstone, *J. Amer. Chem. Soc.* **64**, 1985 (1942).
101. D. K. Banerjee, *J. Indian Chem. Soc.* **47**, 1 (1970).
102. D. K. Banerjee, B. Sugavanam, and G. Nadamuni, *Tetrahedron Lett.* p. 2771 (1968).
103. R. H. Martin and R. Robinson, *J. Chem. Soc., London* p. 1866 (1949).
104. R. H. Lenhard and S. Bernstein, *J. Amer. Chem. Soc.* **77**, 6665 (1955).
105. L. Velluz, G. Nominé, and R. Bucourt, *C. R. Acad. Sci.* **257**, 811 (1963).
106. R. Bucourt and G. Nominé, French Patent, 1,404,413 (1965); *Chem. Abstr.* **63**, 18212 (1965).
107. R. Bucourt, D. Hainaut, J. C. Gasc, and G. Nominé, *Tetrahedron Lett.* p. 5093 (1968).
108. R. Bucourt, D. Hainaut, J. C. Gasc, and G. Nominé, *Bull. Soc. Chim. Fr.* 1920 p. 1920 (1969).
109. K. K. Koshoev, S. N. Ananchenko, and I. V. Torgov, *Khim. Prir. Soedin.* **3**, 108 (1965).
110. V. I. Zaretskii, N. S. Wulfson, and V. L. Sadovskaya, *Tetrahedron Lett.* p. 3879 (1966).
111. G. C. Buzby, Jr., E. Capaldi, G. H. Douglas, D. Hartley, D. Herbst, G. A. Hughes, K. Ledig, J. McMenamin, T. Pattison, H. Smith, C. R. Walk, G. R. Wendt, J. Siddall, B. Gadsby, and A. B. A. Jansen, *J. Med. Chem.* **9**, 338 (1966).
112. A. J. Birch and G. S. R. Subba Rao, *J. Chem. Soc., London* p. 5139 (1965).
113. J. R. Bull and A. Tuinman, *Chem. Commun.* p. 921 (1972).
113a. J. R. Bull and A. Tuinman, *Tetrahedron* **29**, 1101 (1973).
114. G. Stork, H. J. E. Lowenthal, and P. C. Mukharji, *J. Amer. Chem. Soc.* **78**, 501 (1956).
115. L. B. Barkley, M. W. Farrar, W. S. Knowles, H. Raffelson, and Q. E. Thompson, *J. Amer. Chem. Soc.* **76**, 5014 (1954).
116. M. J. T. Robinson, *Tetrahedron* **1**, 49 (1957).
117. G. Stork, S. D. Darling, I. T. Harrison, and P. S. Wharton, *J. Amer. Chem. Soc.* **84**, 2018 (1962).
118. P. C. Mukharji, J. C. Sircar, and V. V. Devasthale, *Ind. J. Chem.* **9**, 515 (1971).
119. P. T. Lansbury, P. C. Briggs, T. R. Demmin, and G. E. DuBois, *J. Amer. Chem. Soc.* **93**, 1311 (1971).
120. P. T. Lansbury and G. E. DuBois, *Chem. Commun.* p. 1107 (1971).
121. P. T. Lansbury, E. J. Nienhouse, D. J. Scharf, and F. R. Hilfiker, *J. Amer. Chem. Soc.* **92**, 5649 (1970).

122. A. J. Birch, J. A. K. Quartey, and H. Smith, *J. Chem. Soc. London* p. 1768 (1952).
123. D. B. Cowell and D. W. Mathieson, *J. Pharm. Pharmacol.* **9**, 549 (1957).
124. N. A. Nelson, J. C. Wollensak, R. L. Foltz, J. B. Hester, J. I. Brauman, R. B. Garland, and G. H. Rasmusson, *J. Amer. Chem. Soc.* **82**, 2569 (1960).
125. D. Lednicer and D. E. Emmert, *J. Org. Chem.* **34**, 1151 (1969).
126. M. P. Rao and S. K. Pradhan, *Synthesis* p. 661 (1971).
127. D. Bertin and J. Perronnet, *Bull. Soc. Chim. Fr.* p. 1422 (1968).

11
B + D → BD → BCD → ABCD

In Chapter 4 syntheses were described in which ring D was attached to an AB moiety, following which ring C was closed producing the steroid. In theory it should be just as feasible to connect B and D by one of the many routes delineated in Chapter 4, close ring C, then construct ring A by one of several available annelation methods. In practice, however, this approach has enjoyed only very limited success.

One of the few successful syntheses began with the triketone 1 (Scheme 11-1), which was obtained by Michael addition of 1,3-cyclohexanedione to methyl vinyl ketone [1]. The next step was to add carbons 11 and 12 by a Grignard reaction with vinylmagnesium bromide. To protect oxo groups at C-3 and C-5 (steroid numbering) from attack by the Grignard reagents, 1 was converted to the corresponding chromone (temporary closure of ring A) by heating in acetic anhydride. The Grignard reaction and subsequent hydrolysis gave 2 in 24% yield [2,3]. Michael addition of 2-methyl-1,3-cyclohexanedione gave BD intermediate 3, which was cyclized (low overall yield) to the D-homosteroid 4 [2,3]. The latter was identical with the product synthesized by an AB+D→ABD→ABCD route [4]. In an effort to obtain a 10-methyl homolog, 5 [5–7], was condensed with 2-methyl-,3-cyclohexanedione and the BD intermediate was cyclized to give 6 in moderate yield [6–8]. All attempts to construct ring A onto 6, however, were unsuccessful [6,7].

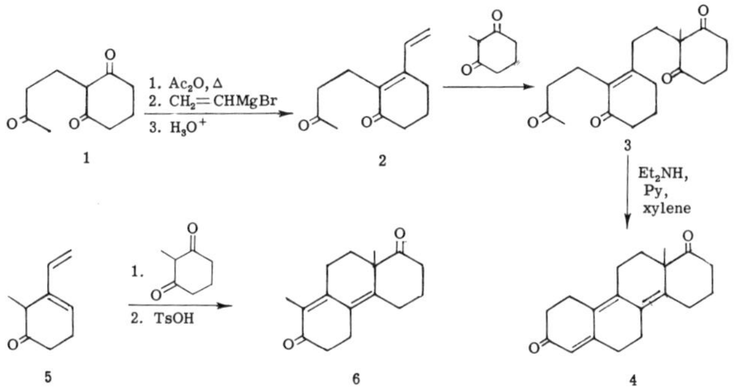

B + D → BD → BCD → ABCD

Scheme 11-1

A 17-deoxy steroid was synthesized from BD intermediate 8 (Scheme 11-2). One method by which 8 was prepared began with the condensation of furfural and p-methoxyacetophenone to give 7, which was hydrolyzed to a diketo acid. Base-catalyzed intramolecular condensation produced 8 in 50% yield from the p-methoxyacetophenone [9,10]. Alternatively, 8 was synthesized in 30% yield from β-bromo-p-methoxyacetophenone via the diester 12 [11]. The D-homolog of 8 also was synthesized [12]. Catalytic hydrogenation of 8, followed by Clemennsen reduction and cyclization gave 9 [13]. Attempts to cyclize without first removing the 17-keto group failed [14]. Reduction of the 11-oxo group, Birch reduction of ring B, and hydrolysis gave 10, which on Robinson annelation gave the steroid 11 [10,14,15].

Another route to BCD intermediates capable of extension to tetracyclic steroids was developed by a joint Indian–American effort [16]. The ketone 13 (Scheme 11-3) was obtained by Stobbe condensation of the isobutyl enol ether of 2-methyldihydroresorcinol [7] with di-t-butyl

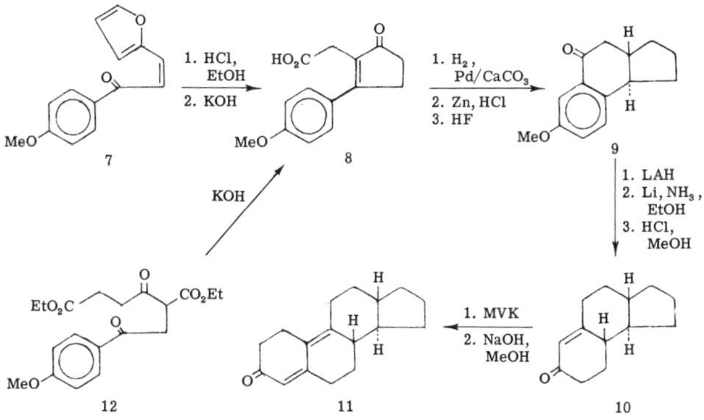

Scheme 11-2

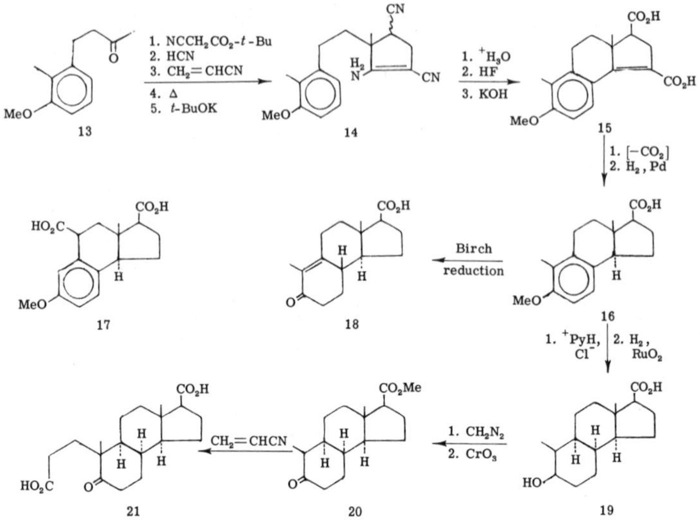

11 B+D → BD → BCD → ABCD

Scheme 11-3

succinate followed by heating with palladium to give 3-(3-methoxy-2-methylphenyl)propanoic acid; the latter was converted via its acid chloride to **13**. Condensation of **13** with *t*-butyl cyanoacetate, addition of HCN, cyanoethylation, pyrolysis, and Thorpe cyclization led to a mixture of isomers **14**. Acid hydrolysis, cyclization (HF), and alkaline hydrolysis gave **15**, which was decarboxylated and hydrogenated to **16** in 41% yield from **13**. An 11-carboxy-10-demethyl analog **17** also was synthesized. Birch reduction of **16** and hydrolysis gave **18**, identical to that prepared from CD intermediates, but the yield was low. Although **18** was convertible to a steroid, an alternative route was sought because of the low yield of **18** and because its annelation produced a mixture of isomers [17]. Consequently, **16** was demethylated and the phenol was hydrogenated over ruthenium oxide to the hydroxy acid **19**. Esterification with diazomethane and oxidation gave **20**, which was cyanoethylated to **21**. The latter is an 8-epimer of a known compound which had been converted into steroids, but efforts to epimerize **21** at C-8 were unsuccessful (a successful epimerization in a comparable instance has been reported [18]).

The BD intermediate **23** (Scheme 11-4) was synthesized by alkylation of the Birch reduction product of resorcinol dimethyl ether with the bromide **22** [19]. Angular methylation, ring closure and hydrogenation gave the BCD intermediate **24**. The D ring was contracted by oxidation and esterification to the diester **25**, Dieckmann cyclization and acid hydrolysis to **26** [20], which had been obtained earlier via BC intermediates [21]. The tricyclic intermediates **27** and 9(11)-dihydro **27** were prepared by unique routes, but the yields were low and the syntheses were not extended to steroids [22].

The first example of a steroid containing the indole ring system was synthesized from a BD intermediate [23]. Demethoxy **8** (Scheme 11-2) was converted via **28** and **29** into **30** (Scheme 11-5). Diazotization of **30** and reduction with stannous chloride gave the corresponding hydrazine in 57% yield; the latter was converted into **31** in 98% yield by reaction with ethyl pyruvate. Ring closure in alcoholic hydrogen chloride produced a 70% yield of **32**, which was hydrolyzed and decarboxylated to afford a mixture from which **33** was isolated [23].

References

1. I. N. Nazarov and S. I. Zav'yalov, *Izv. Akad. Nauk SSSR, Otd. Khim. Nauk* p. 207 (1957).
2. S. I. Zav'yalov, G. V. Kondrat'eva, and L. F. Kudryavtseva, *Med. Prom. SSSR* **15**, 56 (1961).
3. S. I. Zav'yalov, G. V. Kondrat'eva, and L. F. Kudryavtseva, *Izv. Akad. Nauk SSSR, Otd. Khim. Nauk* p. 529 (1961).
4. N. N. Gaidamovich and I. V. Torgov, *Izv. Akad. Nauk SSSR, Otd. Khim. Nauk* p. 1803 (1961).
5. I. N. Nazarov, I. V. Torgov, and G. P. Verkholetova, *Dokl. Akad. Nauk SSSR* **112**, 1067 (1957).
6. S. N. Ananchenko and I. V. Torgov, *Izv. Akad. Nauk SSSR, Otd. Khim. Nauk* p. 1649 (1960).
7. A. Eschenmoser, J. Schreiber, and S. A. Julia, *Helv. Chim. Acta* **36**, 482 (1953).
8. CIBA British Patent 745,758 (1956); *Chem. Abstr.* **51**, 15585f (1957); Swiss Patent 309,919 (1955); *Chem. Abstr.* **51**, 10583e (1957).
9. R. Robinson and W. M. Todd, *J. Chem. Soc., London* p. 1743 (1939).
10. G. S. Grinenko and V. I. Maksimov, *Med. Prom. SSSR* **15**, 50 (1961).
11. G. S. Grinenko and V. I. Maksimov, *Zh. Obshch. Khim.* **28**, 528 (1958).
12. M. Guha, U. Rakshit, and D. Nasipuri, *J. Indian Chem. Soc.* **37**, 267 (1960).
13. G. S. Grinenko and V. I. Maksimov, *Zh. Obshch. Khim.* **28**, 532 (1958); V. I. Maksimov and G. S. Grinenko, *ibid.* p. 2179.
14. G. S. Grinenko and V. I. Maksimov, *Dokl. Akad. Nauk SSSR* **112**, 1059 (1957); V. I. Maksimov and G. S. Grinenko, *Zh. Obshch. Khim.* **28**, 2182 (1958); G. S. Grinenko and V. I. Maksimov, *ibid.* **30**, 574 (1960).
15. V. I. Maksimov and G. S. Grinenko, *Zh. Obshch. Khim.* **29**, 2056 (1959).
16. D. K. Banerjee, H. N. Khastgir, J. Dutta, E. J. Jacob, W. S. Johnson, C. F. Allen, B. K. Bhattacharyya, J. C. Collins, Jr., A. L. McCloskey, W. T. Tsatsos, W. A. Vredenburgh, and K. L. Williamson, *Tetrahedron Lett.* p. 76 (1961).
17. R. B. Woodward, F. Sondheimer, D. Taub, K. Heusler, and W. M. McLamore, *J. Amer. Chem. Soc.* **74**, 4223 (1952).
18. G. Stork, S. D. Darling, I. T. Harrison, and P. S. Wharton, *J. Amer. Chem. Soc.* **84**, 2018 (1962).
19. A. J. Birch, H. Smith, and R. E. Thornton, *J. Chem. Soc., London* p. 1338 (1957); A. J. Birch and H. Smith, *ibid.* p. 1882 (1951).
20. C. S. Rao and D. K. Banerjee, *Tetrahedron* **19**, 1611 (1963).
21. W. E. Bachmann and D. G. Thomas, *J. Amer. Chem. Soc.* **64**, 94 (1942).
22. P. C. Dutta, *J. Indian Chem. Soc.* **26**, 102 (1949).
23. V. N. Sladkov, V. F. Shner, L. M. Alekseeva, K. F. Turchin, O. S. Anisimova, Yu. N. Sheinker, and N. N. Suvorov, *Dokl. Akad. Nauk SSSR* **198**, 605 (1971).

12
CD → BCD → ABCD

12.1 CD Intermediates

The starting CD intermediates for the various total syntheses outlined in this chapter and in Chapter 7 are often the same or quite analogous. Historically, a large effort has been expended in the preparation of such material, both in the context of steroid synthesis and, recently, as a goal unto itself. A discussion of the pathways developed for the synthesis of CD intermediates has been included at this point such that later, when the individual total synthetic schemes are presented, it will be unnecessary to outline the preparation of the starting materials, except in those cases where unique pathways have been developed.

Most of the early synthetic plans started with appropriate naphthalene derivatives which in turn gave rise to D-homosteroids. For example, Johnson's estrone synthesis started with decalin-1,5-dione (3, Scheme 12-1) which had earlier been prepared from 1,5-dihydroxynaphthalene (58%) [1]. By modifying the previously employed conditions to include Raney nickel hydrogenation at elevated temperatures and pressures in a basic ethanolic solution, it was possible to obtain a mixture of diols, 2 (stereochemistry undetermined), which, without purification, could be oxidized to a mixture of *cis* and *trans* diketones 3 [2]. The yield for this conversion (1→3) was 55% and, for synthetic purposes, the mixture

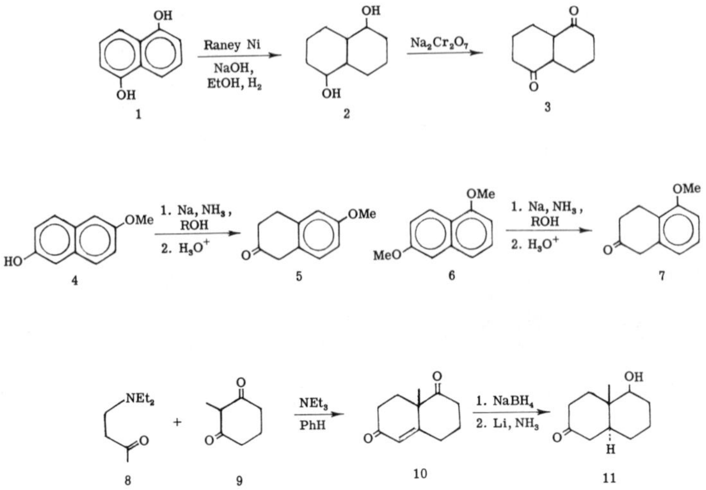

Scheme 12-1

could be used as such or, by heating with acid, could be converted entirely into the thermodynamically more stable *trans* isomer.

Disubstituted naphthalenes 4 and 6 have also served as precursors for CD intermediates used in total synthesis. With the former compound, selective Birch reduction yielded the methoxyketone 5 [3,] which has also been prepared by an alternate, though related, procedure [4]. Similarly, reduction of 2,5-dimethoxynaphthalene afforded 7 (73%) [5] which, besides serving as the starting material in Robinson's synthesis of epiandrosterone, was a key intermediate in Johnson's hydrochrysene approach (Section 12.2). Other substituted naphthalenes have also served as precursors to CD intermediates, and the procedure illustrated above, i.e., various degrees of Birch reduction, has been a general one.

An obvious limitation embodied in the above approach is the lack of an angular methyl group. Unless 18-norsteroids are the goal, this group must be introduced at some stage, usually after the steroid skeleton has been derived. Most schemes could be considerably shortened if the starting CD intermediate already possessed an angular group, and, although it is possible to alkylate certain preformed hydronaphthalene derivatives [6] such intermediates are preferably made by annelating various alkyl substituted cyclohexane-1,3-diones. Thus, the alkali-sensitive intermediate 10 has been the starting point of a number of synthetic schemes. It may be prepared in a two-step sequence by the alkylation of 9 with methyl vinyl ketone in the presence of catalytic amounts of triethylamine, and subsequent cyclization with aluminum tri-*t*-butoxide

12.1 CD Intermediates

[7,8]. The preferred process, however, is the one-step condensation of 8 and 9 in refluxing benzene containing a trace of pyridine (66%) [9–11]. This annelation reaction has proved quite general, and a variety of 8-substituted derivatives of 10 have been prepared for use in total synthesis by starting with derivatives of 8 (or the analogous vinyl ketones) [12,13]. Further investigations showed 10 could be stereoselectively reduced with borohydride to the 17aβ-alcohol [14,15] and, subsequently, with lithium–ammonia to the *trans* ketol 11 [15,16]. Though the *cis*-fused product can also be obtained [15,17], the more highly desired *trans* products (corresponding to the natural ring fusion) are quite readily prepared; as will be seen shortly, this is not the case when the CD fragment is a hydrindane derivative.

The overall sequence could, of course, be shortened if one were to start with CD fragments possessing 5-membered D rings. In recent years this has indeed been the preferred route as more simplified methods of preparation of hydrindanes have been described (Scheme 12-2). The first breakthrough was the preparation of 5,6,7,8-tetrahydro-8-methylindane-1,5-dione (14) [18,19]. 2-Methylcyclopentane-1,3-dione [20,21] condensed nearly quantitatively with methyl vinyl ketone in hot methanol containing catalytic amounts of potassium hydroxide. The resulting trione 13

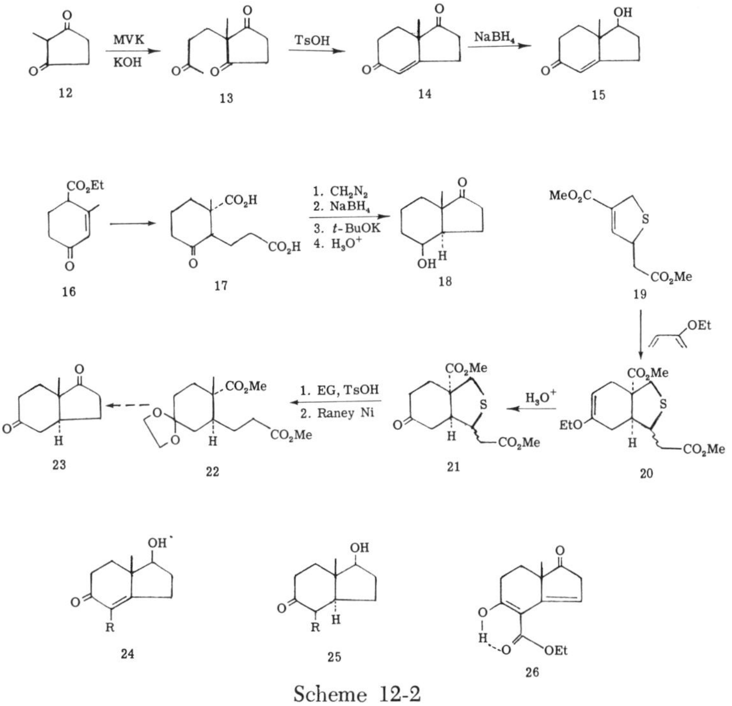

Scheme 12-2

was cyclized in excellent yield by heating in benzene with *p*-toluenesulfonic acid. Other acids (e.g., oxalic) were much less effective for this cyclization. The yield of **14** was greater than 70% from **12**. The annelation could also be accomplished using 4-diethylaminobutan-2-one in place of methyl vinyl ketone, but much less successfully. As with the 6-membered ring analog, the hydrindane **14** could be selectively reduced by a number of metal hydrides to give only the 17β-equatorial configuration **15** (racemic). Also, the optically active form, (+)−**15**, has been obtained by microbiological reduction [18] and by chemical resolution [22]. To this point, the chemistry of hydrindane derivatives closely parallels that of the hydronaphthalene compounds; and intermediates **14** and **15** have served as the starting foci for a great number of total syntheses (see [23] and the references cited therein for an overview of the reactions employing **15**). It has been a relatively simple matter to prepare derivatives of **14** by starting with variously substituted vinyl ketones (or Mannich bases) in place of methyl vinyl ketone. However, the quite vexing problem of preparing *trans* hydrindanes has, until very recently, remained a frustrating synthetic challenge. Thus, catalytic hydrogenation under a variety of conditions as well as lithium–ammonia reduction of **14** or **15** gave only the *cis*-fused product [24,25]. Because of this, the elusive *trans* product has historically been prepared by ring closure of the appropriate cyclohexane diacids [26]. Inhoffen, in his extensive investigations of vitamin D chemistry, converted Hagemann's ester **16** into **18** by the sequence shown in Scheme 12-2 (in fact, both enantiomers of **17** were separated and individually transformed into **18**) [27,28]. In addition, by degradation of natural products [29,30], several more such compounds were obtained which later proved valuable for identification purposes. An elegant elaboration of this general method has recently been reported by Lythgoe *et al.* [31] in the total synthesis of calciferol and its relatives.

A related synthesis, shown in the preparation of **23**, has been recently outlined [32]. This unique approach took advantage of the stereospecificity of the Diels-Alder reaction and gave **20** which, after mild hydrolysis and ketalization was subjected to Raney nickel desulfurization and gave a 54% yield of the diester **22**. This procedure should be amenable to modification and, by employing variously substituted dienes, it should be possible to prepare modified *trans* hydrindanes. Another cyclization procedure for the preparation of *trans*-fused hydrindanes (and decalins) depends on the hydroboration–carbonylation of appropriate dienes (e.g., 1-vinylcyclohexene), and could find some use in steroid synthesis [33].

As indicated above, simple hydrogenation of unsaturated hydrindane derivatives related to **14** and **15** have invariably led to the less desirable *cis* products [34]. However, under appropriate conditions, such catalytic hydrogenations may be made to give predominantly, or exclusively, the natural *trans*-fusion. Two different reports in 1963 indicated that **24** (R = $CH_2CH_2CO_2H$ [35] and R = $CH_2CH_2C_6H_4OCH_3$-*p* [36]) could be selectively hydrogenated over palladium-on-charcoal to yield appreciable quantities of **25**. Additional work has detailed some possible reasons for such selectivity, and it was suggested that for **24** (R = CO_2Et) the enol **26** may be the critical intermediate. Further investigation uncovered the interesting fact that **24** (R = $CH_2CH_2CO_2H$) gave 75% of the desired *trans* product, whereas **24** (R = CH_2CO_2H) could be hydrogenated much less readily to the *trans* hydrindane [37]. With this latter compound, the 17β-alcohol moiety was necessary for proper reduction; if the alcohol function was replaced by a ketone, the reduction, under a variety of conditions, gave mostly the undesired *cis* product [38]. Thus, reduction of the alcohol over 10% palladium-on-calcium carbonate (or palladium-on-charcoal) gave predominantly the *trans* material. On the other hand, if R = CO_2Et, the ketone need not be reduced to the alcohol, but could be converted quantitatively to the *trans* hydrindane. Clearly, more work needs to be done to delineate all the forces operating in this reduction, but the potential for steroid total synthesis is obvious (see, for example, Sections 12.6 and 12.7).

12.2 Hydrochrysene Approach

A major portion of the work devoted to total synthesis by the CD→BCD→ABCD route is that of W. S. Johnson and his collaborators in their development of the hydrochrysene approach. Considered in its entirety, this work, extending for the most part from 1953 to the mid 1960's, represents a monumental effort. For the purposes of the present discussion it will be divided into five parts represented in Scheme 12-3 by (1) the preparation of **28** from 5-methoxy-2-tetralone; (2) the hydrogenation of **28** to give various isomers of **29**; (3) the reduction of the aromatic nucleus; (4) the angular methylation and ring contraction of **30** to give the steroid nucleus **31**; and (5) functionalization of **31** to produce a multitude of steroids including aldosterone.

The initial objective of the route [39] was the preparation of a highly unsaturated precursor **28** which could then, hopefully, be stereoselectively reduced to produce the natural *trans-anti-trans* form of **29**. To this end, the tetralone **27** (see Section 12.1) was annelated with the methio-

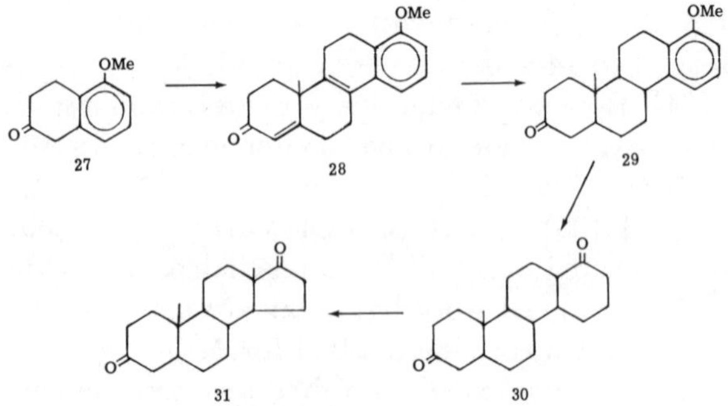

Scheme 12-3

dide of 1-diethylamino-3-pentanone to produce a tautomeric mixture (33 and 34; Scheme 12-4) [40,41]. Interestingly, an isomeric form of 33/34 had previously been employed in the total synthesis of epiandrosterone [42]. The common anion derived from 33/34 by treatment with sodium methoxide was reacted with either methyl vinyl ketone or 1-diethylamino-3-butanone methiodide to give 28. Under milder conditions a mixture of epimeric ketols 35 (the mixture could be separated) was produced [43] which, by further treatment with methoxide, was converted to 28. (This ketol mixture was initially formulated as a mixture of 5α- and 5β-hydroxy compounds, 36. Later work [44] showed this as well as some previous, similar formulations to be incorrect.) For preparative purposes, however, it was not necessary to isolate the tautomeric mixture 33/34. Thus, after the tetralone 27 and metallic sodium were dissolved in anhydrous methanol, the Mannich base 32 was added, followed by more methanol and sodium and, finally, methyl vinyl ketone. The product (34%) was the key, unsaturated steroid precursor 28.

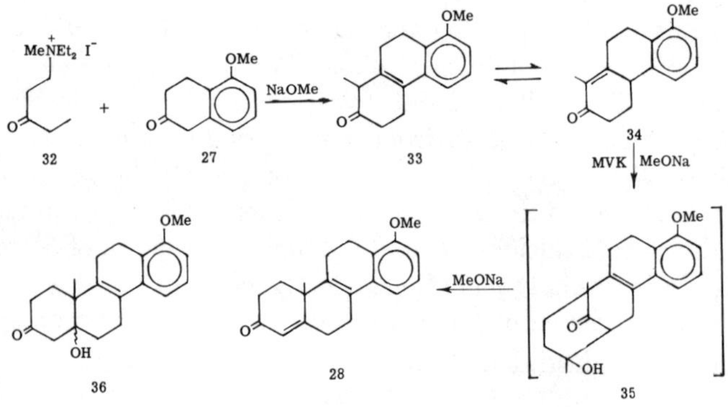

Scheme 12-4

12.2 Hydrochrysene Approach

The next effort was directed toward exploring the selective reduction of the double bonds of **28** [43,45]. For preparation of the AB-*trans* series it was necessary to first prepare the enol ether **37** (Scheme 12-5) which was then reduced with 6% palladium-on-strontium carbonate to give a

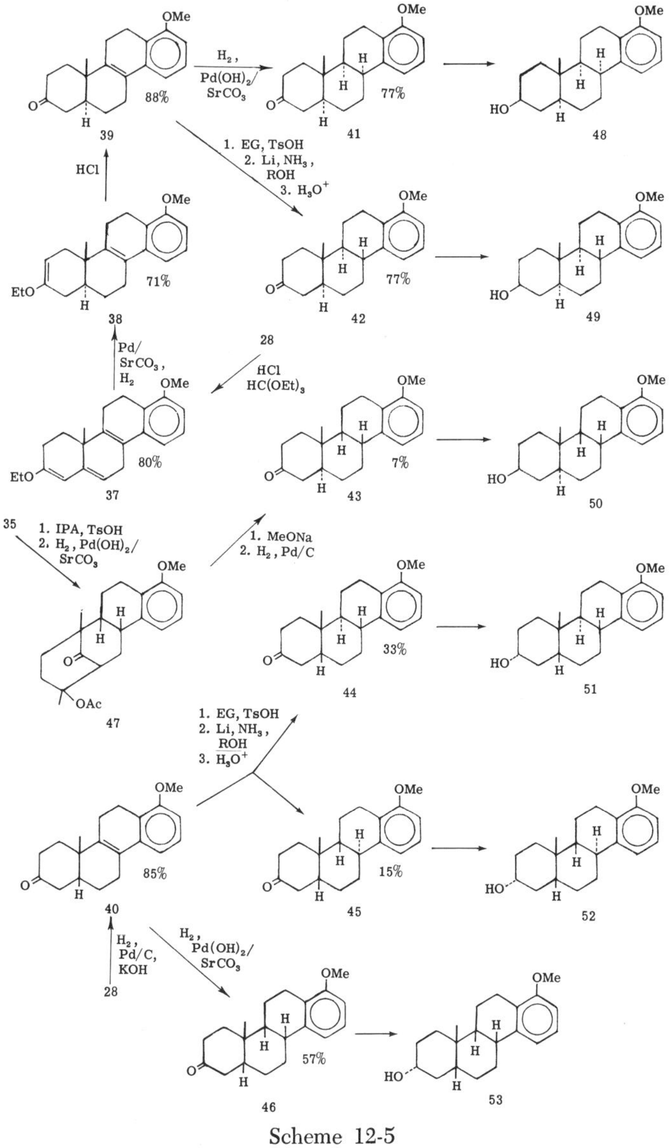

Scheme 12-5

71% yield of **38** (which probably contained some of the isomeric Δ³-olefin). Acid hydrolysis provided **39** (50% yield overall from **28**). Entry into the AB-*cis* series (**40**) was provided in 85% yield simply by palladium-on-charcoal catalyzed hydrogenation of **28** in benzene–ethanol containing a trace of potassium hydroxide.

In the *trans* series it was found that the *trans-anti-cis* ketone **41** could be produced (77%) by hydrogenation of **39** over palladium hydroxide–strontium carbonate. Only a small amount (3%) of the isomeric **42** was thereby produced. Alternatively, **41** could be produced directly from **37** by allowing the reduction (**37→38**) to include the styrene bond. If the ketone **39** was first converted to the ethylene ketal and then reduced according to the Wilds-Nelson procedure (lithium–ammonia with an ether cosolvent, and addition of the alcohol last) [46], a 77% yield of the *trans-anti-trans* ketone **42** resulted. This material could also be prepared via a second route, i.e., the lithium–ammonia–alcohol reduction of **38** followed by acid hydrolysis.

The third member of this series, **43**, was less readily prepared [43]. The intermediate ketol **35** (either isomer could be used) was acetylated with isopropenyl acetate-*p*-toluenesulfonic acid and hydrogenated over palladium hydroxide–strontium carbonate. Cyclization was effected with sodium methoxide to provide the Δ⁴-derivative of **43**. Hydrogenation with 5% palladium-on-charcoal containing a trace of potassium hydroxide allowed isolation of small amounts of ketone **43** in addition to much larger amounts of ketone **46**. A better procedure for the preparation of **43** consisted of first converting the ketoacetate **47** to the Δ⁴-3-ketone, followed by lithium–ammonia–alcohol reduction to provide the 3β-alcohol **50** which could be oxidized to **43** in high yield.

The first entry into the AB-*cis* series, **46**, was isolable from the stereoselective catalytic hydrogenation of **40**. This *cis-syn-cis* material was also obtained above in the preparation of **43**. Ketones **44** and **45** were prepared with some difficulty by first ketalizing **40** followed by the Wilds-Nelson reduction. The resulting saturated ketals were separated and hydrolyzed to give **44** and **45**, 33 and 15% overall yields, respectively. Alternate approaches to these materials were also described [43,45].

In addition to the six tetrahydro derivatives, **41–46**, described above, the four possible Δ⁴-3-ketodihydro derivatives were also identified. Furthermore, with the exception of **45**, ketones **41–46** were reduced (lithium aluminum hydride) to give, stereoselectively, the corresponding alcohols, **48–53**. Ultimately it was possible to effect the conversion of **28** into the acetate of the "natural" *trans-anti-trans* alcohol **49** in 43% yield by metal-in-ammonia reduction (either sodium or lithium could be em-

12.2 Hydrochrysene Approach

ployed) followed by acetylation of the crude product with isopropenyl acetate. This remarkable achievement (four new asymmetric centers are stereoselectively introduced) climaxed this phase of the synthetic scheme.

Following this effort, methods for the reduction of the D ring were investigated [47–49]. Catalytic hydrogenation was first studied, but it soon became apparent that this method was not suitable. The hydroxyphenol (54, R = R' = H; Scheme 12-6) was prepared by demethylation with methylmagnesium iodide, and then subjected to hydrogenation over platinum oxide in acetic acid. Approximately 30% of the resulting material was that arising from hydrogenolysis of the 17a-hydroxy group; additionally there were obtained two isomeric perhydro compounds, dl-3β,17aα-dihydroxy-14-iso-18-nor-D-homoandrostane and dl-3β,17aβ-dihydroxy-13-iso-18-nor-D-homoandrostane, in 21% and 9% yield, respectively. Further studies with Raney nickel gave similarly disappointing results. It was determined that the unwanted hydrogenolysis could be minimized by employing ruthenium oxide as the catalyst; and, in fact, when the *trans-anti-cis* analog of 54 (R = Ac, R' = H) was thus reduced, a crystalline isomer of the *trans-anti-cis-syn-cis* configuration was isolated in over 50% yield. Rather than pursuing this course, however, further investigations were directed toward metal-in-ammonia reductions of the aromatic nucleus.

Interestingly, the Wilds–Nelson reduction procedure readily reduced 1-methoxytetralin [46], but failed totally when applied to 54 (R = H, R' = Me) [48]. Further work showed, however, that the reaction would proceed if increased concentrations of alcohol were employed. Thus, reduction of 54 (R = H, R' = Me) by metallic lithium in ethanol–ammonia (2:3) provided a mixture of the isomers 55 and 56 with the former predominating. These steroid precursors could be further reduced in good yield using 10% palladium-on-charcoal. With 55 it was necessary

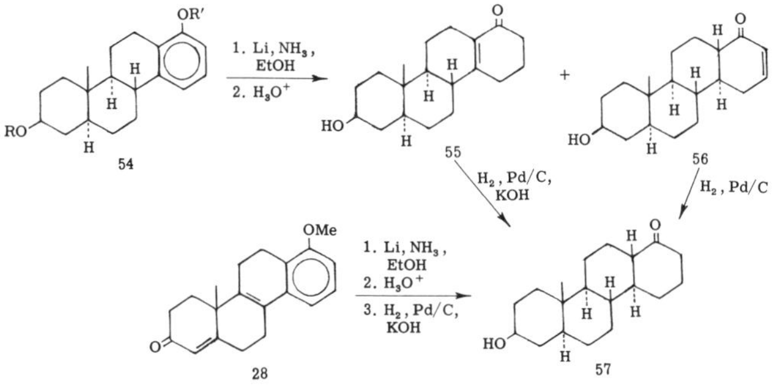

Scheme 12-6

to carry out the hydrogenation in the presence of small amounts of potassium hydroxide in order to obtain the desired CD-*trans* ring fusion.

These reduction studies culminated in the establishment of conditions whereby the highly unsaturated **28** could be stereoselectively reduced (six new asymmetric centers) to **57** in a yield of 29% without purification of any intermediates [48]. The ketone **28** was reduced by lithium in a dioxane–ethanol–ammonia solution under carefully controlled conditions. This was followed by hydrolysis with dilute hydrochloric acid and, finally, hydrogenation over 10% palladium-on-charcoal in an ethanolic solution containing trace amounts of potassium hydroxide. Thus, by the methods outlined in Schemes 12-4 and 12-6, 5-methoxy-2-tetralone was converted to *dl*-18-nor-D-homoepiandrosterone (**57**) in an overall yield of 10%—no mean achievement!

The final phase of Johnson's total synthesis (and probably the least satisfying) was the introduction of the C-13 angular substituent and ring contraction [48–52], illustrated in Scheme 12-7. Condensation of **57** with furfuraldehyde proceeded nearly quantitatively, as did formation of the tetrahydropyranyl ether, to produce **58** (R=tetrahydropyranyl, R'=H, Ar=furanyl). Angular methylation was effected with methyl iodide and potassium *t*-butoxide in *t*-butanol. After acid hydrolysis to remove the tetrahydropyranyl protecting group, there were isolated the epimeric keto alcohols **58** (R=H, R'=Me, Ar=furanyl; 24%) and **61** (R=H, R'=Me, Ar=furanyl; 52%). (The problem thus illustrated— preferential formation of the undesirable CD-*cis* ring fusion—has occupied the attention of many workers [53–55] and remains only partially solved.) Johnson later determined that the $\Delta^{9(11)}$-analog of **58** (R=tetra-

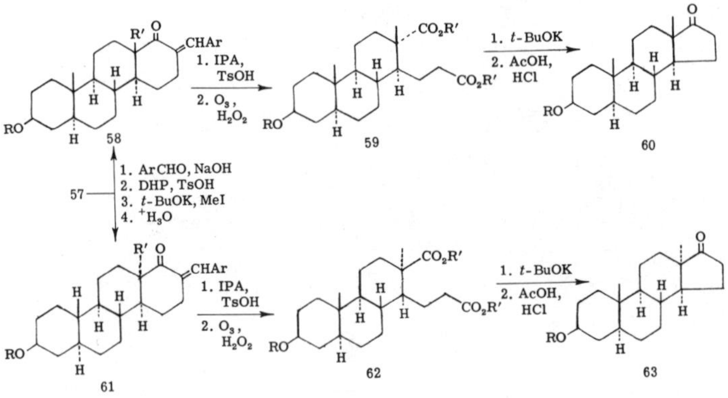

Scheme 12-7

12.2 Hydrochrysene Approach

hydropyranyl; R′=H, Ar=furanyl) was methylated under similar conditions to give the desired CD-*trans* product in 69% yield [56]. Presumably, by eliminating the axial hydrogen at C-11 the steric hindrance to *trans* approach is lessened.

Ring contraction of **58** and **61** was initiated by acetylation of the 3β-hydroxy group followed by ozonization and esterification of the resulting diacids to afford **59** and **62** (R=Ac, R′=Me) in good (ca. 75%) yields. Surprisingly, attempted Dieckmann cyclization with sodium methoxide in methanol was unsuccessful, as was a similar effort using sodium hydride [57]. However, potassium *t*-butoxide in benzene effected the desired conversion smoothly; the resulting β-keto esters were readily hydrolyzed and decarboxylated to provide good yields of *dl*-epiandrosterone (**60**) and *dl*-13-isoepiandrosterone (**63**). Based on intermediate **57**, this represented yields of roughly 28% of the 13-iso material and 13% of the "natural" steroid; based on 5-methoxy-2-tetralone, the figures were 3% and 1%, respectively.

This, then, completed the basic plan of total synthesis. Modifications were later introduced and some yields improved and a multitude of steroid derivatives were subsequently realized, but the essence is illustrated in Schemes 12-3–12-7.

With a procedure at hand for elaboration of the natural (racemic) steroid skeleton, a significant effort was directed toward establishing methods for introducing oxygen substituents at C-11 [51,58,59]. As outlined in Schemes 12-8 and 12-9, the goal was reached, but the reactions often yielded several components and/or provided the intended product in lowered amounts. Lead tetraacetate selectively attacked the C-12 carbon [60] to give epimeric mixtures of 12-acetoxy derivatives of **67**, **69**, and **72** in very good yields. However, the olefin-forming elimination reaction did not proceed so readily and required heating for several hours in acetic acid. Olefin **68** was oxidized with a variety of peracids (performic, peracetic, perbenzoic, monoperphthalic) to give, presumably (it was never isolated), the corresponding 11β,12β-epoxide. Pyrolysis of this crude material gave a 49% yield of the 11-keto compound **71**; alternatively, metal–ammonia–alcohol reduction gave a separable mixture of the 11α- (11%) and 11β- (31%) hydroxy compounds **66** and **65**. The triol **64** was obtained by treating the crude epoxide with acetic acid followed by basic hydrolysis.

When the *trans-syn-cis* olefin **70**, obtained from **69**, was epoxidized with perbenzoic acid and treated with methanolic hydrochloric acid, the resulting hydroxyketone **71** (R=H), possessed the *trans-anti-trans* con-

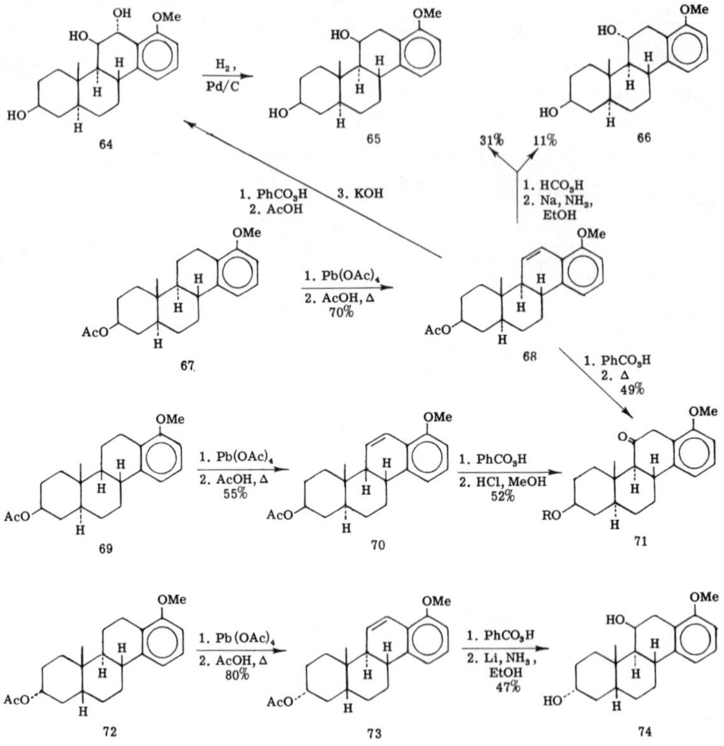

250   12 CD → BCD → ABCD

Scheme 12-8

figuration. Thus, ketonization at C-11 provided sufficient labilization of the C-9 hydrogen to permit isomerization at that point. This isomerization was facile enough that attempts to isolate the *trans-syn-cis* ketone failed.

Brief studies concerning functionalization in the AB-*cis* series were also carried out. By the depicted sequence in Scheme 12-8, the diol **74** was prepared in moderate yield; but similar attempts in the *cis-syn-cis* series were much less encouraging. Apparently in this case, elimination and extraneous acetoxylation were complicating factors.

By combining several of the previously outlined procedures, Johnson was able to prepare *dl*-3β,11β-dihydroxyandrostan-17-one (**80**, Scheme 12-9) and the 13-isosteroid **79**. Formation of the 11-olefin, peroxidation, and reduction were combined, without purification of any of the intermediates, to give a dihydroxy ketone (**75**, R = R′ = H, CHAr replaced by H$_2$) in 15% overall yield. The usual angular methylation sequence was

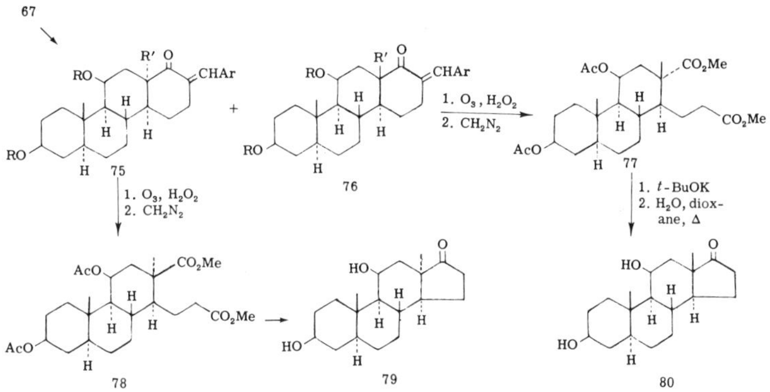

12.2 Hydrochrysene Approach

Scheme 12-9

performed after first preparing the furfurylidene derivative and protecting the hydroxy groups as the tetrahydropyranyl ethers. Methylation provided both the CD-*cis* and CD-*trans* epimers in addition to some O-methylated product which presumably arose because the bulky C-11 substituent sterically hindered C-13 methylation. Acetylation gave a 40% yield of a mixture (3:1) of **75** and **76** (R = Ac; R' = Me, Ar = furanyl) which could be separated by fractional crystallization. These products were then individually ozonized and the resulting diacids esterified with diazomethane to give **77** and **78**. Dieckmann cyclization was effected with potassium *t*-butoxide in alcohol-free benzene. However, the usual acid hydrolysis and decarboxylation was replaced by pyrolysis in aqueous dioxane to avoid the possibility of elimination of the 11β-hydroxy group. The natural and isosteroids (**80** and **79**, respectively) were thereby obtained in roughly 0.6% and 1.6% yields from **67**.

With the foregoing work, Johnson's hydrochrysene approach to total synthesis was developed to the stage where the possible natural (racemic) derivatives that could be prepared were almost unlimited. Indeed, *dl*-testosterone was very early synthesized from **28** [52,61]. Later work modified the procedure and produced *l*-testosterone and *dl*-isotestosterone [62] in addition to *dl*-18-nortestosterone [63,64]. Other natural products to be prepared included *dl*-progesterone [65,66], *dl*-cholesterol [66,67], conessine [53,66], and veratramine [68–70]. These syntheses represent an enormous effort but, in the interest of space, cannot be elaborated here. Johnson's synthesis of aldosterone, however, is detailed in Scheme 12-10 [71,72].

The furfurylidene ketone **81** was readily prepared as outlined in the foregoing discussion; however, the investigator's ingenuity became appar-

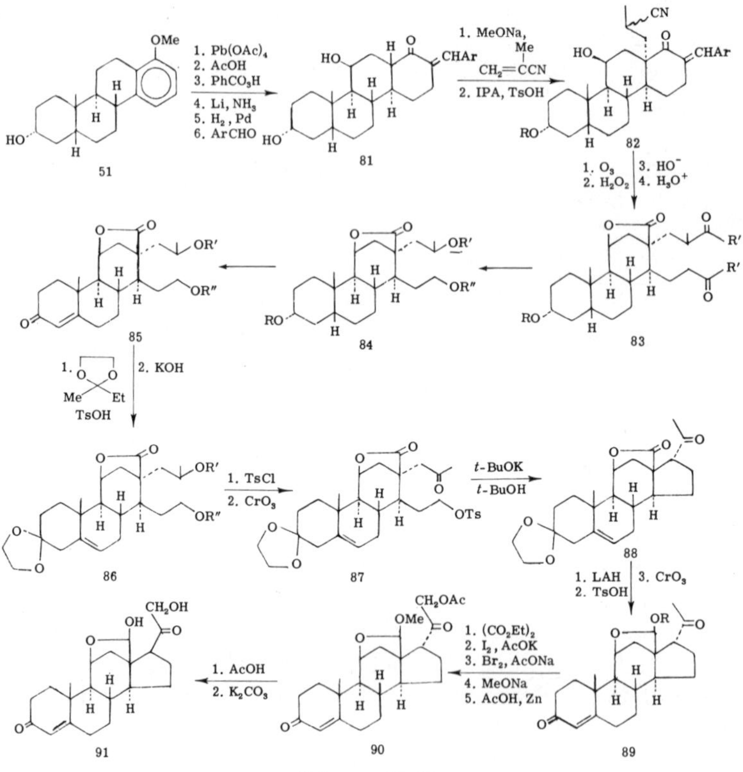

Scheme 12-10

ent in the succeeding transformation. Fully aware that alkylation (i.e., KOtBu-MeI) of systems of this type gave preferentially the undesired α-attack, Johnson elected to take advantage of this propensity and introduce an alkyl group that could ultimately become part of the D ring, while the existing carbonyl carbon (C-17a) would be elaborated into the C-18 of aldosterone. Previous model studies had illustrated the feasibility of this approach [73], and so, in the present case, 81 was condensed with methacrylonitrile using as a base sodium methoxide in methanol. Under these conditions, alkylation occurred essentially exclusively from the α-side to provide an excellent yield of 82 (R=H) as an epimeric mixture (at the potential C-20 carbon). These epimers could be separated at this and latter stages, and in some instances they reacted somewhat differently. However, because this asymmetric center was ultimately ablated, it did not detract from the overall synthesis. In the present discussion only the mixture will be considered. Acetylation,

12.2 Hydrochrysene Approach

accomplished with isopropenyl acetate, was followed by ozonolysis and hydrogen peroxide oxidation. Basic hydrolysis and acidification gave the lactone **83** (R = H; R' = OH) in a 64% overall yield from **81**. It was next necessary to shorten the side chains to produce a material suitable for cyclization. Hunsdiecker decarboxylation gave poor (10–15%) yields; a second approach involving conversion to the dialdehyde, then to the bisenamine which was to be ozonized, was also unsuccessful. The final answer was the Baeyer-Villiger degradation of the diketone **83** (R = Ac; R' = Me). To prepare this material, the 3-acetate was converted to the diacid dichloride (**83**; R = Ac, R' = Cl). This was transformed into the diketone by treatment with ketene dimethylacetal followed by hydrolysis [74]; an alternative procedure involved treatment first with sodium dibenzyl malonate followed by hydrogenolysis and decarboxylation. By this latter method [75], the epimeric mixture of **83** (R = Ac; R' = CH$_3$) was isolated in a 54% yield. Trifluoroperacetic acid and disodium hydrogen phosphate were combined for the oxidation to produce **84** (R = R' = R'' = Ac) in yields up to 83%.

To introduce the α,β-unsaturated ketone system into ring A, the triacetate was selectively saponified to the dihydroxyacetate **84** (R = R'' = H; R' = Ac). Oxidation was effected with N-bromoacetamide to give, after reacetylation, the diacetate of dihydro-**85** (ca. 60%). The Δ^4-unsaturation was introduced by bromination in acetic acid followed by elimination with lithium bromide in dimethylformamide; **85** was thus produced in 52% yield along with 28% of the Δ^1-ketone. This latter material could be reconverted to dihydro-**85** and thence recycled. Ketalization and saponification of **85** gave a nearly quantitative yield of **86** (R' = R'' = H). It was anticipated that ring-D cyclization could be effected by the procedure outlined earlier by Sarett [76]. The keto sulfonate **87** was therefore needed and was prepared in 86% yield by selective tosylation and subsequent oxidation with chromium trioxide in pyridine. At this point the asymmetry at the potential C-20 carbon was eliminated. Cyclization was achieved with potassium t-butoxide in t-butanol, but only in moderate yields (30–54%); and for the overall conversion of **86** (R' = R'' = H) to **88** the yield was 43%. Unfortunately, this cyclization proceeded to give the 17α-epimer. Partial equilibration to the 17β-epimer could be effected under various conditions including methoxide catalysis. At this point, the work converged with that of Wettstein and his collaborators who had obtained both epimers and converted the 17β-analog of **88** to aldosterone [77–79]. Johnson, however, developed an alternative procedure utilizing the 17α-epimer. Two equivalents of lithium aluminum hydride stereoselectively reduced both carbonyl groups of **88** in 82%

yield. Hydrolysis of the ketal moiety, etherification of the lactol, and oxidation of the C-20 alcohol produced **89** (R=CH$_3$). The 21-acetoxy group was introduced by a previously reported method [80] involving the formation of the glyoxolate adduct, treatment with bromine, sodium methoxide, and zinc–acetic acid. In this way, **90** was formed in 20% yield. Mild acid hydrolysis regenerated the lactol and, finally, *dl*-aldosterone (**91**) was isolated by treatment with dilute potassium carbonate, a method known [79] to effect C-17 epimerization and, in this case, simultaneous hydrolysis of the acetate.

12.3 Modifications of the Hydrochrysene Synthesis

Attesting to the soundness of the hydrochrysene approach were the later reports from two independent research groups detailing the preparation and reactions of **93**, an isomer of Johnson's key intermediate. Thus, in a manner analogous to that shown in Scheme 12-4, 6-methoxy-2-tetralone (**5**) was condensed first with the methodide of 1-diethylaminopentan-3-one and then with methyl vinyl ketone (or 1-diethylaminobutan-3-one methiodide) to provide **93** in approximately 50% yield (Scheme 12-11) [81–83]. Under mild conditions the ketol **92** could be isolated but need not be for preparative purposes. However, this material provided a ready route (ca. 70%) to the *cis-syn-cis* product **95** by acetylation, hydrogenation of the styrene double bond, methoxide catalyzed rearrangement, and hydrogenation of the Δ^4-unsaturation. The metal–ammonia reduction of **93** proved highly sensitive to reaction con-

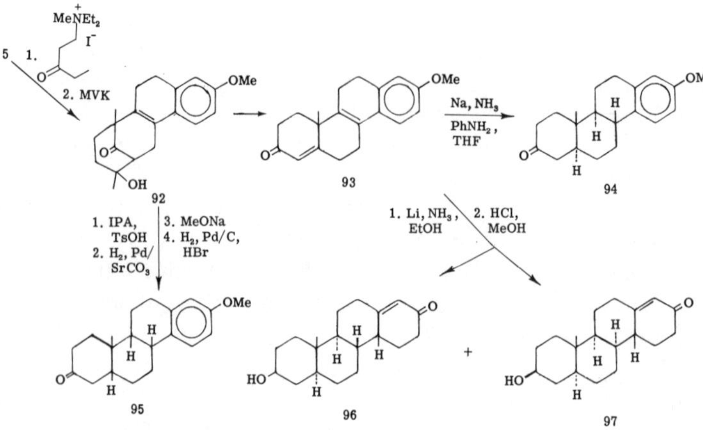

Scheme 12-11

12.3 Modifications of the Hydrochrysene Synthesis

ditions. Kutney *et al.* [84] obtained unsatisfactory results when applying the Birch reduction conditions of Johnson, or variations thereof, and found substantial improvements only when they turned to other amine solvents. Conversely, the Japanese group [85] was able to perform the reduction in ammonia, although the preferred procedure was somewhat complex. The natural *trans-anti-trans* material **94** was ultimately prepared in nearly 70% yield by reduction of **93** using sodium in ammonia–aniline–tetrahydrofuran. The lithium–ammonia reduction of **93** gave rise to an inseparable mixture which was hydrolyzed in refluxing methanolic hydrochloric acid. From this hydrolysate was isolated **96** (55%) and **97** (19%) as well as trace amounts of other components.

To effect angular methylation, Nagata introduced a hydrocyanation procedure which he has since developed and used extensively [86]. Initially, it was found that treatment of **96** with potassium cyanide in methanol gave only minimal amounts of the desired 13-cyano compound, presumbaly because of subsequent *in situ* basic hydrolysis [55]. By adding ammonium chloride to offset this developing basicity, the intended products **98** and **99** (Scheme 12-12) could be isolated, after ketalization and acetylation, in 40 and 24% yield, respectively. Further efforts to improve the stereoselectivity of this reaction [87–91] led to the discovery that the combination of triethylaluminum and hydrogen cyanide

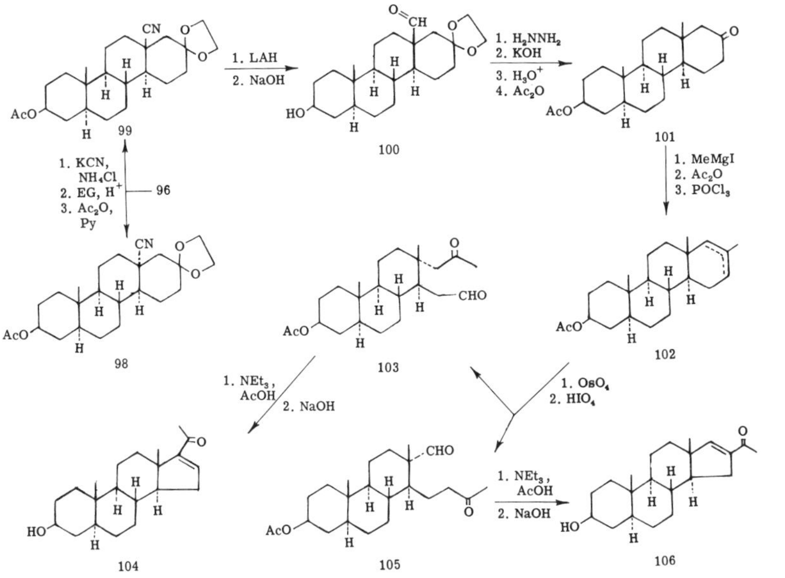

Scheme 12-12

(or, preferably, diethylaluminum cyanide) in aprotic solvents effected addition of cyanide ion to **96** in yields as high as 80% with greater than 90% stereoselectivity for the 13β-cyano epimer.

The next objective, conversion of the cyano moiety into an angular methyl group, was accomplished by lithium aluminum hydride reduction of **99** to the corresponding imine followed by basic hydrolysis to the aldehyde **100** [82,87,92]. Huang-Minlon reduction, deketalization, and acetylation gave the keto acetate **101** in a 73% yield from **99**. Ring contraction was initiated by treating **101** with methylmagnesium iodide, and gave the 17α-methyl-17β-hydroxy analog which was dehydrated with phosphorus oxychloride in pyridine to give **102**, containing approximately equal amounts of the two olefins. Ring opening could be performed with either ozone or (preferably) osmium tetroxide–periodic acid. When osmium tetroxide was employed there resulted four intermediate *cis* diols each of which could be isolated, with the two resulting from α-attack (16α,17α- and 17α,17aα-diol) predominating. Periodic acid cleavage provided **103** and **105**, the former being obtained from the 16,17-diols and the latter from the two 17,17a-diols. These same products were obtained directly from the olefinic mixture via ozonization, but in much poorer yields. Cyclization was achieved by refluxing **103** and **105** in xylene containing triethylammonium acetate to give, after saponification, *dl*-3β-hydroxy-5α-pregn-16-en-20-one (**104**) and the isomeric **106**, respectively.

An analogous series of reactions was carried out in the AB-*cis* series (**96**: AB *cis*, 3α-OH) [89]. In this case, a 9(11)-unsaturation was introduced before the hydrocyanation sequence. The ultimate product (**104**:AB *cis*) was attained in less than 0.1% yield starting from **93**; this disappointing yield exemplifies the major drawback in this synthetic plan. However, the merit of the stereoselective introduction of the cyano group, first developed in this sequence, has been proved in several subsequent natural product syntheses, and the scope of the reaction has recently been detailed [92–94].

12.4 C-Nor-D-homosteroids

Primarily because of the interest in the veratrum alkaloids, several schemes have been presented delineating the total synthesis of C-nor-D-homosteroids which can serve as convenient precursors to these alkaloids. Interestingly, Johnson's key intermediate **28** and the isomeric **93** have both been converted to this steroid skeleton by very similar routes. Kutney *et al.* [84] were able to prepare **108** from the acetate **107** by

12.4 C-Nor-D-homosteroids

employing *t*-butyl chromate (Scheme 12-13). Previous efforts to functionalize C-12 with lead tetracetate or *N*-bromosuccinimide had met with failure, the only isolable material in both instances being the ring-C aromatized product. It was thought that the C-8 position was activated enough by the *p*-methoxy group to cause acetoxylation at this point, followed by elimination to the olefin. The sequence repeated at C-12 would then provide the naphthalene chromophore. Direct oxidation of **107** with chromium trioxide gave very small quantities of the ketone; the bulkier chromate oxidant ultimately used gave a 22% yield of **108**. This material was converted to the olefin **109** in an unspecified yield by sodium borohydride reduction followed by phosphorus pentoxide dehydration [95]. This olefin as well as isomers derived from **28**, have all been converted to the corresponding C-nor-D-homo derivatives by a sequence first outlined by Johnson [68,96]. Conversion to the dialdehyde **110** was achieved by periodic acid cleavage of the 11β,12β-diol (or, with other isomers, by direct ozonization). This material was cyclized to the aldol **111** by refluxing with sodium hydroxide in aqueous methanol. The

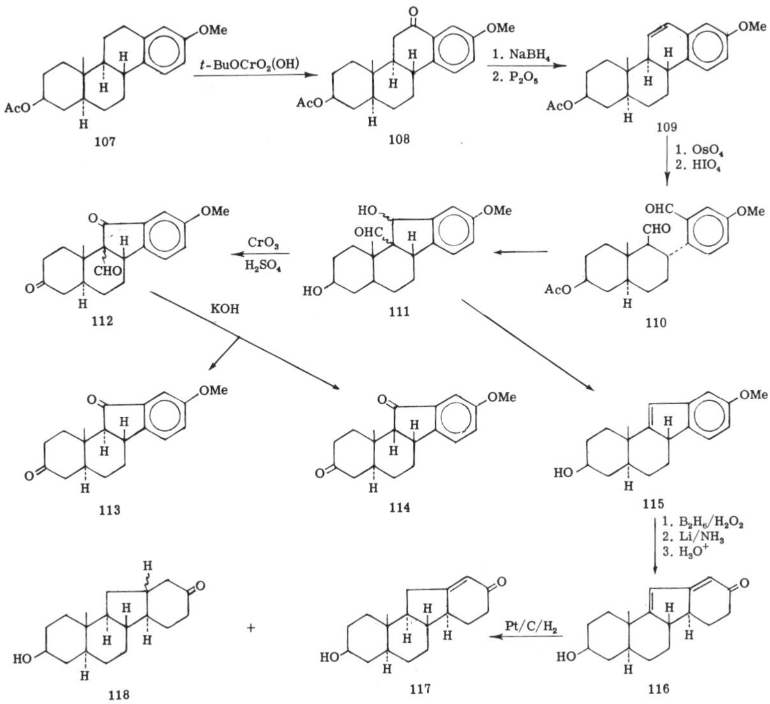

Scheme 12-13

configurations at C-12 and C-9 were not determined, although seemingly only one isomer was produced. This material could be oxidized under mild conditions to provide the diketo aldehyde 112, which, because of its instability, was deformylated to a mixture of 113 and 114. (In this case the ratio of 113/114 was not specified; however, starting with the 17a-methoxy derivative instead of 107, the analogous isomers were found in a 9:1 ratio which, under equilibrating conditions, gave a 9:6 ratio [96]. In the AB-*cis* series the equilibrium ratio was reversed to the extent that the isomer corresponding to 114 predominated 7:3 [97].)

In an alternate reaction sequence [68,98] 111 was diacetylated and converted to the olefin 115, which gave, on catalytic hydrogenation of the $\Delta^{9(11)}$-bond, only the 9β-epimer. Hydroboration of 115 (68%) followed by Birch reduction and dehydration (48%) gave 116, which underwent hydrogenation to give a mixture of 9α-compounds 117 and 118. The latter could be converted to the former by a standard bromination–dehydrobromination sequence. Further transformations of 117 were aimed at alkaloid synthesis and are beyond the scope of this discussion.

Two other synthetic routes to C-nor-D-homosteroids are illustrated in Schemes 12-14 and 12-15. In the first [70], Hagemann's ester 119 was alkylated in a rather novel manner [99] by employing ethyl β-ethoxy-α-bromocrotonate as the alkylating agent. Hydrolysis of 120 then gave the β-keto ester, readily cyclizable to 121. Aromatization concurrent with decarboxylation was effected by treating with 10% palladium-on-charcoal to produce the indanone ester 122, which was alkylated in the form of its pyrrolidine enamine with ethyl vinyl ketone. Subsequent acid hydrolysis and cyclization with 85% phosphoric acid produced the tricyclic keto ester 123 along with minor amounts of the β,γ-unsaturated product (Δ^8). A completely analogous product lacking the aromatic substituents had

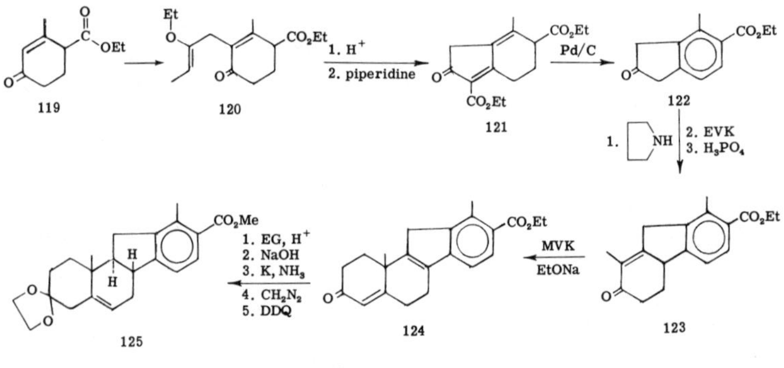

Scheme 12-14

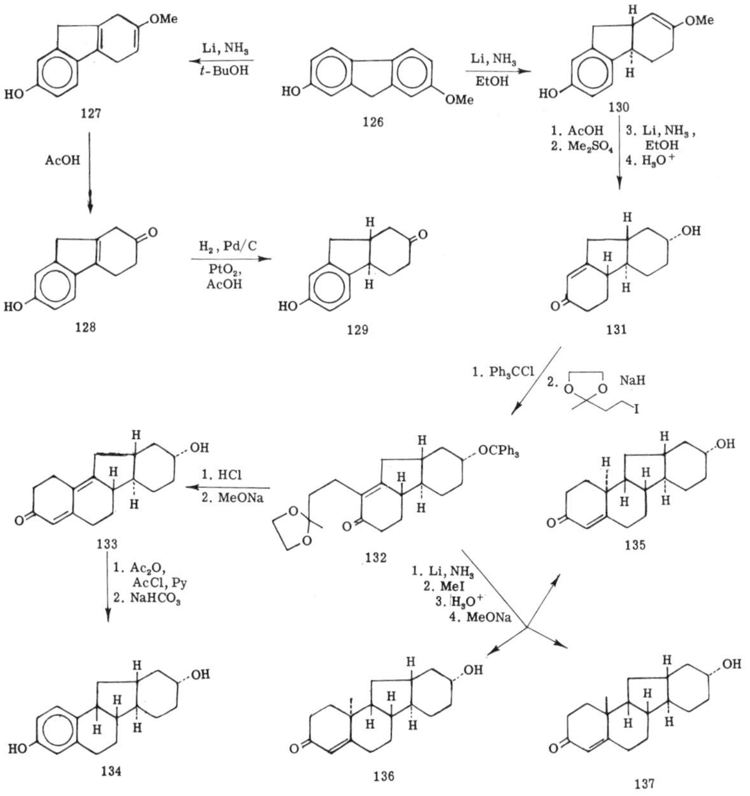

12.4 C-Nor-D-homosteroids

Scheme 12-15

previously been prepared in poor yield by an alternate reaction sequence starting with indanone [100]. Alkylation of **123** with methyl vinyl ketone provided **124** which, after conversion first to the ethylene ketal and then to the sodium salt of the acid, was reduced with potassium in ammonia. The aromatic ring was reduced to the dihydro form in this sequence; however, rearomatization with dichlorodicyanoquinone provided **125** in an overall 50% yield from the ketal acid. This racemic material was resolved in the form of its 1-α-[1-naphthyl]ethylamine salt and provided material identical to that from the degradation of natural veratramine.

Scheme 12-15 illustrates an intriguing approach to C-nor-D-homosteroids which would seem to hold much potential. Although not truly a CD→BCD sequence, it is included in this chapter nonetheless. The synthetic plan was based on the hope that certain fluorene derivatives could be reduced in a stereoselective manner to provide appropriate BCD intermediates which could then be elaborated to tetracyclic materials by standard procedures. Initial efforts were directed toward exploring the Birch reduction of 2-hydroxy-7-methoxyfluorene **126** [101]. Depending on the conditions employed, various reduction products were obtained. With 6 g-atoms of lithium in ammonia-t-butanol, a 73% yield of **127** was isolated, whereas more drastic conditions (90 g-atoms of lithium in ammonia–ethanol) provided **130** (50%). Amounts of lithium between 6 and 90 g-atoms provided mixtures of **127** and **130**, and some evidence indicated that **127** was not a major intermediate in the formation of **130**. Hydrolysis of **127** with 95% acetic acid yielded ketone **128** which, on catalytic hydrogenation, gave only the *cis*-fused isomer **129**. Hydrolysis of **130** gave the expected *trans*-fused ketone, which was methylated and converted to **131** (70%) by further reduction (lithium–ammonia–tetrahydrofuran–ethanol) and hydrolysis with dilute hydrochloric acid. This material, after etherification of the hydroxyl, was alkylated with 1-iodo-3,3-ethylenedioxybutane to give the ketal **132** (50%). Acid hydrolysis afforded the diketo alcohol, and this was cyclized with sodium methoxide giving **133** in 78% yield (nearly 10% overall from **126**) [102,103]. Aromatization was then performed by one of two methods: heating with acetic anhydride, acetyl chloride, and pyridine, or isomerization over 10% palladium-on-charcoal in refluxing ethanol. From either procedure the product was **134** (in the former instance hydrolysis was, of course, necessary).

In another sequence, **132** was reductively methylated using the procedure reported by Stork and McMurry [104] to give a mixture from which three tricyclic compounds were isolated. These materials, after hydrolysis, were cyclized to **135**, **136**, and **137** with the latter predominating.

12.5 Woodward Synthesis

One of the earliest total synthesis of a nonaromatic steroid was developed by Woodward and his associates in the early 1950's, and subsequently modified by a Monsanto research group. Never refined to the degree of the Johnson synthesis [the scheme initially involved some twenty steps and yielded *dl*-3-oxo-21-nor-4,9(11),16-pregnatrien-20-al (**155**) in approximately 1% yield from starting 4-methoxytoluquinone], the route nonetheless embodied much unique chemistry and paved the way for further approaches.

12.5 Woodward Synthesis

The necessary starting CD intermediate **143** was prepared as shown in Scheme 12-16 [105,106] via a Diels-Alder reaction of butadiene and 4-methoxytoluquinone. The product **139** (86%) was exclusively the *cis*-fused material, but could be converted quantitatively to *trans*-**139** by simple crystallization from aqueous base. Lithium aluminum hydride reduction to the methoxyglycol followed by treatment with acid gave ketol **140**. Removal of the hydroxy group was effected by heating the acetate of **140** in zinc–acetic anhydride solution. The *trans* bicyclic ketone **143** was thereby prepared in 45% yield from **139**.

An alternate route from **139** to **143** was later described which, by including a resolution step, afforded the optically pure enantiomorphs [107]. Zinc–acid reduction of **139** could be performed to give **141** (R=H) selectively. Tosylation, reduction, and acid treatment then gave racemic **143**; however, employing *d*-camphor-10-sulfonyl chloride led to resolvable sulfonate diastereomers, which in turn provided (+)−**143** and (−)−**143** in yields of 55–60% from **139**.

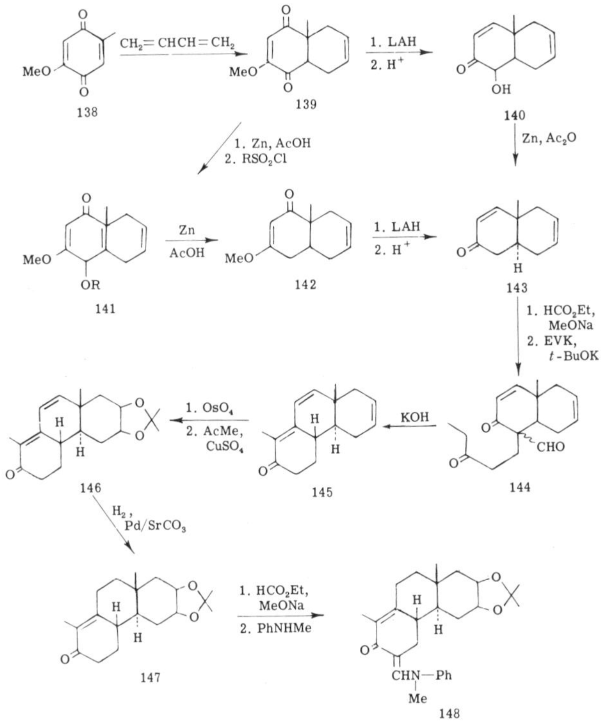

Scheme 12-16

Elaboration of ring B followed previous procedures [108] and involved condensation of **143** with ethyl formate. The resulting hydroxymethylene ketone was reacted with ethyl vinyl ketone (the analogous Mannich base methiodide proved unsatisfactory) to give **144**. Elimination of the formyl group and ring closure was evoked by potassium hydroxide in aqueous dioxane. A single stereoisomer, **145**, was thereby formed in 55–60% yield from **143**.

At this stage in the original scheme, Woodward felt it expedient to protect the isolated double bond in ring D in the form of the acetonide **146**. However, the Monsanto group did not feel this was necessary [109]. The reactions outlined in Scheme 12-16 and discussed below are those of Woodward; but essentially the same sequence and reaction conditions were used, and the same results obtained, when the Δ^{16}-bond was left intact.

Using osmium tetroxide to oxidize **145**, a mixture of *cis-α* and *cis-β* glycols was obtained with the former predominating by a 4:1 ratio. Later work showed the β-analog could be made the major product (28:1) by employing silver acetate–iodine in moist acetic acid [110]. However, for preparative purposes, it was not necessary to separate these glycols nor the resulting mixture of acetonides **146** (for the present, only the β-isomer will be shown). Selective hydrogenation over 2% palladium-on-strontium carbonate in dry benzene proceeded well (85%) to give **147**. Very inferior results were obtained using other catalyst systems or other solvents. The α,β-unsaturated ketone was then condensed with ethyl formate, and the resulting product was further reacted with methylaniline in methanol to provide **148** in which one of the carbonyl-activated positions had been blocked. As shown in Scheme 12-17, condensation with acrylonitrile in *t*-butanol–benzene–water with added Triton B, followed by basic hydrolysis gave material in which cyanoethylation had occurred exclusively at C-10. Little stereoselectivity was evident, however, as shown by the nearly 2:1 preponderance of the C-10 α-methyl configuration over the natural C-10 β-methyl. This lack of stereochemical control in the alkylation step was one of the major shortcomings of the overall synthetic plan (the Monsanto group found 22% of the natural product and 45% of the C-10 α-methyl compound in their series [109]). In fact, further efforts were directed exclusively to this problem (*vide infra*).

Separation of **149** and **152** was readily effected and, by refluxing the keto acids in acetic anhydride containing a catalytic amount of sodium acetate, the enol lactones **150** and **153** were isolated. Interestingly, the natural isomer **153** reacted readily with methylmagnesium bromide to

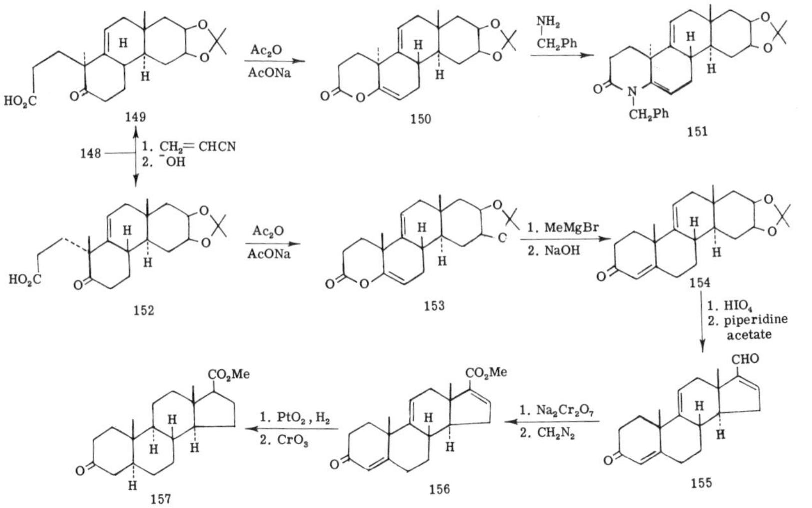

12.5 Woodward Synthesis

Scheme 12-17

produce, after base treatment, the sought for D-homosteroid **154** in good (58%) yield. Conversely, **150** underwent the same reaction very sluggishly giving a 10% yield of the C-10 α-methyl epimer. The novel 4-aza-D-homosteroid **151** was prepared in poor yield in an incidental finding.

With the desired tetracyclic compound **154** possessing the correct stereochemistry at the four quarternary carbons available, efforts were next directed toward contraction of the D ring. To this end it was found that treatment of the foregoing with periodic acid in aqueous dioxane, followed by cyclization with catalytic amounts of piperidine acetate in refluxing benzene produced **155** in 65% yield. Further procedures converted **155** into a racemic mixture of the known natural methyl 3-keto-$\Delta^{4,9(11),16}$-etiocholatrienate (**156**). From this racemic mixture was separated, after borohydride reduction to the 3-alcohols, the dextrorotatory isomer (by repeated preferential precipitation with digitonin [111]). Subsequent oxidation of this 3β-alcohol gave the optically active (+)−**156**. (Barkley et al. prepared this same active material by starting with their optically pure **145**, thus bypassing the later resolution [109]). Finally, this material was hydrogenated over platinum oxide and then oxidized with chromium trioxide in acetic acid. From the resulting mixture there was isolated (2–3%) the known methyl d-3-ketoetiocholanate (**157**). Since this naturally derived material had been interrelated to numerous other steroids, this completed the formal total synthesis of many steroids including 3β-cholestanol [112] and cortisone [113–115].

Further efforts to improve the overall scheme were only partially successful. A substantial investigation was aimed at better methods for elaborating ring A. It was ascertained that, by first contracting the D ring of the tricyclic intermediate, reactivity at C-10 was enhanced. Furthermore, introduction of the C-10 methyl group after cyanoethylation led to much improved stereochemical results. Scheme 12-18 summarizes these developments [116].

When the hydroxymethylene ketone **158** (drawn as the aldehyde) was condensed with methyl 5-oxo-6-heptenoate (**159**) in the presence of benzyltrimethylammonium butoxide, and the resulting product treated with aqueous base, tricyclic acid **160** was isolated in greater than 50% yield. The usual methylation (methyl iodide–potassium *t*-butoxide) sequence was performed on 11,12-dihydro-**160** but proceeded very poorly. Ring contraction was then initiated by converting **160** to **161** by previously described reactions. The acetonide was cleaved to the glycol, and ring opening with lead tetracetate gave the expected dialdehyde which was cyclized to **162**. This ring contraction was later improved considerably by subjecting the Δ^{16}-olefin to ozone in the presence of trimethyl phosphite [117].

A second hydrogenation was followed by acetalization and resulted in crystalline **163**, which was submitted to the usual angular methylation conditions (it is worth noting that the reactivity of C-10 was such that

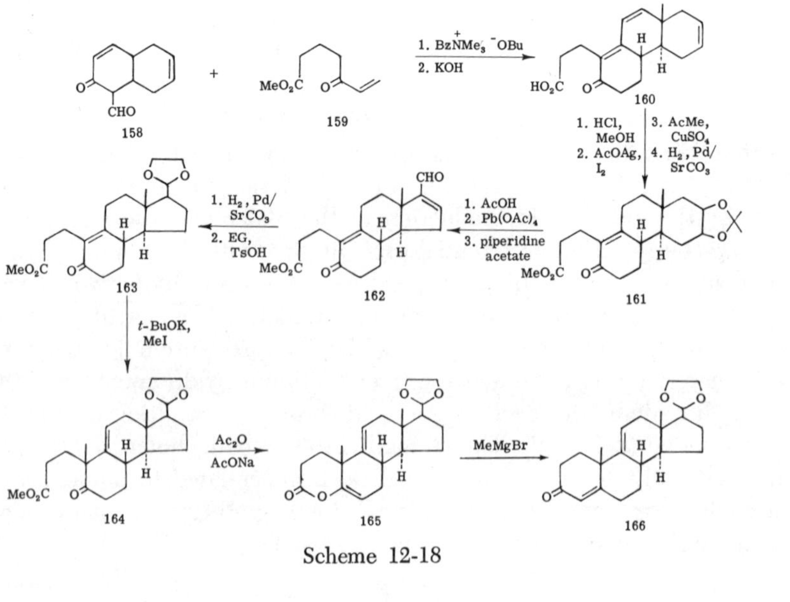

Scheme 12-18

12.5 Woodward Synthesis

the usual precaution of blocking C-6 could be abandoned). The precise amount of β-methyl product obtained was not determined, but was considered to be at least 46% based on the 38% conversion of **163** into **166** [116,118]. This latter figure was a considerable improvement over the previous method; however, the attractiveness of the overall scheme was lessened by the poor yields and the difficulty in handling some of the prior intermediates obtained when working with tricyclic materials substituted with the carboxyethyl function in place of the methyl group. Other attempts were made to improve the synthesis [119,120], but notable progress was not forthcoming.

A brief attempt at total synthesis, which led to a BCD intermediate quite similar to some prepared by the Monsanto group is recorded in Scheme 12-19 [121]. The interesting ketal **167**, prepared in 42% yield starting from 2-methyl-5-keto-1-hexene, was cyclized to **168** (48%) by successive treatment with potassium *t*-butoxide (to cyclize the D ring), strong acid (to effect extensive hydrolysis and ring C cyclization), and methanolic hydrogen chloride (to esterify the C-17 carboxyl group). The dienol acetate of **168** was prepared and, on exposure to sodium borohydride, gave the alcohol **169** (76%) which was felt to be a mixture of materials epimeric at both C-9 and C-17. On reduction of this mixture over 10% palladium-on-charcoal followed by basic hydrolysis and Jones oxidation, a 32% yield of the CD-*trans* hydrindane **170** was isolated in addition to a small amount (18%) of the *cis* acid (both were still mixtures of epimers at C-17). Bromination of **170**, conversion to the 2,4-dinitrophenylhydrazone and cleavage thereof by the Demaecker-Martin procedure [122] afford the α,β-unsaturated ketone **171** in excellent (88%) yield. This material, bearing a close resemblance to Woodward's CD intermediate **143**, was treated accordingly. Thus, condensation with ethyl formate, alkylation with ethyl vinyl ketone, and treatment with base led to the

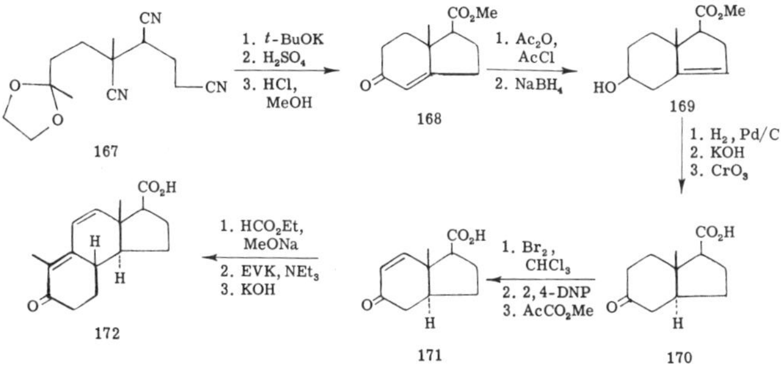

Scheme 12-19

tricyclic keto acid **172** in 25% yield. Esterification and selective hydrogenation gave the methyl ester of 11-dihydro-**172**. Because this material was very similar to the previously described BCD intermediate **145**, the synthesis was not carried further.

12.6 Asymmetric Induction

Of the recent advances in steroid total synthesis, perhaps none is more exciting nor holds more potential than the concept of asymmetric induction. This principle, which has been the subject of various recent reports [123–126], has been exploited to a remarkable degree by the Hoffmann-LaRoche group in the synthesis of several racemic and optically active steroid and 19-norsteroid derivatives. It may be briefly defined as the generation, with some degree of configurational specificity, of a new asymmetric center during the course of a chemical reaction. Such a transformation may be brought about either by employing a dissymmetric catalyst, or by using a reactant already possessing a dissymmetric center which, during the course of the reaction, "induces" asymmetry at a second center. Without doubt, the future will see a large effort devoted to schemes employing asymmetric induction.

The initial observation which kindled the synthetic flame is recorded in Scheme 12-20. Condensation of **173** with **12** by refluxing in pyridine—

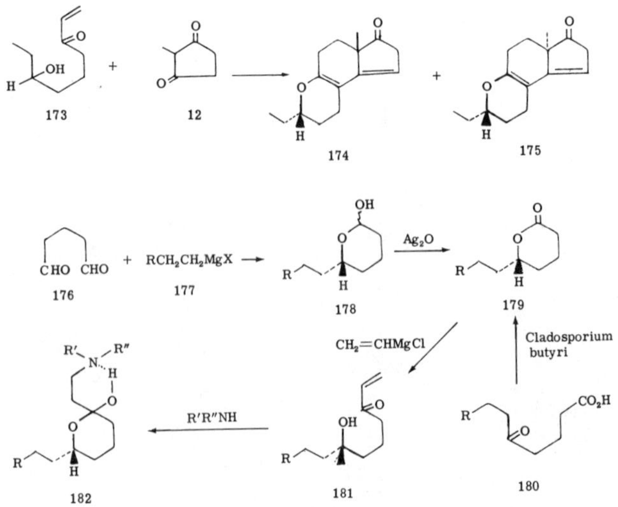

Scheme 12-20

12.6 Asymmetric Induction

toluene provided a good yield of a mixture of **174** and **175** (shown as single enantiomorphs, but were, in this instance, racemic mixtures); however (±)-**174** was by far the predominant isomer [127]. In later work, optically active **173** [7(S)-hydroxy-1-nonen-3-one] was prepared and, on reaction with **12**, led to a 65% yield of crystalline (−)-**174**, contaminated with minor amounts of (−)-**175** [128]. Although these initial reports further detailed the conversion of (+)- and (±)-**174** into natural and racemic 17β-hydroxy-des-A-androst-9-en-5-one, these and similar transformations will be discussed later.

The immediate problem reduced to the preparation of **173** (both optically active and racemic) and various analogs thereof. The racemic material was initialy prepared by a rather tortuous route starting with α-chlorobutyryl chloride [127,129], but this proved unsatisfactory and, ultimately, a novel scheme starting with glutaraldehyde was developed [130]. Various alkyl Grignard reagents (**177**; R=H, n-Pr, $CH_3CH(O$-t-$Bu)CH_2$ 2,2-o-phenylenedioxypropyl) were thus reacted with **176** to provide the lactols **178** (racemic) in yields of 60–75%. These lactols could then be oxidized to the desired lactones **179**, which were in turn reacted with vinyl magnesium chloride at −50° to provide (±)-**181** in very good yield. However, because of the instability of (±)-**181**, it was found expedient to convert it the Mannich base (±)-**182** (R′=R″=Et) (which, interestingly, existed predominantly in the spiroheterocyclic structure due to hydrogen bonding between the amino and hydroxy moieties), by the addition of diethylamine to the reaction mixture. This material was suitable for condensation with **12**. This unique, direct conversion (**179**→**181**→**182**) was accomplished in nearly 75% yield where R=H and, as will be seen, greatly added to the attractiveness of the overall synthesis. It should be pointed out that this reaction sequence was also carried out with **179** (R=cyanomethyl [131], 3,5-dimethyl-4-isoxazolyl [132–134]) although the starting lactones were prepared in quite different manners.

Access was thus provided to a variety of derivatives of **174**, but only as racemic mixtures. Optically active **182** (R=H) could be prepared from active **179** (R=H) which resulted (55%) from the microbiological reduction of δ-oxoheptanoic acid **180** (R=H) [128]. However, this was not a general reaction, and an alternate procedure was desired for obtaining the enantiomorphic forms of **182**. This problem was admirably solved by substituting (+)- or (−)-α-methylbenzylamine for diethyl amine in the conversion of **181** to **182**. Scheme 12-21 depicts a specific instance of what theoretically should be a general method for preparing optically active components. Thus, (±)-**185**, which may be considered

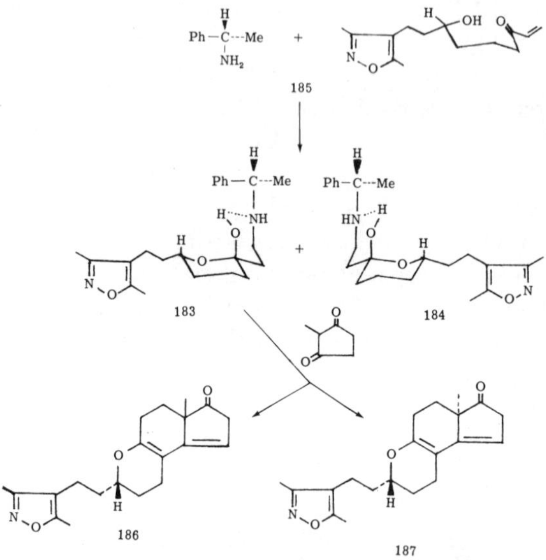

Scheme 12-21

to have been prepared by the route shown in Scheme 12-20, i.e., **176**+**177** (R=3,5-dimethyl-4-isoxazolyl)→**178**→**179**→**181** [in fact, (±)-**185** was prepared in an alternate fashion] combined with (−)-α-methylbenzylamine to give a good yield (ca. 60%) of the separable diastereomers **183** and **184** [134]. When **183** was condensed with 2-methylcyclopentane-1,3-dione, the result was **186** and **187** with the former predominating (72% optical induction). This ratio could be increased substantially by first quaternizing **183** with methyl iodide, reacting this salt with **12**, and then treating the product with *p*-toluenesulfonic acid*; the result was a 46% yield of **186** and **187** with optical induction occurring to the extent of 89%. For synthetic purposes this mixture need not be separated but could be converted to a BCD tricyclic mixture which was then purified.

An analogous resolution was performed starting with **182** (R=H [135], CH$_2$CN [131]), the procedure proving quite general. Further improvements have essentially doubled the yield of the desired **183** by performing a stereochemical inversion cycle on **184** [136] (in actuality, the work

---

* The mechanistic aspects of this reaction have been discussed in some detail [135], but cannot be presented here.

12.6 Asymmetric Induction

was carried out on the analogous compound where the isoxazole group was replaced with a cyanomethyl group [131]).

The plan as thus outlined represents a powerful new methodology for preparing racemic or optically active derivatives of **188** (Scheme 12-22), which can be converted, in unremarkable fashion, into BCD tricyclic steroid intermediates. Thought of in its simplest terms, it involves the preparation of various δ-substituted-δ-lactones, their conversion into Mannich bases (represented by **182**) employing diethylamine for obtaining racemic products or (−)-α-methylbenzylamine for obtaining optically active products, and the condensation of these bases with 2-methylcyclopentane-1,3-dione (the 2-ethyl compound has also been used and, it would seem, any 2-alkyl derivative should react comparably) to give, via a high degree of asymmetric induction, the stereoisomer **188**. Although these latter compounds are usually not obtained completely free of the other isomer, the purity is such that further transformations lead, ultimately, to pure materials.

Scheme 12-22 outlines the general preparation of the key intermediate **193**, and has been performed with a variety of R-groups (R = hydrogen, 2-*t*-butoxypropyl, 3,5-dimethyl-4-isoxazolyl, 2,2-*o*-phenylenedioxypropyl, cyanomethyl) some optically active and some not. For convenience, only one of two possible isomers (see **186** and **187**, Scheme 12-21) is shown in Scheme 12-22; in fact, the reaction sequence was most often performed on the mixture, albeit highly enriched in **188**. Purification was then effected with **193** by simple recrystallization. Reduction (lithium aluminum hydride or sodium borohydride) of **188** gave a nearly quantitative yield of the 17β-alcohol **189** which could be isolated but was conveniently hydrogenated in the crude form to **190**. This hydrogenation (5% palladium-on-charcoal in toluene) was not completely stereospecific as some of the CD-*cis* material was concurrently formed; however, the major product was the desired *trans* isomer (ca. 85%) and, again, for

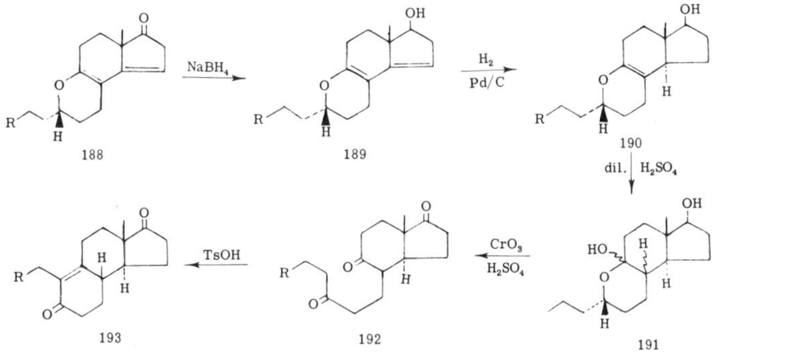

Scheme 12-22

preparative purposes, the crude mixture could be used as such. The enol ether **190** was hydrated with dilute sulfuric acid in acetone to give the hemiketal **191** as a mixture of isomers with unknown configurations at C-8 and C-9. Oxidation to the triketone **192** (configuration at C-8 unspecified) was accomplished with Jones reagent, and the final cyclization was effected either by treatment with p-toluenesulfonic acid in refluxing benzene, or with hydroxide in alcohol. Representative yields for this sequence were (±)-**188**→(±)-**193** (R=cyanomethyl), 46%; and (−)-**188**→(+)-**193** (R=3,5-dimethyl-4-isoxazolyl), 31%. The beauty of this series is illustrated very dramatically with this latter compound. Thus, referring back to Scheme 12-21, a mixture of **186** and **187** was obtained which corresponded to 89% optical induction. Yet, this mixture was converted in 31% yield, without purification of any intermediates, into optically pure (+)-19-(3,5-dimethyl-4-isoxazolyl)-des-A-androst-9-ene-5,17-dione [(+)-**193**; R=3,5-dimethyl-4-isoxazolyl] [133].

At this point there remained only the elaboration of ring A onto **193**. Several methods have been described, each offering some advantages over the previous ones. Initially, (±)-**193** (R=H) was converted to (±)-9β,10α-testosterone by essentially the same methods described in Section 12.4 involving methyl vinyl ketone annelation [135]. The primary effort, however, has been directed toward the total synthesis of (±)- and (+)-estr-4-ene-3,17-dione and some closely related derivatives including Norgestrel (the oral contraceptive Ovral).

Scheme 12-23 illustrates two interesting conversions of BCD intermediates into various steroid derivatives. When (±)-**197** [R=CH(CH$_3$)O-tBu; R′=Et] was treated with p-toluenesulfonic acid in refluxing benzene, the product, obtained in excellent yield, was the rather unstable oxasteroid **194**, as a mixture of C-3 epimers [137]. Analogous treatment of (±)-**197** (R=CN; R′=Me) gave **200** (X=NH, 24%; X=O, 47%). With **194** as the substrate, catalytic hydrogenation was stereoselective from the α-face to give **195**. Attempts to convert this material to (±)-13β-ethylgon-4-en-3,17-dione by hydration–oxidation–cyclization were unsatisfactory, giving only a poor yield of **198**. Treatment with potassium acetylide in ammonia–benzene–ether gave Norgestrel in 23% yield; but a much superior route to **199** was later developed. Ethynylation of **195** followed by reaction with methoxyamine hydrochloride in pyridine gave a good yield of oxime **196**. Oxidation with chromium trioxide in dimethylformamide followed by acid treatment to produce concurrent hydrolysis and cyclization gave the desired product. The overall conversion of **195** to Norgestrel (**199**) by this route was greater than 20% [137].

It was also possible to prepare **199** in an alternate fashion. Hydrogen-

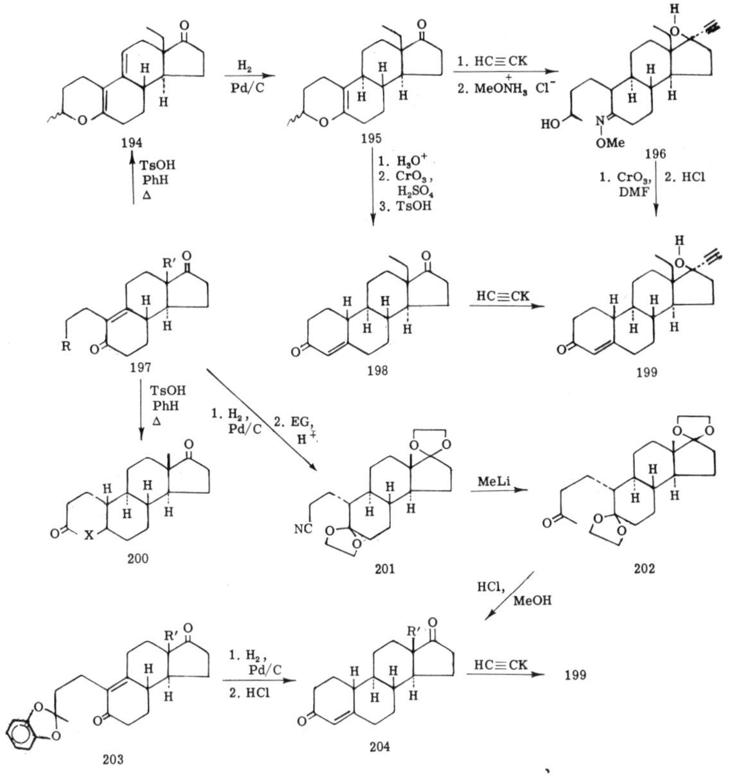

12.6 Asymmetric Induction

Scheme 12-23

ation of (±)-**197** (R=CN; R'=CH$_3$ or CH$_2$CH$_3$) and subsequent ketalization gave (±)-**201** (R'=CH$_3$ or CH$_2$CH$_3$). The cyano derivative was reacted with methyllithium and, following hydrolysis with refluxing methanolic hydrogen chloride, cyclization gave good yields of **204** [R'=CH$_3$ (40%); C$_2$H$_5$ (47%), based on starting **197**]. For **204** (R'=CH$_3$), this represents a 16% overall yield from the Mannich base. A similar sequence of reactions starting with (+)-**197** (R=CN, R'=CH$_3$) gave (+)-estr-4-ene-3,17-dione in 61% yield [131].

Scheme 12-23 also illustrates the use of a third ketone-protecting group, i.e., pyrocatechol [136,138]. Thus, **203** (R'=CH$_3$, C$_2$H$_5$) was easily converted to **204** (R'=CH$_3$, C$_2$H$_5$) in two steps. Catalytic hydrogenation followed by acid hydrolysis–cyclization gave the desired product in over 70% yield (in fact, this represents a 27% yield starting with the Mannich base) [138]. Furthermore, the 13β-ethyl steroid **204** (R'=C$_2$H$_5$) was reacted with potassium acetylide to give Norgestrel (**199**)— a route superior to that previously described. The use of pyrocatechol as a ketone protecting group was shown to be quite advantageous in regard to its acid stability and the crystallinity which it imparted to most of the intermediates.

A fourth masking function extensively investigated by the Hoffmann-LaRoche researchers was the 3,5-dimethylisoxazole group initially exploited by Stork and his collaborators [104, 139] (Scheme 12-24). As with the other protecting groups, the first reaction in this series was hydrogenation (10% palladium-on-charcoal in a 3:1 ethanol–triethylamine solution) of ($\pm$)-205 (R=CH_3, C_2H_5) giving ($\pm$)-206. Under these conditions only one equivalent of hydrogen was absorbed; however, when the hydrogenation solution was made basic with dilute potassium hydroxide, further hydrogenation ensued to provide ($\pm$)-207 (R=CH_3, C_2H_5). Heating the methyl derivative with base yielded ($\pm$)-estr-4-ene-3,17-dione [($\pm$)-210 (R=CH_3)] but only in 45% overall yield. By modifying this procedure, much improved yields could be attained. Thus, ($\pm$)-206 (R=CH_3, C_2H_5) was bisketalized and then hydrogenated as before to give ($\pm$)-208 (R=CH_3, C_2H_5) which was not isolated, but simply refluxed in aqueous base to give ($\pm$)-209 (R=CH_3, C_2H_5). Effecting hydrolysis and cyclization by heating with hydrochloric acid in methanol gave, overall, 80–85% ($\pm$)-210 (R=CH_3) and 70% ($\pm$)-210 (R=C_2H_5) [132]. Starting with optically pure (+)-205 (R=CH_3), the same reaction sequence provided a 66% yield of pure (+)-210 (R=CH_3) [133], a notable achievement indeed.

It is evident that the principle of asymmetric induction is an exciting new act in the steroid chemist's repertoire. Most certainly further developments in the area are imminent; and, although the work discussed in

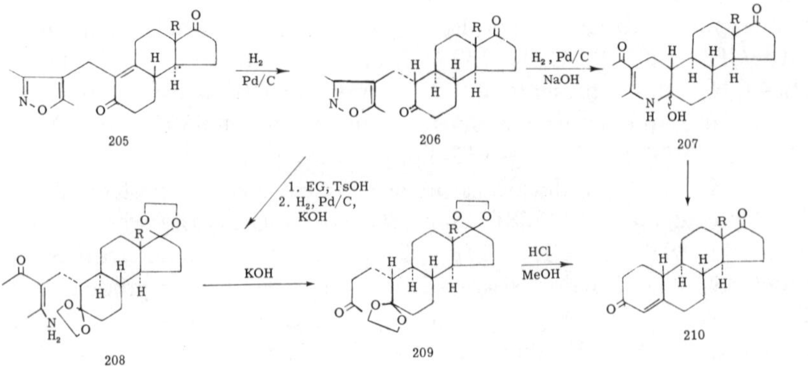

Scheme 12-24

12.7 Hoffman-LaRoche Approach

this section is currently limited to the preparation of the basic estr-4-ene-3,17-dione system, the principle should be applicable to many other steroid skeletons.

12.7 Hoffmann-LaRoche Approach

This section will concern primarily a second approach to total synthesis by investigators at Hoffmann-LaRoche but, as their route is, in many respects, similar to previous ones, attempts will be made to include pertinent efforts by various other groups. The starting material was the indane derivative 211 (Scheme 12-25), which was prepared and used in the optically active and racemic forms [22, 140, 141] (see Section 12.1). This material was converted into the BCD tricyclic intermediate 216 by first alkylating the tetrahydropyranyl ether of 211 with 2-(2-bromoethyl)-2-ethyl-1,3-dioxolane using sodium hydride in dimethylsulfoxide. The result of this reaction was 213 (58%) and 212 (32%), the latter of which could be reconverted to 211. The keto ketal 213 was hydrogenated (10% palladium–barium sulfate), subjected to equilibrating conditions (sodium methoxide to equilibrate the initially formed 8β-alkyl group to the more stable 8α-position), and cyclized by refluxing with

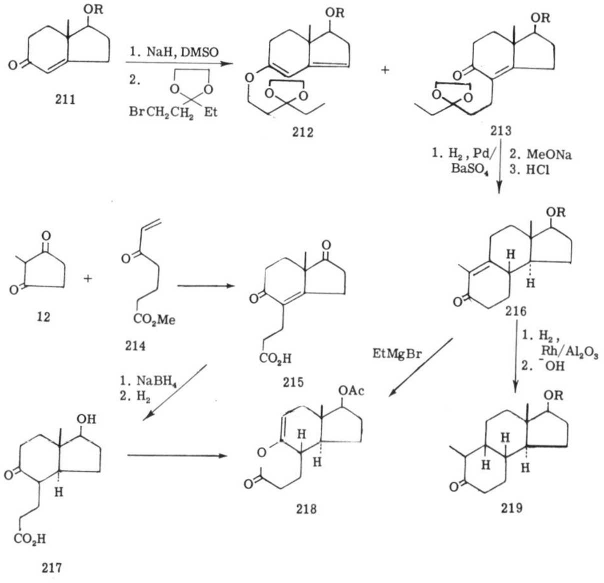

Scheme 12-25

dilute hydrochloric acid to provide **216** (R=H) in a 39% yield. Although the intermediates were never isolated in going from **213** to **216**, it should be appreciated that the hydrogenation must have given appreciable amounts of the CD-*trans* product (see Section 12.1 for a discussion of this problem). The synthesis at this point intersected with many other plans; this precise compound had previously been prepared in an alternate fashion [35, 142, 143]. Thus, methyl 5-oxo-6-heptenoate (**214**) condensed with 2-methylcylcopentane-1,3-dione to give an 88% yield of **215**. Optical resolution followed by borohydride reduction and hydrogenation gave the CD *trans* product **217** in yields of 60–70% [37]. Lactonization to **218** followed by a Fujimoto-Belleau reaction with ethyl magnesium bromide gave optically active **216** (R=H) [144] (in fact, various Grignard reagents could be, and were, used to produce numerous 19-substituted tricyclic steroid precursors; this same type lactone conversion has also been effected with Wittig and phosphonate reagents [145], and a related reaction has been performed with α-sulfonyl carbanions [146]). Other synthesis of BCD intermediates very similar to **216** have been recorded (i.e., **216** with OR replaced by H [147]; **216** with OR replaced by a ketone and without the 19-methyl group [148]; **216** with OR replaced by a ketone [149]; **216** with an 11-ketone and without the 5-ketone [150]; **216** with a C-13 ethyl group, and the analogous D-homo derivative [151]). In fact, the Roussel-UCLAF group has prepared numerous such BCD tricyclics by the BC→BCD route, and a thorough discussion of their work has been presented in Chapter 10.

As indicated (see, for example, Schemes 12-4 and 12-14), such BCD intermediates have often been annelated to provide the steroid skeleton [148, 152, 153] and the uniqueness of the present case lies in their conversion into retrosteroids (9β, 10α). Prior work with naturally obtained derivatives of **216** had established the feasibility of such transformations [154]. Hydrogenation of **216** (R=Ac) over 5% rhodium-on-alumina provided, after saponification, a 60% yield of **219** (R=H) which contained only trace amounts of the isomeric BC-*trans* material [155]. The elaboration of ring A and the side chain was accomplished in one of three ways (Scheme 12-26). Ketalization and oxidation with the Snatzke reagent [156] gave **221** (ca. 85%) which was subjected to a Wittig reaction with ethylidenetriphenylphosphorane [157]. The product obtained was a 12:1 mixture of **222** and the isomeric olefin (a combined yield of greater than 90%). Hydroboration gave predominantly the 20α-ol which was annelated with methyl vinyl ketone and oxidized to give retroprogesterone **225** in 5–10% overall yield from **222**. Altern-

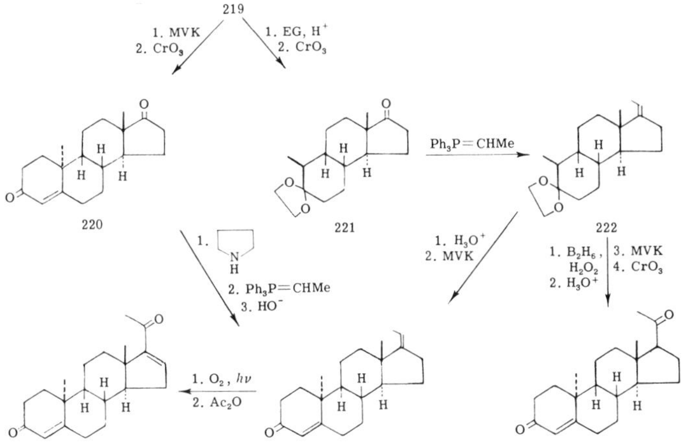

12.8 Miscellaneous Syntheses

Scheme 12-26

atively, **222** was deketalized and annelated to produce **224**, which was subjected to a novel photolytic oxidation [158]. The result was a 55% yield (after dehydration of the initially formed hydroperoxide) of enone **223**. The Δ^{16}-bond in this material provided a means of further functionalization in this series.

A final variation in the synthetic scheme was the conversion of **219** (R=H) to retrotestosterone and then retroandrostenedione (**220**) in a moderate (ca. 25%) yield. This, then, was also converted to **224**. It should be borne in mind that all the compounds described in Schemes 12-25 and 12-26 have been utilized in both racemic and optically active forms. Furthermore, similar sequences have been applied to D-homo- and 18-methyl-D-homo analogs.

12.8 Miscellaneous Syntheses

The work described in this section is that which has been only briefly described and, yet, is sufficiently unusual to warrant individual discussion.

Scheme 12-27 presents the isoxazole annelation procedure for total synthesis reported by Stork and co-workers [104,139,159–162]. In actuality, most of the work reported has dealt with mechanistic details and, though the potential embodied in the sequence would seem to lend itself to further development, few steroids have been prepared by this series of reactions (see, however, Section 12.6). The required isoxazole

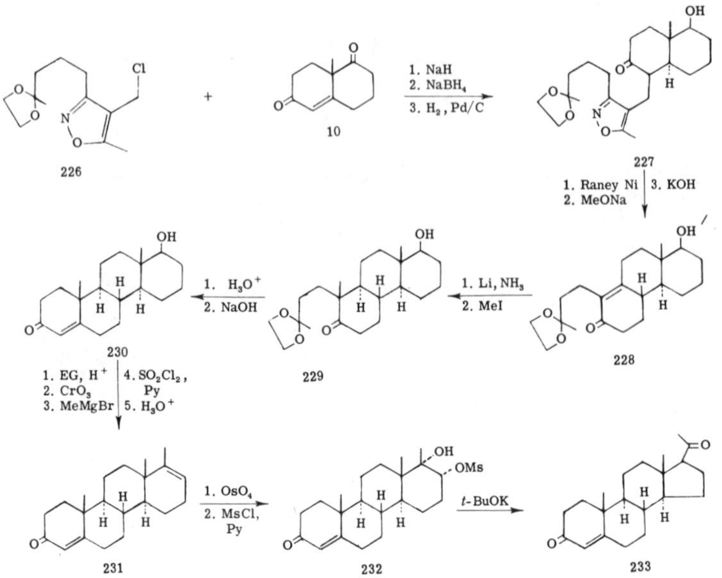

Scheme 12-27

226 was prepared [161] from 6-nitro-2-hexanone and the pyrrolidine enamine of ethyl acetoacetate and reacted with the enolate of 10-methyl-$\Delta^{1,9}$-octalin-2,5-dione to give the expected alkylation product in 55% yield. Sodium borohydride selectively reduced the C-17a ketone to the β-alcohol, and hydrogenation with palladium-on-charcoal in 3:1 ethyl acetate–triethylamine produced the intermediate **227**, which was not purified. This hydrogenation procedure was found to be very sensitive to the reaction conditions used, as variations thereon gave undesirable reduction of the isoxazole ring and/or less of the desired CD *trans* product. Hydrogenolysis with Raney nickel, followed by treatment with methoxide and aqueous base led to **228** in good yield (33% from **10**). The important discovery was made that **228** could be angularly methylated in nearly quantitative yield to the pure β-methyl isomer **229** by employing Stork's previously described lithium–ammonia system [163]. Simple acid hydrolysis and basic cyclization gave *dl*-D-homotestosterone **230** (80%). This material was converted to *dl*-progesterone via intermediate **231**, produced by the usual reactions. Oxidation of **231** with osmium tetroxide gave the *cis*-α glycol which was monomesylated to give **232**. Rearrangement to **233** was then effected by warming with potassium *t*-butoxide in *t*-butanol. Although yields for these latter steps were not described, the overall preparation of **230** would appear attractive.

12.8 Miscellaneous Syntheses

Another interesting new approach to total synthesis [164, 165] has, as its key intermediate, the triketone **235** (Scheme 12-28), prepared from the sodium salt of 2-methylcyclopentane-1,3-dione and **234** by Michael addition and subsequent dehydrohalogenation (Michael addition of divinyl ketone and **12** to afford **235** was not successful). A two-stage condensation (potassium *t*-butoxide in *t*-butanol followed by *p*-toluenesulfonic acid–acetic acid) of **235** and **236** produced the tricyclic **238**. This interesting reaction presumably proceeded through intermediate **237** though it was not isolated. Other derivatives of **236** were found to react in a like manner; however, when the *t*-butyl ester was replaced by a methyl ester, B-ring aromatization leading to salicylate derivatives became a problem. Acid hydrolysis of **238** and base-catalyzed cyclization gave **240**, a compound previously obtained by an alternate total synthesis (see Chapter 4, Section 4.4) [166,167]. The excellent yields (34% of **240** from **12**) obtained speak well for future applications of this method.

Further investigation aimed at introducing the C-19 methyl group revealed that the 3,17-diketal of **239** could be methylated to produce the natural **242** and the α-methyl isomer in yields of 43 and 32%, respectively [168]. Further conversions of the former to **241** and **243** were also reported. The overall synthetic scheme, still in its infancy, will no doubt be further exploited.

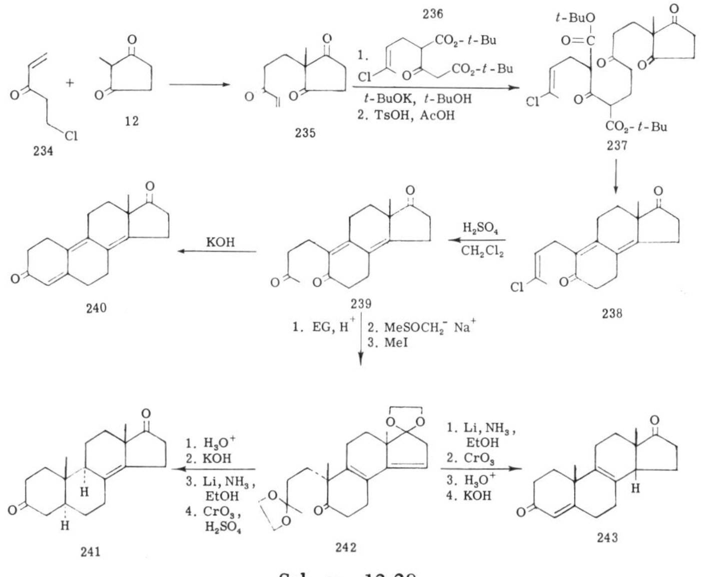

Scheme 12-28

Finally, outlined in Scheme 12-29 are two more, novel syntheses reported in only minimal detail. In the first [169], the optically active *trans* hydrindane **244** [35, 143] underwent a Leuckart reaction with ammonium bicarbonate to provide the expected tricyclic lactam as a mixture of the 9α- (72%) and 9β- (26%) isomers. These compouds were easily separated by taking advantage of their rather marked solubility differences. The 9α-derivative was converted to the 17β-formate (63%) and treated with triethyloxonium fluoroborate to provide **245**. This material was not purified but was treated with diethyl 3-ketoglutarate and triethylamine to give the crystalline **246** (21%). The steroid skeleton was completed by condensing **246** with malonic acid dichloride. The result, after hydrolysis, was the 10-aza compound **247** (23%). An alternate method for effecting this same conversion involved the use of ketene in place of the diacid dichloride. Decarboxylation was performed by heating **247** in formic acid (79%), and further treatment with excess triethyloxonium fluoroborate gave **248** in 53% yield in addition to a

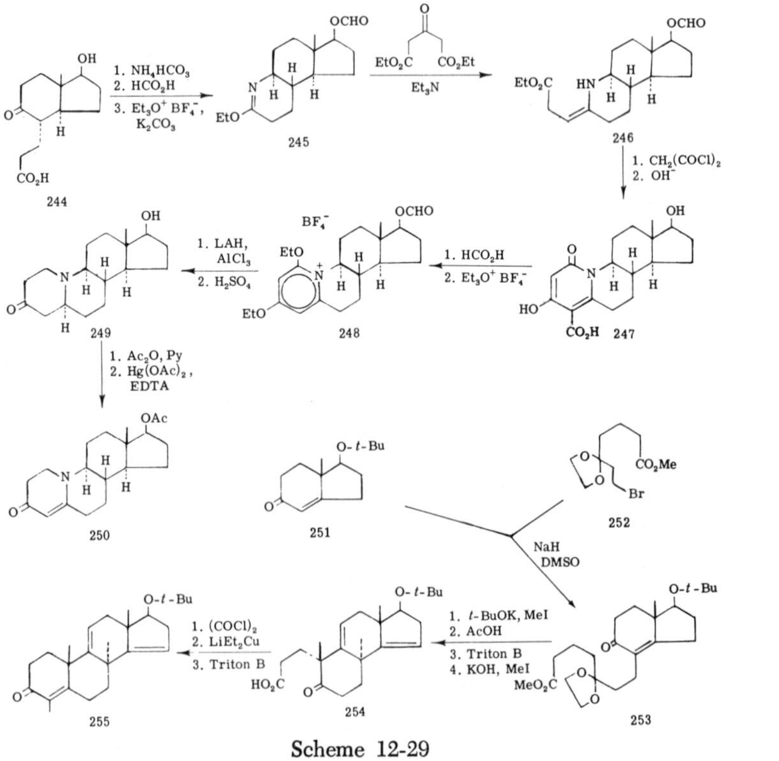

Scheme 12-29

nonaromatic product (the 1-ethoxy group replaced with a ketone). Reduction with lithium aluminum hydride in the presence of aluminum chloride followed by mild acid hydrolysis produced (82%) **249** and the 5β-epimer. Repeated crystallization from ethyl acetate ultimately gave **249** in pure form. The acetate of (+)-10-aza-19-nortestosterone (**250**) was obtained (20%) from **249** by acetylation and subsequent oxidation with mercuric acetate and the sodium salt of ethylenediaminetetraacetic acid. An alternate synthetic sequence for the preparation of 10-aza-19-nortestosterone derivatives has been presented in Chapter 8, Section 8.4.

Also shown in Scheme 12-29 is the preparation of the 4,8-dimethyltestosterone derivative **255** [170]. This compound, obtained in a 5% overall yield from **251** and **252**, was synthesized in an effort to gain entrance into the field of steroidal antibiotics, i.e., the fusidane series. Alkylation of the enone **251** [141] with the bromide **252** gave **253** in 50% yield. Alkylation with methyl iodide produced the 8α-methyl substance; deketalization and cyclization led to the expected tricyclic enone, and introduction of the C-10 angular methyl group was accomplished with potassium hydroxide, t-butanol, and methyl iodide to give **254**. Although the configuration of this product was established, the degree of stereoselectivity of the methylation reactions was not specified. Elaboration of the A ring was performed by treating the acid chloride of **254** with lithium diethylcuprate. The resulting ethyl ketone was cyclized with Triton B thereby providing **255**.

References

1. B. J. F. Hudson and R. Robinson, *J. Chem. Soc., London* p. 691 (1942).
2. W. S. Johnson, C. D. Gutsche, and D. K. Banerjee, *J. Amer. Chem. Soc.* **73**, 5464 (1951).
3. N. A. Nelson, R. S. P. Hsi, J. M. Shuck, and L. D. Kahn, *J. Amer. Chem. Soc.* **82**, 2573 (1960).
4. R. L. Kidwell and S. D. Darling, *Tetrahedron Lett.* p. 531 (1966).
5. J. W. Cornforth and R. Robinson, *J. Chem. Soc., London* p. 1855 (1949).
6. A. J. Birch, J. A. K. Quartey, and H. Smith, *J. Chem. Soc., London* p. 1768 (1952).
7. P. Wieland and K. Miescher, *Helv. Chim. Acta* **33**, 2215 (1950).
8. N. L. Wendler, H. L. Slates, and M. Tishler, *J. Amer. Chem. Soc.* **73**, 3816 (1951).
9. C. A. Friedmann and R. Robinson, *Chem. Ind. (London)* p. 777 (1951).
10. S. Swaminathan and M. S. Newman, *Tetrahedron* **2**, 88 (1958).
11. M. S. Newman and A. B. Mekler, *J. Amer. Chem. Soc.* **82**, 4039 (1960).

12. N. K. Chandhuri and P. C. Mukharji, *Sci. Cul.* 19, 463 (1954).
13. I. N. Nazarov and S. I. Zav'yalov, *J. Gen. Chem. USSR* 23, 1793 (1953).
14. J. D. Cocker and T. G. Halsall, *J. Chem. Soc., London* p. 3441 (1957).
15. C. B. C. Boyce and J. S. Whitehurst, *J. Chem. Soc., London* p. 2680 (1960).
16. A. J. Birch, E. Pride, and H. Smith, *J. Chem. Soc., London* p. 4688 (1958).
17. R. H. Jaeger, *Tetrahedron* 2, 326 (1958).
18. W. Acklin, V. Prelog, and A. P. Prieto, *Helv. Chim. Acta* 41, 1416 (1958).
19. C. B. C. Boyce and J. S. Whitehurst, *J. Chem. Soc., London* p. 2022 (1959).
20. J. J. Panhouse and C. Sannie, *Bull. Soc. Chim. Fr.* [4] 1036 (1955).
21. H. Shick, G. Lehman, and G. Hilgetag, *Chem. Ber.* 102, 3238 (1969).
22. Z. G. Hajos, D. R. Parrish, and E. P. Oliveto, *Tetrahedron* 24, 2039 (1968).
23. D. J. Crispin, A. E. Vanstone, and J. S. Whitehurst, *J. Chem. Soc., London* p. 10 (1970).
24. C. B. C. Boyce and J. S. Whitehurst, *J. Chem. Soc., London* p. 4547 (1960).
25. W. Acklin and V. Prelog, *Helv. Chim. Acta* 42, 1239 (1959).
26. T. M. Dawson, P. S. Littlewood, B. Lythgoe, T. Medcalfe, M. W. Moon, and P. M. Tomkins, *J. Chem. Soc., London* p. 1292 (1971).
27. H. H. Inhoffen and E. Prinz, *Chem. Ber.* 87, 684 (1954).
28. H. H. Inhoffen, S. Schütz, P. Rossberg, O. Berges, K.-H. Nordsiek, H. Plenio, and E. Hödolt, *Chem. Ber.* 91, 2626 (1958).
29. H. H. Inhoffen, G. Quinkert, S. Schütz, D. Kampe, and G. F. Domagk, *Chem. Ber.* 90, 664 (1957).
30. H. H. Inhoffen, G. Quinkert, S. Schütz, G. Friedrich, and E. Tober *Chem. Ber.* 91, 781 (1958).
31. I. J. Bolton, R. G. Harrison, and B. Lythgoe, *J. Chem. Soc., London* p. 2950 (1971); and preceeding publications in this series.
32. G. Stork and P. L. Stotter, *J. Amer. Chem. Soc.* 91, 7781 (1969).
33. H. C. Brown and E. Negishi, *Chem. Commun.* p. 594 (1968).
34. K. H. Baggaley, S. G. Brooks, J. Green, and B. T. Redman, *J. Chem. Soc., London* pp. 2671 and 2678 (1971).
35. L. Velluz, G. Nominé, G. Amiard, V. Torelli, and J. Cérêde, *C. R. Acad. Sci.* 257, 3086 (1963).
36. H. Smith, G. A. Hughes, and B. J. McLoughlin, *Experientia* 19, 178 (1963).
37. G. Nominé, G. Amiard, and V. Torelli, *Bull. Soc. Chim. Fr.* [6] p. 3664 (1968).
38. G. S. Grinenko, E. V. Popova, V. I. Maksimov, and L. M. Alekseeva, *Zh. Org. Khim.* 6, 732 (1970).
39. W. S. Johnson, *J. Amer. Chem. Soc.* 78, 6278 (1956).
40. W. S. Johnson, B. Bannister, B. M. Bloom, A. D. Kemp, R. Pappo, E. R. Rogier, and J. Szmuszkovicz, *J. Amer. Chem. Soc.* 75, 2275 (1953).
41. W. S. Johnson, J. Szmuszkovicz, E. R. Rogier, H. I. Hadler, and H. Wynberg, *J. Amer. Chem. Soc.* 78, 6285 (1956).
42. H. M. E. Cardwell, J. W. Cornforth, S. R. Duff, H. Holtermann, and R. Robinson, *J. Chem. Soc., London* p. 361 (1953).
43. W. S. Johnson, J. Ackerman, J. F. Eastham, and H. A. DeWalt, Jr., *J. Amer. Chem. Soc.* 78, 6302 (1956).
44. W. S. Johnson, J. J. Korst, R. A. Clement, and J. Dutta, *J. Amer. Chem. Soc.* 82, 614 (1960).

References

45. W. S. Johnson, E. R. Rogier, J. Szmuszkovicz, H. I. Hadler, J. Ackerman, B. K. Bhattacharyya, B. M. Bloom, L. Stalmann, R. A. Clement, B. Bannister, and H. Wynberg, *J. Amer. Chem. Soc.* 78, 6289 (1956).
46. A. L. Wilds and N. A. Nelson, *J. Amer. Chem. Soc.* 75, 5360 (1953).
47. W. S. Johnson, E. R. Rogier, and J. Ackerman, *J. Amer. Chem. Soc.* 78, 6322 (1956).
48. W. S Johnson, B. Bannister, and R. Pappo, *J. Amer. Chem. Soc.* 78, 6331 (1956).
49. R. Pappo, B. M. Bloom, and W. S. Johnson, *J. Amer. Chem. Soc.* 78, 6347 (1956).
50. W. S. Johnson, D. K. Banerjee, W. P. Schneider, C. D. Gutsche, W. E. Shelberg, and L. J. Chinn, *J. Amer. Chem. Soc.* 74, 2832 (1952).
51. W. S. Johnson, R. Pappo, and W. F. Johns, *J. Amer. Chem. Soc.* 78, 6339 (1956).
52. W. S. Johnson, B. Bannister, R. Pappo, and J. E. Pike, *J. Amer. Chem. Soc.* 78, 6354 (1956).
53. J. A. Marshall and W. S. Johnson, *J. Amer. Chem. Soc.* 84, 1485 (1962).
54. W. S. Johnson, D. S. Allen, Jr., R. R. Hindersinn, G. N. Sausen, and R. Pappo, *J. Amer. Chem. Soc.* 84, 2181 (1962).
55. W. Nagata, T. Terasawa, T. Aoki, and K. Takeda, *Chem. Pharm. Bull* 9, 783 (1961).
56. W. S. Johnson and D. S. Allen, Jr., *J. Amer. Chem. Soc.* 79, 1261 (1957).
57. D. K. Banerjee and P. R. Shafer, *J. Amer. Chem. Soc.* 72, 1931 (1950).
58. W. S. Johnson, R. Pappo, and A. D. Kemp, *J. Amer. Chem. Soc.* 76, 3353 (1954).
59. W. S. Johnson, A. D. Kemp, R. Pappo, J. Ackerman, and W. F. Jones, *J. Amer. Chem. Soc.* 78, 6312 (1956).
60. W. S. Johnson, J. M. Anderson, and W. E. Shelberg, *J. Amer. Chem. Soc.* 66, 218 (1944).
61. W. S. Johnson, B. Bannister, R. Pappo, and J. E. Pike, *J. Amer. Chem. Soc.* 77, 817 (1955).
62. W. S. Johnson, W. A. Vredenburgh, and J. E. Pike, *J. Amer. Chem. Soc.* 82, 3409 (1960).
63. W. S. Johnson and K. V. Yorka, *Tetrahedron Lett.* p. 11 (1960).
64. K. V. Yorka, W. L. Truett, and W. S. Johnson, *J. Org. Chem.* 27, 4580 (1962).
65. W. S. Johnson, J. F. W. Keana, and J. A. Marshall, *Tetrahedron Lett.* p. 193 (1963).
66. W. S. Johnson, J. A. Marshall, J. F. W. Keana, R. W. Franck, D. G. Martin, and V. J. Bauer, *Tetrahedron, Suppl.* 8, Part II, 541 (1966).
67. J. F. W. Keana and W. S. Johnson, *Steroids*, 4, 457 (1964).
68. P. W. Schiess, D. M. Baily, and W. S. Johnson, *Tetrahedron Lett.* p. 549 (1963).
69. W. S. Johnson, H. A. P. de Jongh, C. E. Coverdale, J. W. Scott, and U. Burckhardt, *J. Amer. Chem. Soc.* 89, 4523 (1967).
70. W. S. Johnson, J. M. Cox, D. W. Graham, and H. W. Whitlock, Jr., *J. Amer. Chem. Soc.* 89, 4524 (1967).
71. W. S. Johnson, J. C. Collins, R. Pappo, and M. B. Rubin, *J. Amer. Chem. Soc.* 80, 2585 (1958).

72. W. S. Johnson, J. C. Collins, Jr., R. Pappo, M. B. Rubin, P. J. Krop, W. F. Johns, J. E. Pike, and W. Bartmann, *J. Amer. Chem. Soc.* **85**, 1409 (1963).
73. W. S. Johnson, D. G. Martin, R. Pappo, S. D. Darling, and R. A. Clement, *Proc. Chem. Soc., London* p. 58 (1957).
74. S. M. McElvain and G. R. McKay, Jr., *J. Amer. Chem. Soc.* **78**, 6086 (1956).
75. R. E. Bowman, *J. Chem. Soc., London* p. 325 (1950).
76. W. F. Johns, R. M. Lukes, and L. H. Sarett, *J. Amer. Chem. Soc.* **76**, 5026 (1954).
77. J. Schmidlin, G. Anner, J.-R. Billeter, and A. Wettstein, *Experientia* **11**, 365 (1955).
78. J. Schmidlin, G. Anner. J.-R. Billeter, K. Heusler, H. Ueberwasser, P. Wieland, and A. Wettstein, *Helv. Chim. Acta* **40**, 1438 (1957).
79. J. Schmidlin, G. Anner, J.-R. Billeter, K. Heusler, H. Uberwasser, P. Wieland, and A. Wettstein, *Helv. Chim. Acta* **40**, 2291 (1957).
80. J. A. Hogg, P. F. Beal, A. H. Nathan, F. H. Lincoln, W. P. Schneider, B. J. Magerlein, A. R. Hanze, and R. W. Jackson, *J. Amer. Chem. Soc.* **77**, 4436 (1955).
81. J. P. Kutney, W. McCrae, and A. By, *Can. J. Chem.* **40**, 982 (1962).
82. W. Nagata, T. Terasawa, S. Hirai, and K. Takeda, *Tetrahedron Lett.* p. 27 (1960).
83. W. Nagata, S. Hirai, T. Terasawa, I. Kikkawa, and K. Takeda, *Chem. Pharm. Bull.* **9**, 756 (1961).
84. J. P. Kutney, J. Winter, W. McCrae, and A. By, *Can. J. Chem.* **41**, 470 (1963).
85. W. Nagata, S. Hirai, T. Terasawa, an K. Takeda, *Chem. Pharm. Bull.* **9**, 769 (1961).
86. W. Nagata, M. Yoshioka, and T. Okumura, *J. Chem. Soc., London* p. 2365 (1970).
87. W. Nagata, *Tetrahedron* **13**, 278 (1961).
88. W. Nagata, T. Terasawa, S. Hirai, and K. Takeda, *Tetrahedron* **13**, 295 (1961).
89. W. Nagata, T. Terasawa, and T. Aoki, *Chem. Pharm. Bull* **11**, 819 (1963).
90. W. Nagata, M. Yoshioka, and S. Hirai, *Tetrahedron Lett.* p. 461 (1962).
91. W. Nagata and M. Yoshioka, *Tetrahedron Lett.* p. 1913 (1966).
92. W. Nagata, T. Wakabayashi, M. Narisada, Y. Hayase, and S. Kamata, *J. Amer. Chem. Soc.* **93**, 5740 (1971).
93. W. Nagata, *Nippon Kagaku Zasshi* **90**, 837 (1969).
94. W. Nagata, M. Yoshioka, and T. Terasawa, *J. Amer. Chem. Soc.* **94**, 4672 (1972).
95. J. P. Kutney, A. By, T. Inaba, and S. Y. Leong, *Tetrahedron Lett.* p. 2911 (1965).
96. W. S. Johnson, N. Cohen, E. R. Habicht, Jr., D. P. G. Hamon, G. P. Rizzi, and D. J. Faulkner, *Tetrahedron Lett.* p. 2829 (1968).
97. D. M. Bailey, D. P. G. Hamon, and W. S. Johnson, *Tetrahedron Lett.* p. 555 (1963).
98. J. P. Kutney, J. Cable, W. A. F. Gladstone, H. W. Hanssen, E. J. Torupka, and W. D. C. Warnock, *J. Amer. Chem. Soc.* **90**, 5332 (1968).
99. D. W. Stoutmire, Ph.D. Thesis, University of Wisconsin, Madison, 1957.
100. B. K. Bhattacharyya, A. K. Bose, A. Chatterjee, and B. P. Sen, *J. Indian Chem. Soc.* **41**, 479 (1964).

101. J. Fried and N. A. Abraham, *Tetrahedron Lett.* p. 3505 (1965).
102. M. J. Green, N. A. Abraham, E. B. Fleischer, J. Case, and J. Fried, *Chem. Commun.* p. 234 (1970).
103. J. Fried, M. J. Green, and G. V. Nair, *J. Amer. Chem. Soc.* **92**, 4136 (1970).
104. G. Stork and J. E. McMurry, *J. Amer. Chem. Soc.* **89**, 5464 (1967).
105. R. B. Woodward, F. Sondheimer, D. Taub, K. Heusler, and W. M. McLamore, *J. Amer. Chem. Soc.* **73**, 2403 (1951).
106. R. B. Woodward, F. Sondheimer, D. Taub, K. Heusler, and W. M. McLamore, *J. Amer. Chem. Soc.* **74**, 4223 (1952).
107. A. J. Speziale, J. A. Stephens, and Q. E. Thompson, *J. Amer. Chem. Soc.* **76**, 5011 (1954).
108. C. H. Shunk and A. L. Wilds, *J. Amer. Chem. Soc.* **71**, 3946 (1949).
109. L. B. Barkley, M. W. Farrar, W. S. Knowles, H. Raffelson, and Q. E. Thompson, *J. Amer. Chem. Soc.* **76**, 5014 (1954).
110. R. B. Woodward and F. V. Brutcher, Jr., *J. Amer. Chem. Soc.* **80**, 209 (1958).
111. R. B. Woodward, F. Sondheimer, and D. Taub, *J. Amer. Chem. Soc.* **73**, 3547 (1951).
112. R. B. Woodward, F. Sondheimer, and D. Taub, *J. Amer. Chem. Soc.* **73**, 3548 (1951).
113. R. B. Woodward, F. Sondheimer, and D. Taub, *J. Amer. Chem. Soc.* **73**, 4057 (1951).
114. L. B. Barkley, M. W. Farrar, W. S. Knowles, and H. Rafflelson, *J. Amer. Chem. Soc.* **75**, 4110 (1953).
115. L. B. Barkley, M. W. Farrar, W. S. Knowles, and H. Rafflelson, *J. Amer. Chem. Soc.* **78**, 4111 (1956).
116. L. B. Barkley, W. S. Knowles, H. Rafflelson, and Q. E. Thompson, *J. Amer. Chem. Soc.* **76**, 5017 (1954).
117. W. S. Knowles and Q. E. Thompson, *J. Org. Chem.* **25**, 1031 (1960).
118. G. Stork, H. J. E. Loewenthal, and P. C. Mukharji, *J. Amer. Chem. Soc.* **78**, 501 (1956).
119. W. S. Knowles and Q. E. Thompson, *J. Amer. Chem. Soc.* **79**, 3212 (1957).
120. Q. E. Thompson, *J. Org. Chem.* **23**, 622 (1958).
121. M. Chaykovsky and R. E. Ireland, *J. Org. Chem.* **28**, 748 (1963).
122. J. Demaecker and R. H. Martin, *Nature (London)* **173**, 266 (1954).
123. Y. Izumi, *Angew. Chem.* **83**, 956 (1971).
124. U. Eder, G. Sauer, and R. Wiechert, *Angew. Chem.* **83**, 492 (1971).
125. Z. G. Hajos and D. R. Parrish, German Patent 2,102,623 (1971); *Chem. Abstr.* **75**, 129414r (1971).
126. J. D. Morrison and H. S. Mosher, "Asymmetric Organic Reactions." Prentice-Hall, Englewood Cliffs, New Jersey, 1971.
127. G. Saucy, R. Borer, and A. Fürst, *Helv. Chim. Acta* **54**, 2034 (1971).
128. G. Saucy and R. Borer, *Helv. Chim. Acta* **54**, 2121 (1971).
129. G. Saucy, W. Koch, M. Müller, and A. Fürst, *Helv. Chim. Acta* **53**, 964 (1970).
130. M. Rosenberger, D. Andrews, F. DiMaria, A. J. Duggan, and G. Saucy, *Helv. Chim. Acta* **55**, 249 (1972).
131. N. Cohen, B. Banner, R. Borer, R. Mueller, R. Yang, M. Rosenberger, and G. Saucy, *J. Org. Chem.* **37**, 3385 (1972).
132. J. W. Scott and G. Saucy, *J. Org. Chem.* **37**, 1652 (1972).
133. J. W. Scott, R. Borer, and G. Saucy, *J. Org. Chem* **37**, 1659 (1972).

134. J. W. Scott, B. L. Banner, and G. Saucy, *J. Org. Chem* **37**, 1664 (1972).
135. G. Saucy and R. Borer, *Helv. Chim. Acta* **54**, 2517 (1971).
136. M. Rosenberger, A. J. Duggan, R. Borer, R. Müller, and G. Saucy, *Helv. Chim. Acta* **55**, 2663 (1972).
137. M. Rosenberger, T. P. Fraher, and G. Saucy, *Helv. Chim. Acta* **54**, 2857 (1971).
138. M. Rosenberger, A. J. Duggan, and G. Saucy, *Helv. Chim. Acta* **55**, 1333 (1972).
139. G. Stork, S. Danishefsky, and M. Ohashi, *J. Amer. Chem. Soc.* **89**, 5459 (1967).
140. Z. G. Hajos, D. R. Parrish, and E. P. Oliveto, *Tetrahedron Lett.* p. 6495 (1966).
141. Z. G. Hajos, R. A. Micheli, D. R. Parrish, and E. P. Olivto, *J. Org. Chem.* **32**, 3008 (1967).
142. J. Warnant and B. Goffinet, French Patent 1,359,675 (1963).
143. L. Velluz, J. Valls, and G. Nominé, *Angew. Chem.* **77**, 185 (1965).
144. J. Weill-Raynal, *Synthesis* **1**, 49 (1969).
145. C. A. Henrick, E. Böhme, J. A. Edwards, and J. H. Fried, *J. Amer. Chem. Soc.* **90**, 5926 (1968).
146. T. Komeno, S. Ishihara, and H. Itani, *Tetrahedron* **28**, 4719 (1972).
147. P. C. Mukharji, *J. Indian Chem. Soc.* **24**, 91 (1947).
148. O. I. Fedorova, G. S. Grinenko, and V. I. Maksimov, *Dokl. Akad. Nauk SSSR* (Eng. Trans.) **171**, 1154 (1966).
149. P. Wieland, G. Anner, and K. Miescher, *Helv. Chim. Acta* **36**, 646 and 1803 (1953).
150. U. S. Patent 2,793,233 (1957); *Chem. Abstr.* **52**, 441 (1958).
151. J. N. Gardner, B. A., Anderson, and E. P. Oliveto, *J. Org. Chem.* **34**, 107 (1969).
152. P. Wieland, H. Ueberwasser, G. Anner, and K. Miescher, *Helv. Chim. Acta* **36**, 1231 (1953).
153. H. Ueberwasser, P. Wieland, G. Anner, and K. Miescher, *Angew. Chem.* **66**, 81 (1954).
154. M. Uskokovic, J. Iacobelli, R. Phelion, and T. Williams, *J. Amer. Chem. Soc.* **88**, 4538 (1966).
155. R. A. Micheli, J. N. Gardner, R. Dubuis, and P. Buchschacher, *J. Org. Chem.* **34**, 1457 (1969).
156. G. Snatzke, *Chem. Ber.* **94**, 729 (1961).
157. A. M. Krubiner, G. Saucy, and E. P. Oliveto, *J. Org. Chem.* **33**, 3548 (1968).
158. W. P. Schneider, D. E. Ayer, and J. Huber, *Abstr. Int. Cong. Horm. Steroids, 2nd, 1966* p. 24 (1967).
159. G. Stork, *Pure Appl. Chem.* **9**, 931 (1964).
160. M. Ohashi, H. Kamachi, H. Kakisawa, and G. Stork, *J. Amer. Chem. Soc.* **89**, 5460 (1967).
161. G. Stork and J. E. McMurry, *J. Amer. Chem. Soc.* **89**, 5461 (1967).
162. G. Stork and J. E. McMurry, *J. Amer. Chem. Soc.* **89**, 5463 (1967).
163. G. Stork, P. Rosen, N. Goldman, R. V. Coombs, and J. Tsuji, *J. Amer. Chem. Soc.* **87**, 275 (1965).
164. S. Danishefsky and B. M. Migdalof, *J. Amer. Chem. Soc.* **91**, 2807 (1969).

165. S. Danishefsky, L. S. Crawley, D. M. Soloman, and P. Heggs, *J. Amer. Chem. Soc.* **93**, 2356 (1971).
166. T. B. Windholz, J. H. Fried, H. Schwam, and A. A. Patchett, *J. Amer. Chem. Soc.* **85**, 1707 (1963).
167. D. B. R. Johnson, D. Taub, and T. B. Windholz, *Steroids* **8**, 365 (1966).
168. S. Danishefsky, P. Solomon, L. Crawley, M. Sax, S. C. Yoo, E. Abola, and J. Pletcher, *Tetrahedron Lett.* p. 961 (1972).
169. D. Bertin and J. Perronnet, *Bull. Soc. Chim. Fr.* [1] p. 117 (1969).
170. W. G. Dauben, G. Ahlgren, T J. Leitereg, W. C. Schwarzel, and M. Yoshioko, *J. Amer. Chem. Soc.* **94**, 8593 (1972).

Appendix
Supplementary References

Chapter 1 Designing Total Syntheses

Section 1.1

Total synthesis of heterocyclic steroidal systems. H. O. Huisman, *in* "Steroids," (W. Johns, ed.), MTP International Review of Science: Organic Chemistry, Vol. 8, p. 235. Butterworths, London, 1973.

Architecture of a molecule. K. Wiedhaup, *Chem. Weekbl.* **67**, s8 (1971).

Total synthesis of toad poisons. Recent advances in synthesis of bufadienolides. Y. Kamano, *Yushi Kagaku Kyokaishi* **31**, 212 (1973).

Chemistry of vitamin D and its related compounds. II. Total synthesis of vitamin D_3 and synthesis of vitamin D_3 active metabolites. T. Kobayashi, *Vitamins* **47**, 381 (1973).

The chemistry of steroidal contraceptives. P. D. Klimstra and F. B. Colton, *in* "Contraception. The Chemical Control of Fertility" (D. Lednicer, ed.), 100. Dekker, New York, 1969.

Section 1.3

Cycloalkanecarboxylic acid synthesis by intramolecular chloro olefin annelation. P. T. Lansbury and R. C. Stewart, *Tetrahedron Lett.* p. 1569 (1973).

α-Silylated vinyl ketones. A new class of reagents for the annelation of ketones. G. Stork and B. Ganem, *J. Amer. Chem. Soc.* **95**, 6152 (1973).

Conjugate addition–annelation. A highly regiospecific and stereospecific synthesis of polycyclic ketones. R. K. Boeckman, Jr., *J. Amer. Chem. Soc.* **95**, 6867 (1973).

Appendix: Supplementary References

Section 1.4

Enzyme-analogue built polymers and their use for the resolution of racemates. G. Wulff, A. Sarhan, and K. Zabrocki, *Tetrahedron Lett.* p. 4329 (1973).

Chapter 2 Biogenetic-like Steroid Synthesis

Section 2.2

Direct formation of the steroid nucleus by a nonenzymic biogenetic-like cyclization. Preparation of the cyclization substrate. R. L. Markezich, W. E. Willy, B. E. McCarry, and W. S. Johnson, *J. Amer. Chem. Soc.* **95**, 4414 (1973).

Direct formation of the steroid nucleus by a nonenzymic biogenetic-like cyclization. Cyclization and proof of structure and configuration of products. B. E. McCarry, R. L. Markezich, and W. S. Johnson, *J. Amer. Chem. Soc.* **95**, 4416 (1973).

Acetylenic bond participation in biogenetic-like olefinic cyclizations in nitroalkane solvents. Synthesis of the 17-hydroxy-5β-pregnan-20-one system. D. R. Morton, M. B. Gravestock, R. J. Parry, and W. S. Johnson, *J. Amer. Chem. Soc.* **95**, 4417 (1973).

Nonenzymic biogenetic-like olefinic cyclizations. Stereospecific cyclization of dienic acetals. A. van der Gen, K. Wiedhaup, J. J. Swoboda, H. C. Dunathan, and W. S. Johnson, *J. Amer. Chem. Soc.* **95**, 2656 (1973).

Acetylenic bond participation in biogenetic-like olefinic cyclizations in nitroalkane solvents. A facile total synthesis of *dl*-testosterone benzoate. D. R. Morton and W. S. Johnson, *J. Amer. Chem. Soc.* **95**, 4419 (1973).

A stereospecific total synthesis of estrone via a cationic olefinic cyclization. P. A. Bartlett and W. S. Johnson, *J. Amer. Chem. Soc.* **95**, 7501 (1973).

Concerning the mechanism of a nonenzymic biogenetic-like olefinic cyclization. P. A. Bartlett, J. I. Brauman, W. S. Johnson, and R. A. Volkmann, *J. Amer. Chem. Soc.* **95**, 7502 (1973).

Intermediates in the total synthesis of 16-dehydroprogesterone. W. S. Johnson, U. S. Patent 3,741,987 (1973); *Chem. Abstr.* **79**, 79061h (1973).

Total synthesis of steroids. W. S. Johnson, German Patent 2,234,018 (1973); *Chem. Abstr.* **78**, 159987j (1973).

Cyclisations radicalaires XXIV—Cyclisation oxydante du phenyl-13-trimethyl-2,6,10-tridecatriene-2,6,10 par le peroxyde de benzoyle. J. Y. Lallemand, M. Julia, and D. Mansuy, *Tetrahedron Lett.* p. 4461 (1973).

Chapter 3 AB → ABC → ABCD

Section 3.5

Synthese d'une trans-anti-trans perhydrobenz(e)indenone-2, un intermediaire potentiel dans la synthese des C-nor-D-homosteroides. E. Brown and M. Ragault, *Tetrahedron Lett.* p. 1927 (1973).

Section 3.6.3

Heterocyclic nitrogen compounds. VII. Heterocyclic steroids and analogous compounds. Synthesis of 11,14-diazaestra-1,3,5(10),6,8-pentaene derivatives. V. Nacci, G. Stefancich, G. Filacchioni, R. Giuliano, and M. Artico, *Farmaco, Ed. Sci.* **28**, 49 (1973).

Aza steroids. V. Introduction of 11-hydroxy and 11-amino groups. R. E. Brown, H. V. Hansen, D. M. Lustgarten, R. J. Stanaback, and R. I. Meltzer, *J. Org. Chem.* **33**, 4180 (1968).

Chapter 4 AB + D → ABD → ABCD

Section 4.2.2

Total synthesis of steroids. III. Reaction of 3-methoxy-8,14-secoestra-1,3,5(10),9(11)-tetraene-14,17-dione and its derivatives with 2,3-dichloro-5,6-dicyano-p-benzoquinone (DDQ). An improved method of preparation of racemic 14-dehydroequilenin methyl ether. A. R. Daniewski, M. Guzewska, and M. Kocor, *Bull. Acad. Pol. Sci., Ser. Sci. Chim.* **21**, 91 (1973).

Asymmetric synthesis of 19-norsteroids. R. Pappo, R. B. Garland, C. J. Jung, and R. T. Nicholson, *Tetrahedron Lett.* p. 1827 (1973).

Improved synthesis of optically active steroids. T. Asako, K. Hiraga, and T. Miki, *Chem. Pharm. Bull.* **21**, 107 (1973).

Total synthesis of estrone via estriol dimethyl ether. T. Asako, K. Hiraga, and T. Miki, *Chem. Pharm. Bull.* **21**, 697 (1973).

Secogonatetraen-14-ones. T. Miki and K. Hiraga, Japanese Patent 72/32,977 (1972); *Chem. Abstr.* **78**, 30089p (1973).

Racemic 3,17β-estradiol. T. A. Serebryakova, A. V. Zakharychev, M. A. Nekrasova, S. N. Ananchenko, and I. V. Torgov, French Patent 2,136,891 (1973); *Chem. Abstr.* **79**, 42755m (1973).

13-Alkylgona-1,3,5(10),8,14-pentaen-17-ones. H. Smith and G. A. Hughes, German Patent 1,593,441 (1973); *Chem. Abstr.* **78**, 84648p (1973).

Stereochemistry of the ethoxyethynyl carbinol-αβ-unsaturated ester conversion. M. H. Tankart and J. S. Whitehurst, *J. Chem. Soc., Perkin Trans.* **1**, 615 (1973).

Selektive elektrochemische Reduktionen in der Steroidreihe. K. Junghans, *Chem. Ber.* **106**, 3465 (1973).

Section 4.2.3

(±)-6,6-Difluoronorgestrel, a new synthetic hormonal steroid. A. L. Johnson, *J. Med. Chem.* **15**, 360 (1972).

Synthesis of 3-methoxy-6-methyl-B-norestra-1,3,5(10),8,14-pentaen-17β-ol and its 3-deoxy analog. S. R. Ramadas and K. P. Sujeeth, *Chem. Ind. (London)* p. 381 (1973).

Appendix: Supplementary References 289

(±)-3-Methoxy-13-allyl-8,14-secogona-1,3,5(10),9(11)-tetraene-14,17-dione. K. Yoshioka, T. Asako, K. Hiraga, and T. Miki, Japanese Patent 72/46,592 (1972); *Chem. Abstr.* **78**, 84649q (1973).

13-Carbocyclic 3-oxo-4-gonenes. A. A. Patchett and T. B. Windholz, U. S. Patent 3,725,439 (1973); *Chem. Abstr.* **79**, 32190j (1973).

Section 4.2.4

Darstellung und Strukturuntersuchung diastereomerer 4-oxa-steroide. G. Sauer, U. Eder, and G.-A. Hoyer, *Chem. Ber.* **105**, 2358 (1972).

Mechanistic studies on the total synthesis of heterocyclic steroids: Attempts of synthesis of 4,6-diazasteroids. J. J. Artús, J-J. Bonet, and A. E. Peña, *Tetrahedron Lett.* p. 3187 (1973).

13-Aminogonanes and their N-acyl and N-alkyl derivatives. T. B. Windholz, D. B. R. Johnston, and A. A. Patchett, U. S. Patent 3,714,208 (1973); *Chem. Abstr.* **78**, 111608j (1973).

Section 4.3

Bridgehead nitrogen heterocycles by imide reduction. J. C. Hubert, W. N. Speckamp, and H. O. Huisman, *Tetrahedron Lett.* p. 4493 (1972).

Darstellung und Strukturuntersuchung diastereomerer 4,8,14-Androstatrien-3,17-dione. J. Ruppert, U. Eder, and R. Wiechert, *Chem. Ber.* **106**, 3636 (1973).

Section 4.7

A Wittig reaction involving a novel rearrangement: Confirmation by X-ray crystallography. E. G. Brain, F. Cassidy, and A. W. Lake, *Chem. Commun.* p. 497 (1972).

Potentially carcinogenic cyclopenta[a]phenanthrenes. Part VI. 1,2,3,4-Tetrahydro-17-ketones. M. M. Coombs and T. S. Bhatt, *J. Chem. Soc., Perkin Trans.* 1, 1251 (1973).

Chapter 5 AB + D → ABCD

Section 5.2

Darstellung von hexahydrochrysen-derivaten. G. Stubenrauch, K. Reiff, U. Schumacher, and W. Tochtermann, *Tetrahedron Lett.* p. 1549 (1973).

Benzologe cyclohexadien-1,2-diol-derivative. K. Reiff, U. Schumacher, G. Stubenrauch, and W. Tochtermann, *Tetrahedron Lett.* p. 1553 (1973).

Section 5.3

Phosphasteroids. I. Cycloaddition reactions of phospholenes with various dienes. Y. Kashman and O. Awerbouch, *Tetrahedron Lett.* p. 3217 (1973).

Thiocarbonyl ylides. Photogeneration, rearrangement and cycloaddition reactions. A. G. Schlotz and M. B. DeTar, *166th Nat. Meet., Amer. Chem. Soc., Chicago* ORGN 106 (1973).

7-Azasteroid analogues. I. S. F. Dyke, P. A. Bather, A. B. Garry, and D. W. Wiggins, *Tetrahedron* **29**, 3881 (1973).

Chapter 7 A + CD → ACD → ABCD

Section 7.1

Lactones for steroid synthesis. V. Torelli and M. Vignau, French Patent 2,145,812 (1973); *Chem. Abstr.* **79**, 79048j (1973).

Section 7.2

Ethylenedioxy derivatives of substituted naphthalenone compounds. M. Los, U. S. Patent 3,714,195 (1973); *Chem. Abstr.* **78**, 148131z (1973).

D-Homoestra-1,3,5(10)-trienes and D-homo-19-norpregna-1,3,5(10)-trien-20-ynes. M. Los, U. S. Patent 3,723,533 (1973); *Chem. Abstr.* **79**, 5515n (1973).

Steroid-like compounds. M. Los, U. S. Patent 3,736,345 (1973); *Chem. Abstr.* **79**, 42756n (1973).

Total synthesis of steroids. IV. New Path of total synthesis of 19-norsteroidal system. A. R. Daniewski, *Bull. Acad. Pol. Sci., Ser. Sci. Chim.* **21**, 17 (1973).

8-Methoxycarbonyl-3,17-dioxo-13-methyl-1-gona-4,9-diene. K. Sakai and S. Amemiya, Japanese Patent 73/02,529 (1973); *Chem. Abstr.* **78**, 148128d (1973).

Section 7.6

Azasteroids. X. Synthesis of 3,4-dihydro-8-methoxybenzo[c]phenanthridin-1(2H)-one. S. V. Kessar, N. Parkash, and G. S. Joshi, *J. Chem. Soc., Perkin Trans.* 1, 1158 (1973).

Synthesis of the methyl ether of *dl*-6-azaestradiol. E. V. Popova and G. S. Grinenko, *Zh. Obshch. Khim.* **43**, 668 (1973).

Synthesis of quinolizine derivatives. XXV. Azasteroids. III. Synthesis of 17-acetoxy-10-azaandrostan-7-one and related compounds. M. Akiba and S. Ohki, *Chem. Pharm. Bull.* **21**, 40 (1973).

Intramolecular Diels-Alder reactions: Construction of aza- and diazasteroid type skeletons. H. W. Gschwend, *Helv. Chim. Acta* **56**, 1763 (1973).

Chapter 8 A + D → AD → ABCD

Section 8.1

Synthesis of some novel 16-methyl analogs of estrone and 19-nor testosterone. D. K. Banerjee and P. R. Srinivasan, *Indian J. Chem.* **10**, 891 (1972).

Steroid total synthesis. X. Optically active estrone precursors and racemic equilenin methyl ether. N. Cohen, B. L. Banner, J. F. Blount, M. Tsai, and G. Saucy, *J. Org. Chem.* **38**, 3229 (1973).

Experiments directed toward the total synthesis of terpenes. XVIII. The convergent, stereoselective total synthesis of the unsymmetrical pentacyclic triterpene alnusenone. R. E. Ireland, M. I. Dawson, S. C. Welch, A. Hagenbach, J. Bordner, and B. Trus, *J. Amer. Chem. Soc.* **95**, 7824 (1973).

Appendix: Supplementary References

Section 8.3

Synthesis of diazasteroids. II. Synthesis of 2,3-dimethoxy-8,11-diazagona-1,3,5(10),9(11)-tetraene. M. Nagata, M. Goi, K. Matoba, T. Yamazaki, and R. M. Castle, *J. Heterocyl. Chem.* **10**, 21 (1973).

Chapter 9   BC → ABC → ABCD

Section 9.5.4

Synthesis Aldosteron-ähnlicher corticosteroide. Racemisches 11-dehydroaldosteron (rac.-18-oxo-11-dehydrocorticosteron). J. Schmidlin, *Helv. Chim. Acta* **54**, 2460 (1971).

Chapter 10   BC → BCD → ABCD

Section 10.1

Steroidal analogs of unnatural configuration. VII. Total synthesis of 17β-hydroxy-9-methyl-9β-10α-estr-4-en-3-one, a novel retrotestosterone isomer. J. R. Bull and A. Tuinman, *Tetrahedron* **29**, 1101 (1973).

Section 10.3

Gonatriene derivatives. J. Warnant and J. Jolly, German Patent 2,212,589 (1972); *Chem. Abstr.* **78**, 30085j (1973).

Section 10.8

Gon-9-enes and gona-9,11-dienes, J. Prost-Marechel and G. Tomasik, French Patent 2,125,640 (1972); *Chem. Abstr.* **78**, 136529s (1973).

Chapter 11   B+D → BD → BCD → ABCD

A,B-Indolosteroids, IV. Synthesis of *trans*-3a-methyl-1,2,3a,4,5,9b-hexahydro-3*H*-benz[*e*]indene-3,5-dione, a tricyclic starting compound. V. N. Sladkov, V. F. Shner, O. S. Anisimova, V. P. Pakhomov, N. S. Kuryatov, and N. N. Suvorov, *Zh. Org. Khim.* **9**, 810 (1973).

Chapter 12   CD → BCD → ABCD

Section 12.1

The stereocontrolled synthesis of *trans*-hydrindan steroidal intermediates. Z. G. Hajos and D. R. Parrish, *J. Org. Chem.* **38**, 3239 (1973).

Section 12.6

New total synthesis of 19-norsteroids. G. Saucy, M. Rosenberger, and R. Borer, *Proc. Int. Congr. Horm. Steroids, 3rd, 1970.* p. 117 (1971).

Section 12.7

Stereocontrolled total synthesis of 19-nor steroids. Z. G. Hajos and D. R. Parrish, *J. Org. Chem.* **38**, 3244 (1973).

Benz[e]indenes. R. A. Micheli, French Patent 2,129,392 (1972); *Chem. Abstr.* **78**, 136,531s (1973).

Section 12.8

Bicyclic polyketone intermediates. G. Saucy, U. S. Patent 3,714,262 (1973); *Chem. Abstr.* **78**, 111613g (1973).

Author Index

Numbers in parentheses are reference numbers and indicate that an author's work is referred to although his name is not cited in the text. Numbers in italics show the page on which the complete reference is listed.

A

Abola, E., 277 (168), *285*
Abraham, N. A., 260 (101, 102), *283*
Abraham, S., 35 (2), *46*
Abramson, H. N., 169 (46, 47), 170 (47), *174*
Acevedo, R., 218 (93), *231*
Achini, R. S., 39 (24), *47*
Ackerman, J., 10 (46), 22 (46, 105), *30, 32*, 244 (43), 245 (43, 45), 246 (43, 45), 247 (47), 249 (59), *280, 281*
Acklin, W., 241 (18), 242 (18, 25) *280*
Adam, G., 93 (77), *121*
Ahlgren, G., 279 (170), *285*
Akalaer, A. N., 152 (41), *159*
Akhrem, A. A., 1, *29*, 71 (157), *83*, 110 (159), 117 (196), *124, 125*
Akiba, M., *290*
Alberola, A., 156 (54), *160*
Albrum, H. E., 161 (12), 166 (12), *173*
Alburn, H. E., 27 (134), *34*
Alderova, E., 138 (22, 23), *141*
Aleksandrova, G. V., 104 (117), *123*
Alekseeva, L. M., 237 (23), *238*, 243 (38), *280*
Allen, C. F., 111 (170), *124*, 235 (16), *238*
Allen, D. S., Jr., 16 (74), *31*, 248 (54), 249 (56), *281*
Allen, W. M., 208, *230*
Al'Safar, N., 92 (62), *121*
Ambelang, T., *14*
Amemiya, S., 150, *159, 290*
Amiard, G., 25 (119), *33*, 204 (19), *229*, 243 (35, 37), 274 (35, 37), 278 (35), *280*
Ananchenko, S. N., 9 (38), 12 (59), 19 (59), 22 (59), *30, 31*, 58 (53), *80*, 87 (1, 5, 9, 10, 13, 17), 88 (1, 20, 21), 89 (1, 5, 22, 27, 28, 29, 30), 90 (1, 9, 10, 20, 22, 35, 36, 38, 39, 40, 41), 91 (10, 20, 44, 45, 46, 47, 48, 49, 52, 53, 54, 55, 56), 92 (30, 61, 62), 104 (117, 118), *118, 119, 120, 121, 123*, 131 (27), *132*, 220 (109), 222 (109), *232*, 234 (6), *238, 288*
Anderson, B. A., 274 (151), *284*
Anderson, J. M., 66 (120), *82*, 249 (60), *281*
Anderson, R. J., 36 (8), 39 (8, 24), 40 (8), *47*
Andrews, D., 267 (130), *283*
Anisimova, O. S., 152 (41), *159*, 237 (23), *238, 291*
Anner, G., 2 (18, 24), 4, 5, 25 (120), *29, 33*, 58, 60, 61, 63 (87, 90), 65, *80, 81*, 129, (3), *131*, 187, 189 (61, 67, 74, 75, 76), 192 (66, 67, 68, 79, 80), 194 (99, 100) *198, 199, 200*, 253 (77, 78, 79), 254 (79), 274 (149, 152, 153) *282, 284*
Aoki, T., 10 (43), *30*, 194 (107, 108), *200*, 248 (55), 255 (55, 89), 256 (89), *281, 282*
Ap Simon, J. W., 13 (63), *31*
Archer, S., 43 (40), 44 (40), 45 (40), *48*
Arigoni, D., 35 (5), 39 (5), *47*, 194 (99, 100, 103), *200*
Arison, B., 94 (81), 101 (81), *122*
Arth, G. E., 11 (52), 25 (120), *31, 33*, 65 (113), *82*, 182 (27, 28, 33, 34, 36, 37, 39), 184 (33), 185 (37, 39), 186 (33, 34), *197, 198*
Artico, M., 107 (151), *124, 288*
Artús, J. J., *289*
Arunachalan, T., 71 (168), 72 (168), *83*
Asako, T., 21 (99), *32*, 87 (7, 14), 88 (7, 14), 90 (14), 92 (66, 70), 93 (76), *118, 119, 121, 288, 289*
Awarebouch, O., *289*
Axelrod, E. H., 39 (28), *47*
Axelrod, L. R., 71 (156), *83*, 93 (74, 75), *121*
Ayer, D. E., 275 (158), *284*

B

Babbe, D., 64 (103), *81*, 99 (98), *122*
Babcock, J. C., 213 (72), *231*
Baburao, K., 107 (147), 108 (147), *124*, 158 (60), *160*
Bachmann, W. E., 2 (17), 4, 8 (17), 13 (60), 16, 17 (17, 87), 18 (17), 26 (17), *29, 31, 32*, 49, 51, 53, 55 (22), 58, 60 (22), 61, 65 (22), 59 (1), 78 (2), 79, 89 (26), 116 (186), 117 (200), *119, 125*, 201, 203 (3), *228*, 237 (21), *238*
Baeur, V. J., 1 (3), *29*
Bagchi, P., 146, *159*
Baggaley, K. H., 243 (34), *280*
Bagli, J. F., 71 (161), *83*
Baily, D. M., 41 (31), *47*, 251 (68), 257 (68), 258 (68, 97), *281, 282*
Baker, P., 13 (63), *31*
Balasubramanian, S. K., 16 (78), *31*, 203 (13, 14), 218 (13, 96), *229, 232*
Banerjee, A., 180, *197*
Banerjee, D. K., 9 (41), 15 (71), 16 (78), *30, 31*, 53, 54 (20), 58 (54, 55, 62), 61 (54), 79, *80*, 91 (43), 111 (170), 115 (181, 182), *120, 124, 125*, 129 (4), *131*, 136 (13, 14), *140*, 143 (1, 3), 144 (3), 146, *158, 159*, 201, 202, 203 (1, 13, 14), 215 (87), 218, 219 (87, 101, 102), 220 (101), 221, 222 (101, 102), *228, 229, 231, 232*, 235 (16), 237 (20), *238*, 239 (2), 248 (50), 249 (57), *279, 281, 290*
Banerjee, S., 70 (153, 154), *83*
Banes, D., 89 (23), *119*
Banner, B. L., 267 (131, 134), 268 (131, 134), 271 (131), *283, 284, 290*
Bannister, B., 2 (23), 5, 10 (46), 22 (46), *29, 30*, 244 (40), 245 (45), 246 (45), 247 (48), 248 (48, 52), 251 (52, 61), *280, 281*
Barcza, S., 10 (50), *30*, 102 (106), *122*
Bardhan, J. C., 51, 79, 103 (109), *122*, 138 (18), *141*
Bardoneschi, R., 194 (94), *200*
Barends, R. J. P., 76 (181), 77 (182), *83, 84*

Barkley, L. B., 12 (56), *31*, 223 (115,) *232*, 262 (109), 263, 264 (116), 265 (116), *283*
Barnes, R. A. 103 (108), *122*, 139, *141*, 158 (63), *160*
Bartlett, P. A., *287*
Bartlett, W. R., 42 (35), 43 (41), 44 (41), 45 (41, 49), *47, 48*
Bartmann, W., 251 (72), *282*
Barton, D. H. R., 2 (22, 29), 3 (22, 33), 5, *29, 30*, 67 (126), 68, *82*, 194 (101, 102, 106), *200*
Baty, J. D., 77 (193), *84*
Bauer, V. J., 251 (66), *281*
Bayer, H., 103 (111), 111 (111), *122*
Beachem, M. T., 139 (36), *141*
Beal, P. F., 187 (58a), *198*, 206 (35), *230*, 254 (80), *282*
Beaton, J. M., 194 (101, 102, 106), *200*
Beck, L. W., 56, *79*
Bell, R. A., 41 (31), *47*
Bellet, P., 27 (133), *33*
Bendas, H., 218 (97), *232*
Benington, F., 78 (195), *84*
Benson, H. D., 78 (198), *84*
Bergel'son, L. D., 131 (26, 30, 37), *132, 133*
Berges, O., 2 (25), 5, 19 (25), 25 (25), *29*, 242 (28), *280*
Bergmann, E., 116 (187, 189), *125*
Bergmann, F., 116 (189), *125*
Bergmann, W., 116 (187), *125*
Berlin, K. D., 49 (9) 55 (23), 78 (23), *79*, 118 (204), *125*
Berndt, H. D., 16 (76), *31*
Bernstein, S., 215 (80, 81, 82), 220 (104), *231, 232*
Bertin, D., 58 (66), *80*, 172 (61), *175*, 201 (8), 203 (8), 204 (8, 17), 206 (29, 36), 207 (45), 208 (45), 214 (79), 228 (127), *229, 230, 231*, 233, 278 (169), *285*
Beslin, P., 68, *82*
Betrus, B. J., 169 (42), *174*
Beverung, W. N., 109 (153, 154), *124*, 158 (61), *160*

Author Index

Beyler, R. E., 11 (52), 25 (120), *31*, *33*, 65 (113), *82*, 182 (27, 28, 29, 31, 32, 33, 34), 184 (33), 186 (33, 34), *197*, 213 (74), *231*
Bhatt, M. V., 16 (78), *31*, 53 (18), *79*, 201 (1), 202 (1), 203 (1), *228*
Bhatt, T. S., *289*
Bhattacharjee, M. K., 73, 74 (171, 175), *83*
Bhattacharyya, B. K., 10 (46), 20 (94), 22 (46), *30*, *32*, 63 (96), *81*, 111 (170), *124*, 136 (7, 8, 9), *140*, 180 (21, 22), *197*, 235 (16), *238*, 245 (45), 246 (45), 259 (100), *281*, *282*
Bhattacharyya, P. K., 194 (98), *200*
Bhide, G. V., 71 (165, 166), 77 (188, 189), *83*, *84*, 139 (42), *141*, 158 (62), *160*
Bialy, G., 154 (48), *160*
Biellmann, J. F., 37 (19), *47*
Bierwagen, M. E., 154 (48), *160*
Billeter, F., 178, *197*
Billeter, J. R., 2 (24), 5, 25 (120), 29, *33*, 63 (85, 86, 89), *81*, 181, 187, 189 (61, 67, 68, 74, 75, 76), 192 (66, 67, 68, 79, 80), *197*, *198*, *199*, 253 (77, 78, 79), 254 (79), *282*
Bindseil, A. W., 131 (19), *132*
Biollaz, M., 187 (58c), *198*
Birch, A. J., 22, *33*, 62, 63, 77, *81*, *84*, 93 (78a), 104 (116), 105 (123, 128, 129), 106 (133, 134), 107 (140), 113 (173), 114 (174), *121*, *123*, *125*, 137 (16, 17), *141*, 146, 147 (16, 17), 150 (30), *159*, 171 (52), *174*, 178 (16), *197*, 222, 227 (122), *232*, *233*, 236 (19), *238*, 240 (6), 241 (16), *279*, *280*
Biswas, K. K., 67 (129), *82*
Blank, B., 194 (95, 96), *200*
Blanke, E., 55 (24), *79*
Bloch, K., 35 (1), *46*
Bloom, B. M., 10 (46), 22 (46), *30*, 244 (40), 245 (45), 246 (45), 247 (49), 248 (49), *280*, *281*
Blount, J. F., *290*
Bluestein, B. R., 139 (30, 32), *141*
Boeckman, R. K., Jr., 13, *286*
Böhme, E., 274 (145), *284*

Bogdanowicz, M. J., 14 (70), *31*
Bohme, E., 99 (99), *122*
Bolton, I. J., 242 (31), *280*
Bonet, J-J., *289*
Boots, S. G., 58 (60), 61 (60), *80*
Bordner, J., *290*
Borer, R., 13 (66), 27 (66), *31*, 267 (127, 128, 131, 133), 268 (131, 135, 136), 270 (135), 271 (131, 136), 272 (133), *283*, *284*, *292*
Bose, A. K., 259 (100), *282*
Bottari, F., 117 (199), *125*
Bowers, A., 63 (83), *81*, 218 (93), *231*
Bowman, R. E., 253 (75), *282*
Boyce, C. B. C., 147 (23, 24), *159*, 241 (15, 19), 242 (24), *280*
Boyle, P. H., 25 (118), *33*
Bradlow, H. L., 35 (2), *46*
Brady, S. F., 8 (36), *30*, 42 (36), 43 (42), *47*, *48*
Brain, E. G., *289*
Brauman, J. I., 227 (124), *233*, *287*
Briggs, P. C., 226 (119), *232*
Brizzolara, A., 206 (30), *229*
Brocksom, T. J., 45 (49), *48*
Brooks, S. G., 243 (34), *280*
Brown, E., 69, *82*, *83*, *287*
Brown, E. V., 73, *83*
Brown, H. C., 14 (69), *31*, 242 (33), *280*
Brown, J. J., 215 (80, 81), *231*
Brown, J. N., 107 (146), *124*, 166 (36), 168 (36), *174*
Brown, O. H., 171 (58), *175*
Brown, R. D., 92 (65), 94 (81), 101 (81), *121*, *122*
Brown, R. E., 107 (145), *124*, 166 (30, 31, 32, 33, 34, 35, 36), 167 (32, 33), 168 (30, 33, 34, 35, 36), *174*, *288*
Brownfield, R. B., 26 (124), 27 (124), *33*
Brutcher, F. V., Jr., 262 (110), *283*
Bry, T. S., 213 (74), *231*
Bryson, T. A., 45 (46), *48*
Buchschacher, P., 194 (104), *200*, 274 (155), *284*
Buchta, E., 103 (111), 111 (111), *122*
Bucourt, R., 27 (125), *33*, 165 (29), *174*, 201 (10), 204 (18, 21), 206 (29, 37), 207 (41, 42, 43, 44, 45), 208 (42, 44, 45, 49, 53, 54, 55), 209 (59), 210 (61,

62, 63, 65), 212 (69), 213 (41, 59, 70, 73), 214 (59, 75, 76), 220 (105, 106, 107, 108), 222 (107, 108), *229, 230, 231, 232*
Bukwa, B., 56, *80*
Bull, J. R., 222, 232, *291*
Burckhalter, J. H., 93 (79), *121*, 139 (44, 45), *141*, 169 (46, 47), 170 (47), *174*
Burckhardt, U., 3 (34), *30*, 69 (150), *83*, 251 (69), *281*
Burgstahler, A. W., 35 (6), 36 (6), 39 (6), *47*
Burmistrova, M. S., 20 (97), *32*, 131 (31), *132*
Burnop, V. C. E., 19 (90), *32*, 55 (31), *79*
Butenandt, A., 49, *79*
Butz, E. W. J., 171 (54), *175*
Butz, L. W., 131 (32, 35), *132*, 171 (54, 55, 56, 57), *175*
Buu-Hoï, N. P., 17 (84), *32*, 111 (169), *124*
Buzby, G. C., Jr., 25 (117), 26 (121), *33*, 87 (8), 91 (50), 92 (8), *119, 120*, 161 (3, 5, 9, 10, 13, 16, 18, 23), 162 (9), 163 (5), 164 (10), 165 (5, 23), 166 (5, 16), *173, 174*, 220, *232*
By, A., 254, 255 (84), 256 (84), 257 (95), *282*

C

Cable, J., 258 (98), *282*
Cainelli, G., 194 (99, 100, 103), *200*
Cameron, D. D., 144 (6), 145 (7), *159*
Campbell, J. A., 213 (72), *231*
Campos-Neves, A. S., 67 (126), 68 (126), *82*
Cantrall, E. W., 26 (124), 27 (124), *33*
Capaldi, E., 161 (10), 164 (10), *173*, 220 (111), *232*
Carboni, S., 117 (194, 195, 199), *125*
Cardwell, H. M. E., 2 (20), 3 (20), 4, 16 (20), *29*, 176 (1, 2), *196*, 244 (42), *280*
Carnero, P., 64 (102), 65 (102), *81*
Carol, J., 89 (23), *119*
Carpio, H., 63 (83), *81*
Casey, A. C., 17 (82), *32*, 111 (168), *124*
Cassidy, F., *289*

Castle, R. M., *291*
Cavanaugh, R., 150 (31), *159*
Cérêde, J. 204 (19), 229, 243 (35), 274 (35), 278 (35), *280*
Cereghetti, M., 194 (104, 105), *200*
Chaikoff, I. L., 35 (2), *46*
Chandhuri, N. K., 241 (12), *280*
Chang, C., 17 (88), *32*, 55 (29, 30), *79*
Charney, W., 28 (139), *34*
Chatterjee, A., 70, *83*, 196, *200*, 259 (100), *282*
Chatterjee, S., 16 (78), *31*, 53 (18), *79*, 201 (1), 202 (1), 203 (1),*228*
Chaudhuri, A. C., 115 (179), *125*
Chauvin, M. C. R., 39 (24), *47*
Chaykovsky, M., 265 (121), *283*
Chen, C. H., 118 (204), *125*
Cherkasov, A. N., 27 (132), *33*
Chesnut, R. W., 49 (9), 55 (23), 78 (23), *79*
Cheung, H. C., 39 (27), *47*
Chevallier, J., 64 (102), 65 (102), *81*
Chidester, C. G., 90 (37), *120*
Chinn, L. J., 12 (54), 15 (71), 25 (54), *31*, 58 (55), 61 (55), *80*, 91 (43), *120*, 129 (4), *131*, 143 (2, 3), 144 (3), 146 (8), *158, 159*, 203 (12), 204 (12), 220 (12), 229, 248 (50), *281*
Christiansen, R. G., 17 (80), 20 (95), *32*, 58 (56, 57), 61 (56, 57), *80*, 135 (2, 6), 136 (2, 6), 137 (2), *140*
Christmann, K. F., 45 (50), *48*
Christy, M. E., 117 (201), *125*
Chuang, C. K., 110 (162, 163, 164), *124*
Chuma, K., 204 (23), *229*
Clarke, R. L., 43 (40), 44 (40), 45 (40), *48*
Clarkson, R., 107 (142), *123*
Clayton, R. B., 37 (21), 39 (23), *47*
Clement, R. A., 10 (46), 22 (46), *30*, 244 (44), 245 (45), 246 (45), 252 (73), *280, 281, 282*
Clemo, G. R., 17 (81), *32*, 111 (167), *124*
Close, W. J., 49 (6), 54, 56 (6), 63 (6), *79*
Cocker, J. D., 241 (14), *280*
Codderre, R. A., 58 (67), 61 (67, 70, 72), *81*
Cohen, A., 2 (16), *29*, 49 (4, 5), *79*, 103 (112), *123*
Cohen, L. A., 61 (72), *81*

Cohen, M. P., 110 (158), *124*
Cohen, N., 2, *29*, 257 (96), 258 (96), 267 (131), 268 (131), 271 (131), *282*, *283*, *290*
Cole, J. E., Jr., 16 (73), *31*, 58 (58), 61 (58), *80*, 127 (2), 129 (2), *131*
Cole, W., 2 (17), 4, 8 (17), 16 (17), 17 (17), 18 (17), 26 (17), *29*, 49, 53, 55 (2), 59 (1), 63 (82), 78 (2), *79*, *81*
Collins, J. C., Jr., 111 (170), *124*, 187 (62a), *198*, 235 (16), *238*, 251 (70, 72), *281*, *282*
Collins, J. F., 146 (13), *159*
Collins, R. J., 73, *83*
Colton, F. B., 213 (71), *231*, *286*
Combs, R. V., 157 (57), *160*
Conia, J.-M., 68, *82*
Constantin, J. M., 25 (120), *33*
Cook, C. E., 218 (95), *232*
Cook, J. W., 2 (16), *29*, 49 (4, 5), *79*, 103 (112), *123*
Cookson, R. C., 2 (29), *30*
Coombs, M. M., *289*
Coombs, R. V., 276 (163), *284*
Corey, E. J., 2 (26), 5, *29*, 37 (20), 39, *47*
Cornforth, J. W., 2 (20), 3 (20), 4, 11 (53), 16 (20), 26 (53), *29*, *31*, 66 (119), 67 (127), *82*, 139 (28, 37, 38), *141*, 176, 177 (9, 10, 11), 178, *196*, *197*, 240 (5), 244 (42), *279*, *280*
Cornforth, R. H., 66 (119), *82*, 177 (10), *197*
Corrall, C., 156 (51, 52), *160*
Costerousse, G., 209 (59), 213 (59), 214 (59), *230*
Cottard, A., 64 (102), 65 (102), *81*
Coverdale, C. E., 3 (34), *30*, 69 (150), *83*, 251 (69), *281*
Cowell, D. B., 227 (123), *233*
Cox, J. M., 3 (34), *30*, 69 (148), *83*, 251 (70), 258 (70), *281*
Crabbé, P., 63, 65, *81*, *82*, 187 (58c), *198*
Crandall, J. K., 41 (31, 32), *47*
Crawley, L. S., 277 (165, 168), *285*
Crenshaw, R. R., 74, *83*, 154 (47, 48), 155 (47), *160*

Crispin, D. J., 58 (65), *80*, 87 (6), 88 (6), 104 (121), *118*, *123*, 147 (18, 19, 20), 148 (20), *159*, 163 (26), *174*, 242 (23), *280*
Cross, A. D., 218 (93), *231*
Cross, P. E., 118 (206, 208), *125*
Crowfoot, D. M., 107 (137), *123*
Cruickshank, P. A., 58 (61, 67), 61 (61, 67), *80*, *81*
Cruz, A., 63 (77), 65 (109), *81*, *82*
Curphey, T. J., 37 (15), *47*
Curtis, P. J., *34*

D

Damodaran, K. M., 99 (100), *122*, 215 (87), 219 (87), *231*
Dane, E., 129 (6, 8, 9), *131* (15, 17, 19, 20, 29), *132*
Danieli, N., 2 (27, 28), 3 (27, 28), 6, *30*, 110 (166), 115 (166), *124*, 180 (18), *197*
Daniels, G. H., 117 (197), *125*
Daniels, R., 2 (21), 4, *29*, 56 (40), *80*, 181 (26), *197*
Daniewski, A. R., 12 (55), *31*, *288*, *290*
Danishefsky, S., 12 (58), *31*, *150*, *159*, 272 (139), 275 (139), 277 (164, 165, 168), *284*, *285*
Dann, O., 98 (97), *122*
Darling, S. D., 66 (122), *82*, 224 (117), *232*, 236 (18), *238*, 240 (4), 252 (73), *279*, *282*
Dauben, W. G., 35 (2), *46*, 279 (170), *285*
Daum, S. J., 43 (40), 44 (40), 45 (40), *48*
David, I. A., *31*, 58 (59), 59 (59), 61 (59), *80*, 144 (5), *159*
Davidson, R. S., 151 (37, 38), *159*
Davies, W., 131 (44), *133*
Davis, R. E., 171 (54), *175*
Davis, S. B., 22 (104), *32*, 219 (100), *232*
Dawson, M. I., 44 (44), *48*, *290*
Dawson, T. M., 144 (4), *158*, 242 (26), *280*
de Gee, A. J., 76 (181), *83*
Dehm, H. C., *31*, 58 (59), 59 (59), 61 (59), *80*, 144 (5), 146 (8), *159*
de Jongh, H. A. P., 3 (34), *30*, 69 (150), *83*, 251 (69), *281*

de Jonge, K., 74 (177), 83, 110 (160), 124, 154 (46), 160
de Koning, H., 21 (101), 32, 77 (190), 84, 94 (87), 122
del Key, A., 156 (54), 160
Demaecker, J., 265, 283
de Martino, G., 107 (151), 124
Demmin, T. R., 226 (119), 232
Denot, E., 218 (93), 231
Desaulles, P., 16 (75), 31, 187 (64), 192 (64), 198
Detar, M. B., 289
Devasthale, V. V., 224 (118), 232
de Waard, E. R., 74 (178), 83, 158 (59), 160
DeWalt, H. A., Jr., 22 (105), 32, 244 (43), 245 (43), 246 (43), 280
Dhage, S. P., 194 (98), 200
Diaz, J., 131 (38, 39), 133
Dickinson, R. A., 129 (11), 132
Dietrich, R., 158 (64), 160
DiMaria, F., 267 (130), 283
Dimroth, K., 146, 159
Dixon, J., 144 (4), 158
Djerassi, C., 3 (31), 23, 30, 33, 63 (83), 81, 178, 196 (117), 197, 200, 208 (57, 58), 218, 230, 232
Doering, W. E., 22 (104), 32, 219 (100), 232
Dolak, L. A., 43 (39), 44 (39), 45 (39), 48
Dolar, J., 218 (95), 232
Domagk, G. F., 242 (29), 280
Douglas, G. H., 10 (47), 22 (107), 25 (117), 30, 32, 33, 87 (2, 8, 15), 88 (2), 89 (a, 32), 90 (2), 91 (50), 92 (8, 32), 93 (15, 73), 94 (86), 95 (86), 98 (86), 118, 119, 120, 121, 122, 147 (22), 151 (22), 159, 161 (3, 4, 5, 8, 9, 10, 14, 18), 162 (4, 9), 163 (4, 5), 164 (4, 10), 165 (4, 5), 166 (5), 173, 174, 220 (111), 232
Dreiding, A. S., 205 (27), 229
Dryden, H. L., Jr., 12 (54), 25 (54), 31, 203 (12), 204, (25), 220 (12), 229
DuBois, G. E., 226 (119), 227 (120), 232
Dubois, R. J., 17 (82), 32, 111 (163), 124
Dubuis, R., 274 (155), 284
Ducep, J. B., 37 (19), 47

Duchamp, D. J., 90 (37), 120
Dufay, P., 208 (49), 230
du Feu, E. C., 11 (51), 30
Duff, S. R., 2 (20), 3 (20), 4, 16 (20), 29, 176 (1, 2), 196, 244 (42), 280
Duggan, A. J., 267 (130), 268 (136), 271 (136, 138), 283, 284
Duisberg, C., 156 (55), 160
Dunathan, H. C., 42 (34), 47, 287
Duncan, G. W., 213 (72), 231
Durham, N. N., 49 (9), 55 (23), 78 (23), 79
Dutta, J., 111 (170), 124, 235 (16), 238, 244 (44), 280
Dutta, P. C., 237 (22), 238

E

Eastham, J. F., 22 (105), 32, 244 (43), 245 (43), 246 (43), 280
Eberhardt, H., 158 (64), 160
Eder, K., 131 (15, 20), 132
Eder, U., 27 (126), 33, 98 (95), 122, 163, 174, 266 (124), 283, 289
Edgren, R. A., 25 (117), 33, 87 (8), 91 (50), 92 (8), 119, 120, 161 (3, 5), 163 (5), 165 (5), 166 (5), 173
Edwards, B. E., 71 (156), 83
Edwards, J. A., 23, 33, 64 (103), 81, 99 (98, 99), 122, 274 (145), 284
Eggerichs, T. L., 218 (89), 231
Eglington, G., 18 (89), 32, 115 (180), 125
Ehmann, L., 103 (111, 113), 111 (111), 122, 123
Ehrenstein, M., 208, 230
Elliott, G. H., 19 (90), 32, 55 (31), 79
Emmert, D. E., 90 (37), 120, 227 (125), 233
Eppstein, S. H., 28 (138), 34, 187 (58b), 198
Erhart, K., 154 (46), 160
Erman, W. F., 58 (61), 61 (61), 80
Eschenmoser, A., 35, 36 (7), 39 (5, 7), 40 (7), 47, 234 (7), 235 (7), 238
Eudy, N. H., 218 (92), 231
Euw, J. V., 187 (59, 60), 189 (73), 198, 199
Evers, H., 157 (58), 160

F

Farkas, E., 210 (60), 230
Farrar, M. W., 223 (115), 232, 262 (109), 263 (109, 114, 115), 283
Faulkner, D. J., 45 (49), 48, 257 (96), 258 (96), 282
Fedorova, O. I., 274 (148), 284
Feldman, J., 171 (56), 175
Felix, D., 36 (7), 47
Feurer, M., 182 (36), 197
Fieser, L. F., 2 (15), 7 (15), 23, 29
Fieser, M., 2 (15), 7 (15), 23, 29
Filacchioni, G., 107 (151), 124, 288
Fisher, J., 25 (117), 33, 91 (50), 120, 161 (5), 163 (5), 165 (5), 166 (5), 173
Fitzi, K., 42 (37), 48
Fleischer, E. B., 260 (102), 283
Foell, T., 25 (117), 27 (134), 33, 34, 91 (50), 120, 161 (5, 11, 12), 163 (5), 165 (5), 166 (5, 11, 12), 173
Folkers, K., 187 (55), 198
Foltz, R. L., 227 (124), 233
Fornefeld, E. J., 210 (60), 230
Fraher, T. P., 270 (137), 284
France, D. J., 22 (106), 32, 148 (26, 27), 159
Franck, R. W., 1 (3), 29, 251 (66), 281
Freed, J. H., 39 (26), 47
Fried, J., 28 (136), 34, 206 (33), 220 (33), 229, 260 (101, 103), 283
Fried, J. H., 23, 33, 64 (103), 81, 87 (3, 11) 88 (3), 89 (3), 99 (98, 99), 105 (130), 111, 119, 122, 123, 187 (58c), 198, 213, 231, 274 (145), 277 (166), 284, 285
Friedmann, C. A., 68, 82, 241 (9), 279
Friedrich, G., 242 (30), 280
Froelich, P. M., 17 (82), 32, 111 (168), 124
Frumkis, H., 110 (166), 115 (166), 124
Fürst, A., 267 (127, 129), 283
Fung, V. A., 42 (35), 47

G

Gabisch, E. W., 116 (188), 125
Gaddis, A. M., 131 (32, 35), 132, 171 (54), 175
Gadsby, B. W., 25 (117), 26 (121), 33, 87 (8), 91 (50), 92 (8), 119, 120, 161 (3, 5, 10, 16, 22), 163 (5), 164 (10), 165 (5), 166 (5, 16), 173, 174, 220 (111), 232
Gaidamovich, N. N., 12 (57), 31, 91 (58), 105 (58, 126, 127), 121, 123, 234 (4), 238
Galantay, E., 93 (78), 121
Gamble, D. J. C., 103 (114), 123
Ganem, B., 286
Gardner, J. N., 274 (151, 155), 284
Gardner, J. O., 13
Garland, R. B., 65, 66 (111), 82, 227 (124), 233, 288
Gasc, J. C., 16 (77), 27 (125), 31, 33, 213 (70), 215 (83, 85), 220 (107, 108), 222 (107, 108), 231 232
Gastambide, B., 64 (102), 65, 81
Gastambide-Odier, M., 64 (102), 65, 81
Gatica, J., 196 (117), 200
Gay, R., 64, 81
Geller, L. E., 194 (106), 200
Georgian, V., 194 (95), 200
Gesson, J. P., 63 (80), 81
Ghosh, A. C., 69 (152), 83, 218 (89, 90, 91), 231
Gibian, H., 27 (130), 33, 92 (67), 121
Girard, A., 49 (4), 79
Giuliano, R., 288
Gladstone, W. A. F., 258 (98), 282
Gloss, M., 158 (64), 160
Goffinet, B., 206 (39), 230, 274 (142), 284
Gogte, V. N., 1, 29, 71 (163), 73 (173), 74 (173), 83
Goi, M., 291
Goldberg, M. W., 71 (155), 83, 103 (113), 123
Goldman, N., 157 (57), 160, 276 (163), 284
Goldsmith, D., 37 (14), 47
Gordon, L., 158 (63), 160
Goto, G., 92 (70), 121
Gottesman, R. T., 139 (31), 141
Graber, R. P., 20 (94), 32, 49 (7), 63 (7, 95, 96), 79, 81
Graham, D. W., 3 (34), 30, 69 (148), 83, 251 (70), 258 (70), 281
Granchelli, F. E., 156 (56), 160

Graves, J. M. H., 22 (107), 32, 87 (2), 88 (2), 89 (2), 90 (2), 91 (50), 93 (78a), 118, 120, 121, 147 (22), 151 (22), 159, 161 (4, 7), 162 (4), 163 (4), 164 (4), 165 (4), 166 (7), 173
Gravestock, M. B., 8 (37), 30, 45 (46, 47, 48), 48, 287
Green, J., 243 (34), 280
Green, M. J., 28 (136), 34, 260 (102, 103), 283
Greenspan, G., 27 (134), 34, 161 (11, 12, 22), 166 (11, 12), 173, 174
Grinenko, G. S., 19 (93), 32, 151 (39, 40), 152 (41), 159, 235 (10, 11, 13, 14, 15), 238, 243 (38), 274 (148), 280, 284, 290
Grob, C. A., 186, 198
Gross, E. M., 56 (43), 80
Gschwend, H. W., 290
Guette, J.-P., 116 (184), 125
Guha, M., 235 (12), 238
Guiliano, R., 107 (151), 124
Gulaya, V. E., 27 (131), 33, 90 (42), 120
Gunar, V. I., 104 (119, 120), 123
Gundkötter, M., 158 (64), 160
Gurbaxani, S., 39 (27), 47
Gurvich, I. A., 20 (97), 32, 68 (142, 143), 82, 130 (13), 131 (14, 45, 46), 132, 133
Gut, M., 36 (7), 47, 205 (26), 206 (31, 32), 220 (31, 32), 229
Gutsche, C. D., 9 (41), 15 (71), 30, 31, 52 (15, 16, 17), 58 (54, 55), 61 (54, 55), 79, 80, 91 (43), 120, 129 (4), 131, 136 (15), 139 (15), 140, 143 (1, 3), 144 (3), 158, 201 (2), 228, 239 (2), 248 (50), 279, 281
Guzewska, M., 288

H

Haberland, G., 17 (85, 87), 32, 55, 79
Habicht, E. R., Jr., 58 (60), 61 (60), 80, 257 (96), 258 (96), 282
Hadler, H. I., 10 (46), 22 (46, 104), 30, 32, 244 (41), 245 (45), 246 (45), 280, 281
Haefliger, W., 187 (58c), 198
Hagedoen, K-W., 98 (97), 122
Hagenbach, A., 290
Hainaut, D. 210 (10), 220 (107, 108), 222 (107, 108), 229, 232

Hajos, Z. G., 13 (62), 27 (126), 31, 33, 71, 83, 242 (22), 266 (125), 273 (22, 140, 141), 279 (141), 280, 283, 284, 291, 292
Halsall, T. G., 241 (14), 280
Hamon, D. P. G., 257 (96), 258 (96, 97), 282
Hand, J. J., 22 (106), 32, 148 (26, 27, 28), 159
Hannah, J., 213 (74), 231
Hansen, H. V., 166 (34), 168 (34), 174, 288
Hanson, J. C., 169 (47), 170 (47), 174
Hanssen, H. W., 258 (98), 282
Hanze, A. R., 187 (58a), 198, 206 (35), 230, 254 (80), 282
Hanzlik, R. P., 39 (29), 47
Hara, S., 180 (20), 197
Harbert, C. A., 43 (41), 44 (41), 45 (41), 46 (51), 48
Harnik, M., 2 (21), 4, 21 (98), 29, 32, 56, 57, 58 (47), 80, 110 (166), 115 (166), 124, 181 (26), 197
Harper, S. H., 103 (115), 123
Harris, L. S., 156 (56), 160
Harris, S. A., 187 (55), 198
Harrison, I. T., 224 (117), 232, 236 (18), 238
Harrison, R. G., 242 (31), 280
Hartley, D., 22 (107), 25 (117), 26 (21) 32, 133, 87 (2, 8), 88 (2), 89 (2), 90 (2), 91 (50), 92 (8), 118, 119, 120, 147 (22), 151 (22), 159, 161 (3, 4, 5, 6, 10, 16), 162 (4), 163 (4, 5), 164 (4, 10), 165 (4, 5), 166 (5, 16), 173, 220 (111), 232
Hartman, J. A., 205 (27), 229
Hartshorn, M. P., 23, 33
Haven, A. C., 187 (56), 198
Hawkes, G. H., 92 (71), 121
Hawthorne, J. R., 106, (136), 123
Hayase, Y., 256 (92), 282
Hazra, B. G., 70 (153), 83, 196, 200
Heer, J., 63 (85, 86, 90, 91, 92), 81
Heggs, P., 277 (165), 285
Heidepriem, H., 94 (80), 121
Heinrich, E., 17 (87), 32, 55 (27), 79
Heller, M., 215 (82), 231
Henrick, C. A., 99 (99), 122, 274 (145), 284

Herbst, D. 25 (117), 33, 161 (5, 10, 19), 163 (5), 164 (10), 165 (5), 166 (5), *173*, *174*, 220 (111), *232*

Herbst, D. R., 87 (8), 92 (8), *119*, 161 (3, 21), *173*, *174*

Herr, M. E., 215 (86), *231*

Herrin, T. R., 43 (41), 44 (41), 45 (41), *48*

Hertler, W. R., 2 (26), 5, *29*

Herzog, H. L., 28 (139), *34*

Hester, J. B., 227 (124), *233*

Heusler, K., 2 (19), 4, 12 (19), 16 (75), 18 (19), 25 (120), 27 (19), *29*, *31*, *33*, 176 (4, 5), 187 (64, 65, 66, 67, 68), 189 (67, 74, 75, 76), 192 (64, 65, 66, 67, 68, 79, 80, 81, 82, 83, 84, 86, 87, 88, 89, 90, 91, 92, 93), 194 (99, 100), *196*, *197*, *198*, *199*, *200*, 236 (17), *238*, 253 (78, 79), 254 (79), 261 (105, 106), *282*, *283*

Heusser, H., 35 (3), *46*

Hewett, C. L., 2 (16), *29*, 49 (4, 5), *79*, 103 (112), *123*

Heyl, F. W., 215 (86), *231*

Highet, R. J., *31*, 58 (59), 59 (59), 61 (59), *80*, 144 (5), *159*

Hilfiker, F. R., 56 (41), *80*, 227 (121), *232*

Hilgetag, G., 97 (93), 102 (104), *122*, 241 (21), *280*

Hill, R. K., 26 (123), *33*, 127 (1), *131*

Hindersinn, R. R., 16 (74), *31*, 248 (54), *281*

Hiraga, K., 21 (99), *32*, 87 (7, 14), 88 (7, 14), 90 (14), 92 (66, 70), 93 (76), *118*, *119*, *121*, *288*, *289*

Hirai, S., 23 (111), *33*, 66 (124), 67 (132), *82*, 194 (111, 112, 114), *200*, 254 (82, 83), 255 (85, 88, 90), 256 (82), *282*

Hiraoka, T., 9 (42), *30*, 163 (27), *174*

Hirscheld, H., 2 (25), 5, 19 (25), 25 (25), *29*

Hirschler, H. P., 139 (30, 32), *141*

Hirschmann, R., 52 (17), *79*

Hoch, J., 17 (85), *32*, 49 (8), *79*

Hodes, W., 116 (190), 117 (193), *125*

Hödolt, E., 242 (28), *280*

Hoffman, C. H., 187 (56), *198*

Hofmann, H., 98 (97), *122*

Hogg, J. A., 59, 65, *81*, *82*, 138 (19, 20, 21), *141*, 187, *198*, 206, *230*, 254 (80), *282*

Holly, F. W., 187 (55, 56), *198*

Holmen, R. E., 17 (87), *32*, 51, 55, *79*

Holtermann, H., 2 (20), 3 (20), 4, 16 (20), *29*, 176 (1, 2), *196*, 244 (42), *280*

Holton, R. A., 36 (9), 39 (9), 41 (9), *47*

Hooper, J. W., 13 (63), *31*

Hooz, J., 42 (38), *48*

Hopla, R. E., 36 (9), 39 (9), 41 (9), *47*

Horeau, A., 64 (98, 100, 101), *81*, 92 (64), 116 (184, 185), *121*, *125*, 162 (24, 25), *174*

Horoldt, E., 2 (25), 5, 19 (25), 25 (25), *29*

Hosli, H., 103 (113), *123*

Hoss, O., 131 (19, 20), *132*

Hotta, S., 35 (2), *46*

Howell, F. H., 194 (109, 110), *200*

Hoyer, G.-A., 98 (95), *122*, *289*

Hsi, R. S. P., 240 (3), *279*

Huang, Y. T., 110 (162, 163, 164), *124*

Huber, J., 275 (158), *284*

Hubert, J. C., 97 (91, 92), 105 (125), *122*, *123*, *289*

Hubner, M., 26 (122), *33*

Hudson, B. J. F., 239 (1), *279*

Hughes, G. A., 22 (107), 25 (117), 26 (121), *32*, *33*, 58 (63), *80*, 87 (2, 4, 8, 15), 88 (2, 4), 89 (2, 4, 24, 25, 32), 90 (2, 4), 91 (50), 92 (8, 32), 93 (15), *118*, *119*, *120*, 147, 151 (22), *159*, 161, 162 (4), 163 (4, 5), 164 (2, 4, 10), 165 (4, 5), 166 (5, 7, 16), *173*, 220 (111), *232*, 243 (36), *280*, *288*

Huisman, H. O., 1, 2, 10 (48, 49), 21 (48, 101), *29*, *30*, *32*, 71 (160, 162, 164), 74 (177, 178), 76 (162, 181, 181a), 77 (162, 182, 183, 190), *83*, *84*, 88 (19), 92 (72), 94 (82, 83, 84, 87), 96 (89, 90), 97 (91, 92), 101 (102, 103), 102 (72, 103, 105), 105 (125), 110 (160, 161), 117 (202, 203), *119*, *121*, *122*, *123*, *124*, *125*, 152, 153 (43, 44, 45), 154 (44, 46, 49), 155 (49), 157 (58), 158 (59), *160*, 171 (60), *175*, *286*, *289*

Hunter, J. H., 59, 65, *81*, *82*

I

Iacobelli, J., 274 (154), *284*
Ilton, M. A., 43 (42), *48*
Ilyukhina, T. V., 131 (14), *132*
Inaba, T., 257 (95), *282*
Inhoffen, H. H., 2 (25), 5, 19 (25), 25 (25), 29, 151, *159*, 242 (27, 28, 29, 30), *280*
Ireland, R. E., 17 (80), 20 (95), 32, 44 (45), *48*, 58 (57), 61 (57), *80*, 135 (2), 136 (2, 11), 137 (2), *140*, 265 (121), *283*, *290*
Iriarte, J., 63 (77, 78), 65 (109), *81*, *82*
Irmscher, K., 2 (25), 5, 19 (25), 25 (25), 29, 151 (36), *159*
Ishihara, S., 274 (146), *284*
Itani, H., 274 (146), *284*
Ivanova, L. N., 131 (16), *132*
Ives, D. A. J., 2 (22), 3 (22), 5, *29*
Iwai, I., 9 (42), *30*, 163 (27), *174*
Izumi, Y., 266 (123), *283*

J

Jackson, R. W., 187 (58a), *198*, 206 (35), *230*, 254 (80), *282*
Jacob, E. J., 93 (75), 111 (170), 115 (181, 182), *121*, *124*, *125*, 235 (16), *238*
Jacquesy, J. C., 63 (80), *81*
Jacquesy, R., 63 (80), *81*
Jaeger, R. H., 241 (17), *280*
Jansen, A. B. A., 25 (117), 26 (121), *33*, 87 (8), 92 (8), *119*, 161 (3, 5, 10, 16), 163 (5), 164 (10), 165 (5), 166 (5, 16), *173*, 220 (111), *232*
Jaques, B., 41 (31), *47*
Jautelat, M., 39, *47*
Jeger, O., 35 (3, 5), 39 (5), *46*, *47*, 69 (152), *83*, 194 (99, 100, 103, 104, 105) *200*
Jen, T. Y., 87 (15), 93 (15), *119*, 161 (7, 14, 17), 166 (7), *173*
Jenks, T. A., 154 (48), *160*
Jensen, E. V., 21 (98), *32*, 57 (45, 46), *80*
Jensen, N. P., 42 (38), *48*
Jhina, A. S., 73 (173), 74 (173), *83*
Jilek, J. O., 136 (10), 138 (22, 24), *140*, *141*

Jit, P., 77 (186, 187), *84*
Jogi, R. R., 166 (38), 169 (38), *174*
Johns, W. F., 25 (120), *33*, 65 (113, 114), 66, *82*, 91 (57), *121*, 182 (27, 32, 33, 35, 36, 38), 184 (33, 35, 38), 185 (35), 186 (33), *197*, *198*, 218 (94), *231*, 248 (51), 249 (51, 59), 251 (72), 253 (76), *281*, *282*
Johnson, A. L., 91 (51), *120*, *288*
Johnson, D. B. R., 277 (167), *285*
Johnson, J. A., Jr., 21 (102), *32*, 56 (34, 35), *80*
Johnson, T. L., 13 (60), *31*, 56, 63 (33), *79*
Johnson, W. S., 1, 2 (23), 3 (34), 5, 8 (36, 37), 9 (41), 10 (46), 15 (71), 16 (73, 74), 17 (80), 20 (94, 95), 22 (46, 105), 23 (112), 27 (112), 28 (135), 29, *30*, *31*, *32*, *33*, *34*, 37 (11, 12, 17, 18), 41 (12, 31, 32, 33), 42, 43 (39, 40, 41, 42), 44 (39, 40, 41), 45 (39, 40, 41, 45, 46, 47, 48, 49), 46 (51), *47*, *48*, 49 (7), 52, 54, 55, 58 (54, 55, 56, 57, 58, 59, 60), 61, 63, 66 (120), 68, 69, 79, *80*, *81*, *82*, 83, 91, 111 (170), *120*, *124*, 127 (2), 129 2, 4), *131*, 135, 136 (2, 6, 11, 15), 137 (2), 139 (15), *140*, 143 (1, 2, 3), 144 (3, 5, 6), 145 (7), 146 (8, 11), *158*, *159*, 187 (62a), 196 (118), *198*, *200*, 201, 228, 235 (16), *238*, 239 (2), 243 (39), 244 (40, 41, 43, 44), 245 (43, 45), 246 (43, 45), 247 (47, 48, 49), 248 (48, 49, 50, 51, 52, 53, 54), 249 (51, 56, 58, 59, 60), 251, 252 (73), 257, 258 (68, 70, 96, 97), *279*, *280*, *281*, *282*, *287*
Johnston, D. B. R., 100 (101), 106 (131), *122*, *123*, *289*
Johnston, E. L., 187 (55, 56), *198*
Jolly, J., 214 (77), 215 (77), *231*, *291*
Joly, R., 214 (77), 215 (77), *231*
Jones, A. R., 135 (1), *140*
Jones, E. R. H., 78, *96*, 107 (149), 118 (206, 208), *124*, *125*
Jones, E. T., *31*, 58 (59), 59 (59), 61 (59), *80*, 144 (5), *159*
Jones, G., 77, *84*
Joshel, L. M., 171 (55, 56, 57), *175*
Joshi, G. S., *290*
Juday, R. E., 56, *80*

Julia, M., 43, 48, 287
Julia, S., 43, 48
Julia, S. A., 234 (7), 235 (7), 238
Jundt, W., 186 (50, 52, 53), 198
Jung, C. J., 288
Junghans, K., 288

K

Kagan, H., 92 (64), 121, 162 (24, 25), 174
Kahn, L. D., 240 (3), 279
Kakisawa, H., 275 (160), 284
Kalvoda, J., 194 (99, 100), 200
Kamachi, H., 275 (160), 284
Kamano, Y., 286
Kamata, S., 256 (92), 282
Kampe, D., 242 (29), 280
Kashman, Y., 289
Kasturi, T. R., 71, 72 (168), 83, 99 (100), 122
Kasymov, S. K., 90 (41), 120
Katz, L. E., 78 (199), 84
Kauder, O., 177 (11), 197
Keana, J. F. W., 1 (3), 29, 251 (65, 66, 67), 281
Kelly, R. B., 2 (22), 3 (22), 5, 29
Kemp, A. D., 244 (40), 249 (58, 59), 280, 281
Kerwin, J. F., 194 (95, 96), 200
Kessar, S. V., 77, 84, 94 (85), 105 (124), 118 (207), 122, 123, 125, 166 (38), 169, 174, 290
Khastgir, H. N., 22 (108), 33, 66 (117), 82, 111 (170), 124, 171 (53), 175, 235 (16), 238
Kidwell, R. L., 66 (122), 82, 240 (4), 279
Kieslich, K., 27 (130), 33, 92 (67), 94 (80), 121
Kikkawa, I., 66 (123, 124), 67 (123), 82, 194 (113), 200, 254 (83), 282
King, J. C., 16
King, M. S., 65 (112), 82
Kinnel, R. B., 41 (33), 42 (33), 47
Kiprianov, G. I., 62, 81, 135 (5), 136 (5), 137 (5), 140
Kirk, D. N., 23, 33
Kitahana, Y., 148 (25), 159
Klemm, L. H., 116 (190), 117 (192, 193), 125

Klimstra, P. D., 213 (71), 231, 286
Kloetzel, M. C., 116 (186), 125
Klyne, W., 178, 197
Knabeschuh, L. H., 139 (40), 141
Knowles, W. S., 12 (56), 31, 223 (115), 232, 262 (109), 263 (109, 114, 115), 264 (116, 117), 265 (116, 119), 283
Kobayashi, K., 69 (149), 83
Kobayashi, T., 286
Koch, H. J., 27 (130), 33
Koch, W., 267 (129), 283
Kocor, M., 147 (17), 159, 288
Koebner, A., 114 (175, 176), 117 (176), 125
Köster, H., 176 (8), 197
Kogan, L. M., 27 (131), 33, 90 (42), 120
Kokor, M., 12 (55), 31
Komeno, T., 274 (146), 284
Kon, G. A. R., 96 (88), 103 (88, 110, 114, 115), 122, 123
Kondrat'eva, G. V., 10 (45), 30, 234 (23), 238
Konort, M. D., 139 (33), 141
Konst, W. M. B., 92 (72), 102, 121
Konz, W. E., 36 (9), 39 (9), 41 (9), 47
Korst, J. J., 244 (44), 280
Koshoev, K. K., 87 (5), 89 (5), 90 (35, 36), 118, 120, 220 (109), 222 (109), 232
Kosmol, H., 27 (130), 33, 92 (67), 94 (80), 121
Kotlyarevskii, I. L., 131 (18), 132
Kraay, R. J., 210 (60), 230
Kraychy, S., 2 (21), 4, 29, 56 (40), 80, 181 (26), 197
Kreutzer, A., 2 (25), 5, 19 (25), 25 (25), 29
Krieger, C., 26 (124), 27 (124), 33
Krivoruchko, V. A., 110 (159), 124
Krop, P. J., 251 (72), 282
Krubiner, A. M., 19 (92), 32, 63 (81), 81, 208 (56), 230, 274 (157), 284
Kubela, R., 129 (11), 132
Kucherov, V. F., 131 (14, 16, 36), 132
Kudryavtseva, L. F., 10 (45), 30, 234 (2, 3), 238
Kuehne, M. E., 16
Kumar, A., 77 (184), 84, 105 (124), 118 (207), 123, 125, 166 (38), 169 (38, 44), 174

Kuo, C. H., 87, *119*
Kuok-kin, F., 92 (62), *121*
Kurath, P., 63 (82), *81*
Kurosawa, Y., 27 (129), *33*, 203 (16), *229*
Kuryatov, N. S., *291*
Kushinsky, S., 205 (28), *229*
Kushner, S., 13 (60), *31*, 55 (22), 58 (22), 60 (22), 61 (22), 65 (22), *79*
Kutney, J. P., 254 (81), 255, 256, 257 (95), 258 (98), *282*
Kutsenko, L. M., 62 (74), *81*, 135 (5), 136 (5), 137 (5), *140*
Kuznetsova, A. I., 131 (45, 46), *133*

L

Labler, L., 194 (97), *200*
Lagidze, D. R., 87 (17), 89 (30), 92 (30, 61), *119*, *121*
Lake, A. W., *289*
Lakhvich, F. A., 110 (159), *124*
Lakum, B., 90 (42), *120*
Lallemand, J. Y., *287*
Landesman, H., 205 (30), *229*
Lansbury, P. T., 14 (67), *31*, 56, *80*, 226, 227 (120, 121), *232*, *286*
Lapin, H., 19 (91), *32*, 65 (104), *81*
Lardon, A., 24 (116), *33*, 187 (62), 189 (62), *198*
Laroche, M. J., 64 (102), 65 (102), *81*
Lavit-Lamy, D., 17 (84), *32*, 111 (169), *124*
Lederman, Y., 110 (166), 115 (166), *124*
Ledig, K. W., 25 (117), *33*, 87 (8), 92 (8), *119*, 161 (3, 5, 10), 163 (5), 164 (10), 165 (5), 166 (5), *173*, 220 (111), *232*
Lednicer, D., 23, *33*, 90 (37), *120*, 227 (125), *233*
Leeming, M. R. G., 161 (22), *174*
Leffler, C. F., 218 (92), *231*
Lehmann, G., 97 (93, 94), 102 (104), *122*, 241 (21), *280*
Leigh, H. M., 28 (138), *34*, 187 (58b), *198*
Leitereg, T. J., 279 (170), *285*
Lematre, J., 64 (101), 78 (200), *81*, *84*
Lenard, K., 17 (83) *32*, 169 (45), 170 (45), *174*

Lenhard, R. H., 215 (82), 220 (104), *231*, *232*
Leong, S. Y., 257 (95), *282*
Leonov, V. N., 58 (53), *80*, 87 (10), 88 (20), 90 (10, 20, 38, 39, 41), 91 (10, 20, 44, 47, 48, 49, 52, 53, 54, 55), *119*, *120*, *121*
Levein, P., 22 (104), *32*
Levina, R. Ya., 116 (191), *125*
Levine, P., 219 (100), *232*
Levine, S. G., 218 (91, 92), *231*
Lewis, B. B., 194 (95, 96), *200*
Lewis, H. J., 170 (50), *174*
Li, T., 43 (41), 44 (41), 45 (41, 49), *48*
Limanov, V. E., 87 (10), 89 (27, 28), 90 (10), 91 (10), *119*
Lincoln, F. H., 187 (58a), *198*, 206 (35), *230*, 254 (80), *282*
Linni, T. D., 91 (56), *121*
Linstead, R. P., 19 (90), 22 (104), *32*, 55 (31), *79*, 219 (100), *232*
Lipinski, C. A., 44 (44), *48*
Littlewood, P. S., 144 (4), *158*, 242 (26), *280*
Littlewood, S. M., 151 (37), *159*
Loewenthal, H. J. E., 265 (118), *283*
Logemann, W., 176 (8), *197*
Logothetic, A. L., 107 (150), *124*
Loke, K. H., 145 (7), *159*
Lora-Tamayo, M., 156 (50, 51, 54), *160*
Lord, K. E., 37 (21), *47*
Lorthioy, E., 116 (184), *125*
Los, M., 22 (106), *32*, 148 (26, 27, 28), *159*, *290*
Loven, R., 77 (183), *84*
Lowenthal, H. J. E., 10 (44), 14 (44), *30*, 222 (114), *232*
Lucke, B., 97 (94), 102, *122*
Luke, G. M., 74, *83*, 154 (47, 48), 155 (47), *160*
Lukes, R. M., 25 (120), *33*, 65 (113, 114), 82, 182 (27, 28, 30, 32, 33, 35, 36, 37), 184 (33, 35), 185 (35, 37), 186 (33, 47), *197*, *198*, 253 (76), *282*
Lumb, A. K., 77 (187), *84*, 94 (85), *122*
Lunn, W. H., 42 (37), *48*

Author Index 305

Lustgarten, D. M., 107 (145), *124*, 166 (30, 31, 32, 33, 34, 35), 167 (32, 33), 168 (30, 33, 34, 35), *174*, *288*
Lyall, J., 1, *29*, 71 (158), *83*
Lyster, S. C., 213 (72), *231*
Lythgoe, B., 144 (4), 151, *158*, *159*, 242, *280*
Lyttle, D. A., 28 (138), *34*

M

Ma, C. M., 110 (163, 164), *124*
MacAlpine, G. A., 129 (11), *132*
Macaulay, S., 13 (63), *31*
McCaleb, K. E., 56 (39), *80*, 139 (39), *141*, 181 (25), *197*
McCarry, B. E., 8 (37), *30*, 45 (48), *48*, *287*
McCloskey, A. L., 111 (170), *124*, 235 (16), *238*
MacConnell, J. G., 169 (47), 170 (47), *174*
McCrae, W., 3 (35), *30*, 254 (81), 255 (84), 256 (84), *282*
McElvain, S. M., 253 (74), *282*
McGinnis, N. A., 108 (152), *124*
McKay, G. R., Jr., 253 (74), *282*
McKillip, A., 92 (71), *121*
McLamore, W. M., 2 (19), 4, 12 (19), 18 (19), 27 (19), *29*, 176 (4, 5), *196*, *197*, 236 (17), *238*, 261 (105, 106), *283*
McLoughlin, B. J., 22 (107), 25 (117), *32*, *33*, 87 (2, 8), 88 (2), 89 (2, 25), 90 (2), 91 (50), 92 (8), *118*, *119*, *120*, 147, *151* (22), *159*, 161 (3, 4, 5), 162 (4), 163 (4, 5), 164 (4), 165 (4, 5), 166 (5), *173*, 243 (36), *280*
McMenamin, J., 25 (117), *33*, 87 (8), 91 (50), 92 (8), *119*, *120*, 161 (3, 5, 10), 163 (5), 164 (10), 165 (5), 166 (5), *173*, 220 (111), *232*
McMurry, J. E., 13 (65), *31*, 260, 272 (104), 275 (104, 161, 162), 276 (161), *283*, *284*
McQuillin, F. J., 11 (51), *30*
Magerlein, B. J., 187 (58a), *198*, 206 (35), *230*, 254 (80), *282*
Magnani, A., 194 (95), *200*

Mahishi, N., 115 (181, 182), *125*
Mak, K. T., *14*
Maksimov, V. I., 19 (93), *32*, 151 (39, 40), *159*, 235 (10, 11, 13, 14, 15), *238*, 243 (38), 274 (148), *280*, *284*
Mancera, O., 3 (31), *30*, 196 (117), *200*
Mannhardt, J. H., 116 (183), *125*
Manson, A. J., 218 (97), *232*
Mansuy, D., *287*
Marin, J., 156 (50), *160*
Markezich, R. L., *287*
Marrion, G. F., 145 (7), *159*
Marshall, J. A., 1 (3), *29*, 196 (118), *200*, 248 (53), 251 (53, 65, 66), *281*
Marsili, A., 117 (195, 199), *125*
Martin, D. G., 1 (3), *29*, 251 (66), 252 (73), *281*, *282*
Martin, H., 16 (79), *31*
Martin, J. G., 127 (1), *131*
Martin, R. H., 201 (4), 203 (4), 219, 228, *232*, 265, *283*
Marvel, C. S., 171 (59), *175*
Masamune, T., 69 (149), *83*
Matejka, R. C., 218 (89), *231*
Mathieson, D. W., 227 (123), *233*
Mathieu, J., 1, 12 (54), 25 (54), 27 (133), *29*, *31*, *33*, 58 (66), *80*, 201 (6, 8, 9), 202 (6), 203 (8, 9), 204 (8, 9), 206 (9, 29, 36), 207 (6), 208 (52), 209 (52, 59), 212 (59, 70), 213 (6, 52, 59, 70), 214 (6, 52, 59, 77), 215 (9, 52, 77), 220 (9), *229*, *230*, *31*
Matoba, K., *291*
Mayor, R. K., 94 (85), *122*
Mazur, Y., 2 (27, 28), 3 (27, 28), 6, *30*, 180 (18), *197*
Medcalfe, T., 151 (37), *159*, 242 (26), *280*
Meier, J., 36 (7), *47*
Meister, P. D., 28 (138), *34*, 187 (58b), *198*
Mekler, A. B., 241 (11), *280*
Meltzer, R. I., 107 (145), *124*, 166 (30, 31, 32, 33, 34), 167 (32, 33), 168 (30, 33, 34), *174*, *288*
Ménager, L., 92 (64), *121*, 162 (24, 25), *174*

Meyer, W. L., 67 (131), 82, 144 (6), 145 (7), *159*
Meyer-Delius, M., 114, 117 (198), *125*
Meyers, A. I., 107 (146, 147), 109 (147, 153, 154, 155, 156), 110 (157), *124*, *158*, *160*, 166 (36), 168, 169, *174*
Meystre, C., 189 (72), 192 (72), 194 (99, 100), *199*, *200*
Micheli, R. A., 13 (62), *31*, 273 (141), 274 (155), 279 (141), *284*
Miescher, K., 2 (18), 4, 29, 58, 60, 61, 63, 65, *80*, *81*, 103, (111), 111 (111), *122*, 129 (3), *131*, 178, 181, *197*, 241 (7), 274 (149, 152, 153), *279*, *284*
Migdalof, B. M., 277 (164), *284*
Mihailović, M. I., 194 (103), *200*
Mijović, M. V., 35 (3), *46*
Miki, I., 21 (99), *32*, 87 (7, 14), 88 (7, 14), 90 (14), 92 (66, 70), 93 (76), *118*, *119*, *121*, *288*, *289*
Miles, D. H., 45 (46), *48*
Miller, M., 146 (11), *159*
Miller, R., 103 (108), *122*
Milne, G. M., 39 (24, 28), *47*
Minaeva, I. N., 117 (196), *125*
Miramontes, L., 208 (57, 58), 212 (58), *230*
Mishra, L. K., 17 (81), *32*, 111 (167), *124*
Mitra, R. B., 21 (103), *32*, 73, 74 (171, 174), *83*
Möller, K., 158 (64), *160*
Moiseenkov, A. M., 110 (159), *124*
Mondal, K., 69, *83*, 178, *197*
Montelaro, R. C., 26 (123), *33*
Moon, M. W., 242 (26), *280*
Moore, C., 77 (193), *84*
Moore, J. A., 3 (32), *30*, 180 (19), *197*
Morand, P., 1, *29*, 71 (158), *83*
Morgan, J. G., 49 (9), 55 (23), 78, *79*
Morgan, R. L., 3 (33), *30*
Mori, K., 218 (90), *231*
Morin, R. D., 78 (195), *84*, 117 (200), *125*
Morrison, J. D., 266 (126), *283*
Morton, D. R., *287*
Mosher, H. S., 266 (126), *283*
Mousseron, M., 131 (38), *133*
Müller, M., 267 (129), *283*
Müller, P., 206 (34), *229*

Müller, R., 267 (131), 268 (131, 136), 271 (131, 136), *283*, *284*
Mukharji, P. C., 10 (44), 14 (44), *30*, 68, 82, 222 (114), 224 (118), 232, 241 (12), 265 (118), 274 (147), *280*, *283*, *284*
Mukherjee, B. B., 180 (22), *197*
Mukherjee, S. M., 136 (7), *140*
Mukherji, S. M., 22 (109), *33*, 62, *81*
Muller, G., 194 (94), *200*
Mundra, K. P., 77 (187), *84*
Munoz, G. G., 107 (147), 109 (147, 153), *124*, 158 (60), *160*
Murai, A., 69 (149), *83*
Murphy, J. W., 39 (25), *47*
Murray, H. C., 28 (138), *34*, 187 (58b), *198*
Murthy, P. S. N., 16 (78), *31*, 203 (14), 215 (87), 219 (87), 229, *231*
Myers, R. F., 45 (46), *48*

N

Nacci, V., 107 (151), *124*, *288*
Nadamuni, G., 53 (19), 54 (20), 79, 219 (102), 221 (102), 222 (102), *232*
Nadeau, R., 38 (22), *47*
Nagal, A., 12 (58), *31*, 150 (32, 33, 34), *159*
Nagata, M., *291*
Nagata, W., 10 (43), 23 (111), *30*, *33*, 66, 67 (123, 130, 132, 133, 134, 135), *82*, 194, *200*, 248 (55), 254 (82, 83), 255, 256 (82, 87, 89, 92, 93, 94), *281*, *282*
Nair, G. V., 28 (136), *34*, 260 (103), *283*
Nakuma, K., 204 (24), *229*
Narisada, M., 256 (92), *282*
Nasipuri, D., 67, 82, 115 (179), *125*, 235 (12), *238*
Nathan, A. H., 187 (58a), *198*, 206 (35), *230*, 254 (80), *282*
Nazarov, I. N., 20 (97), *32*, 68 (142, 143), 82, 104 (117, 118), *123*, 130 (12, 13), 131 (18, 22, 25, 26, 27, 28, 30, 31, 33, 34, 36, 37, 45, 46), *132*, *133*, 147 (15), *159*, 234 (1, 5), *238*, 241 (13), *280*
Nédélec, L., 16 (77), 27 (125), *31*, *33*, 213 (70), 215 (83, 84, 85), 218 (84), *231*

Author Index

Neeter, R., 74 (178), 83, 158 (59), 160
Negishi, E., 14 (69), 31, 242 (33), 280
Neher, R., 16 (75), 31, 187 (59, 60, 64), 189 (73, 74, 75, 76), 192 (64), 198, 199
Nekrasova, M. A., 288
Nelson, N. A., 22 (110), 33, 56, 63, 65, 66 (111), 80, 81, 82, 104 (122), 123, 139 (41, 43), 141, 227 (124), 233, 240 (3), 246, 247, 279, 281
Nesty, G. A., 171 (59), 175
Neustaedter, P. J., 45 (45), 48
Nevenzel, J. C., 18 (89), 32, 115 (180), 125
Newhall, W. E., 187 (55, 56), 198
Newkome, G. R., 26 (123), 33
Newman, M. S., 18 (89), 32, 68 (141), 82, 115 (180), 116 (183), 125, 241 (10, 11), 280
Nicholson, R. T., 288
Nienhouse, l. J., 56 (41), 80, 227 (121), 232
Nijhuis, H., 117 (203), 125
Ninomiya, I, 2, 29
Nominé, G., 1, 12 (54), 25 (54, 119), 27 (133), 29, 31, 33, 58 (66), 80, 201 (5, 6, 8, 9), 202 (5, 6, 89), 203 (5, 8, 9), 204 (1, 8, 9, 17, 18, 19, 20, 21), 206 (9, 29, 36, 37), 207 (6, 41, 42, 43, 44, 47), 208 (42, 44, 45, 49, 53, 54, 55), 209 (59), 210 (61, 62, 63, 65), 213 (6, 41, 59, 70, 73), 214 (6, 59, 75, 76), 215 (9), 220 (9, 105, 106, 107, 108), 222 (107, 108), 229, 230, 231, 232, 243 (35, 37), 274 (35, 37, 143), 278 (35, 143), 280, 284
Nordman, C. E., 169 (47), 170 (47), 174
Nordsiek, K.-H., 242 (28), 280
Novak, L., 138 (22, 23), 141
Novello, F. C., 117 (201), 125

O

Oberster, A. E., 213 (74), 231
Ohashi, M., 272 (139), 275 (139, 160), 284
Ohki, S., 290
Oida, S., 148 (25), 159
Oka, K., 180 (20), 197

Okorie, D. A., 45 (47), 48
Okumura, T., 255 (86), 282
Oliveto, E. P., 13 (62), 19 (92), 31, 32, 63 (81), 81, 208 (56), 230, 242 (22), 273 (22, 140, 141), 274 (151, 157), 279 (141), 280, 284
Olson, A. J., 169 (47), 170 (47), 174
Olson, G. L., 8 (36), 30, 42 (36), 47
O'Neill, R. C., 61 (71, 72), 81
Onken, D., 93 (77), 121
Oppolzer, W., 39, 47
Ortez de Montellano, P. R., 37 (20), 47
Osborne, M. W., 166 (31), 174
Owings, F. F., 194 (95, 96), 200
Owyang, R., 41 (31), 47
Ozawa, Y., 27 (127), 33, 203 (16), 204 (23, 24), 229

P

Page, T. F., 78 (195), 84
Paì, N. R., 77 (188), 84, 158 (62), 160
Pakhomov, V. P., 291
Palmer, K. H., 218 (95), 232
Pandit, U. K., 10 (48, 49), 21 (48, 101), 30, 32, 74, 77 (190), 83, 84, 88 (18, 19), 94 (82, 83, 84, 87), 96 (89, 90), 101 (102), 110 (160, 161), 117 (202, 203), 119, 122, 124, 125, 153 (43, 44, 45), 154 (44, 46), 157 (58), 158 (59), 160, 171 (60), 175
Panhouse, J. J., 241 (20), 280
Papa, D., 65 (112), 82
Pappo, R., 2 (10, 23), 5, 16 (74), 29, 31, 68 (144), 82, 89 (34), 119, 187 (62a), 198, 244 (40), 247 (48), 248 (48, 49, 51, 52, 54), 249 (51, 58, 59), 251 (52, 61, 71, 72), 252 (73), 280, 281, 282, 288
Parkash, N., 290
Parker, K. A., 13
Parnes, Z. N., 90 (41), 120
Parrish, D. R., 13 (62), 27 (126), 31, 33, 71 (155), 83, 242 (22), 266 (125), 273 (22, 140, 141), 279 (141), 280, 283, 284, 291, 292
Parry, R. J., 45 (46, 47), 48, 287

Pars, H. G., 156 (56), *160*
Parthasarathy, P. C., 194 (98), *200*
Patchett, A. A., 2 (22), 3 (22), 5, *29*, 87 (3, 11), 88 (3), 89 (3), 92 (65), 94 (81), 100 (101), 101 (81), 105 (130), *118, 119, 121, 123,* 277 (166), *285, 289*
Pattison, T., 161 (10), 164 (10), *173,* 220 (111), *232*
Pattison, T. W., 25 (117), *33,* 87 (8), 91 (50), 92 (8), *119, 120,* 161 (3, 5), 163 (5), 165 (5), 166 (5), *173*
Paul, V., 16 (78), *31,* 203 (14), 215 (87), 219 (87, 96), *229, 231, 232*
Pawson, B. A., 39 (27), *47*
Peak, D. A., 107 (138, 139), *123*
Pechet, M. M., 194 (106), *200*
Peña, A. E., *289*
Perelman, M., 210, *230*
Perronnet, J., 172 (61), *175,* 228 (127), *233,* 278 (169), *285*
Petersen, J. W., 136 (15), 139 (15), *140*
Petersen, M. R., 45 (49), *48*
Peterson, D., 150 (32), *159*
Peterson, D. H., 28, *34,* 187 (58b), *198*
Peterson, J. W., 52 (15, 16), 79, 201 (2), *228*
Phelion, R., 274 (154), *284*
Phillips, D. H. P., 25 (117), *33*
Phillips, P. C., 87 (8), 91 (50), 92 (8), *119, 120,* 161 (3, 5), 163 (5), 165 (5), 166 (5), *173*
Pierdet, A., 201 (8), 203 (8), 204 (8, 18, 20, 21), 207 (41, 42, 43, 44, 45), 208 (42, 44, 45, 49), 209 (59), 213 (41, 59), 214 (59, 79), *229, 230, 231*
Pike, J. E., 2 (23), 5, 23 (112), 27 (112), *29, 33,* 177 (11), *197,* 248 (52), 251 (52, 61, 62, 72), *281 282*
Pillai, C. N., 16 (78), *31,* 53 (18), 79, 201 (1), 202 (1), 203 (1), *228*
Pinkney, P. S. 171 (59), *175*
Pivnitskii, K. K., 187, *198*
Platonova, A. V., 58 (53), *80,* 87 (13), 88 (20), 89 (27, 28), 90 (20, 36, 41), 91 (20, 44), *119, 120*
Plenio, H., 2 (25), 5, 19 (25), 25 (25), *29,* 242 (28), *280*
Pletcher, J., 277 (168), *285*
Podobrazhnykh, S. D., 152 (41), *159*

Poirier, R. H., 78, *84*
Poittevin, A., 194 (94), *200*
Pollman, M. J. M., 153 (43, 44, 45), 154 (44), *160*
Ponsold, K., 26 (122), *83*
Poos, G. I., 11 (52), 25 (120), *31, 33,* 65 (113), *82,* 182 (27, 28, 30, 32, 33, 34, 36, 37, 38, 39), 184 (33, 38), 185 (37, 39), 186 (33, 34), *197, 198*
Popova, E. V., 19 (93), *32,* 151 (39, 40), 152 (41), *159,* 243 (38), *280, 290*
Popp, F. D., 17 (82), *32,* 77, 78 (199), *84,* 111 (168), *124*
Porter, Q. N., 131 (44), *133*
Poselenov, I. A., 110 (159), *124*
Pradhan, S. K., 227 (126), *233*
Prelog, V., 241 (18), 242 (18, 25), *280*
Prema Madyastha, B. R., 194 (98), *200*
Pride, E., 241 (16), *280*
Prieto, A. P., 241 (18), 242 (18), *280*
Prinz, E., 242 (27), *280*
Prost-Marechel, J., *291*
Protiva, M., 136 (10), 138 (22, 23, 24), *140, 141*
Prout, F. S., 56 (38), *80*

Q

Quartey, J. A. K., 105 (128), 106 (134), *123,* 227 (122), *233,* 240 (6), *279*
Quinkert, G., 151 (35), *159,* 242 (29, 30), *280*

R

Raffelson, H., 12 (56), *31,* 223 (115), *232,* 262 (109), 263 (109, 114, 115), 264 (116), 265 (116), *283*
Ragault, M., 69 (146, 147), *82, 83*
Rakshit, U., 235 (12), *238*
Ralhan, N. K., 169 (40, 41, 42), *174*
Ralls, J. W., 2 (21), 4, 29, 56 (39, 40), *80,* 139 (39), *141,* 181 (25, 26), *197*
Ramachandra, R., 99 (100), *122*
Ramadas, S. R., *288*
Ramage, G. R., 170 (49, 50), *174*
Ramani, Gi, 115 (181), *125*
Rampal, A. L., 166 (38), 169 (38, 43), *174*
Rao, C. S., 237 (20), *238*

Rao, K. B., 169 (42), *174*
Rao, M. P., 227 (126), *233*
Rao, P. N., 71, *83*, 93 (74, 75), *121*
Rapala, R. T., 210 (60), *230*
Rapson, W. S., 106 (135), 107 (135, 137), 114 (135), *123*, 146 (12), *159*
Rasmusson, G. H., 227 (124), *233*
Rautenstrauch, C., 129 (9), *132*
Ray, F. E., 171 (58), *175*
Razdan, R. K., 156 (56), *160*
Re, L., 92 (63), 106 (131), *121*, *123*
Redeuilh, G., 169 (44), *174*
Redman, B. T., 243 (34), *280*
Rees, R., 25 (117), 27 (134), *33*, *34*, 91 (50), *120*, 161 (5, 11, 12, 20), 163 (5), 165 (5), 166 (5, 11, 12), *173*, *174*
Reich, H., 176, *197*
Reichstein, T., 24 (116), *33*, 186 (48), 187 (59, 60, 62), 189, *198*, *199*
Reiff, K., *289*
Reine, A. H., 168, *174*
Reineke, L. M., 28 (138), *34*, 187 (58b), *198*
Renfrow, A., 139, *141*
Renfrow, W. B., 139, *141*, 178, *197*
Reus, H. R., 153 (45), *160*
Rich, D. H., 43 (41), 44 (41), 45 (41), *48*
Richter, J. W., 187 (55), *198*
Riegel, B., 56 (38), *80*
Rieke, A. C., 218 (90), *231*
Rijsenbrij, P. P. M., 77 (183), *84*
Riniker, B., 218 (98, 99), *232*
Riniker, R., 218 (98, 99), *232*
Rizzi, G. P., 257 (96), 258 (96), *282*
Roach, L. C., 26 (123), *33*
Robins, P. A., 9 (39), 16 (73), *30*, *31*, 58 (58), 61 (58), *80*, 127 (2), 129 (2, 5), *131*, *132*, 186 (49), *198*
Robinson, F. M., 182 (36), 187 (56), *197*, *198*
Robinson, J. M., 182 (28), *197*
Robinson, M. J. T., 224 (116), *232*
Robinson, R., 2 (20), 3 (20), 4, 11 (51, 53), 16 (20, 79), 17 (86), 26 (53), *29*, *30*, *31*, *32*, 58 (48, 49), 60, 66 (118, 119), 67, 68, *80*, 82, 106 (133, 135, 136), 107 (135, 137, 138, 139, 140), 108 (152), 112 (171, 172), 114, 117 (178), *123*, *124*, *125*, *138*, 139 (37, 38), *141*, 146 (12), *159*, 170, *174*, 176, 177, (9, 10, 11), *196*, *197*, 201, 203, 219, 228, *232*, 235 (9), *238*, 239 (1), 240 (5), 241 (9), 244 (42), *279*, *280*
Rodig, O. R., 56 (43), *80*
Rogers, N. A. J., 118 (205), *125*
Rogier, E. R., 10 (46), 22 (46), *30*, 244 (40, 41), 245 (45), 246 (45), 247 (47), *280*, *281*
Rosen, P., 13 (61), *31*, 157 (57), *160*, 276 (163), *284*
Rosenberger, M., 267 (130, 131), 268 (131, 136), 270 (137), 271 (131, 136, 138), *283*, *284*, *292*
Rosenkranz, G., 3 (31), *30*, 196 (117), *200*, 208 (57, 58), 212 (58), *230*
Ross, F. T., 218 (95), *232*
Rossberg, P., 2 (25), 5, 19 (25), 25 (25), *29*, 242 (28), *280*
Roy, J., 115 (179), *125*
Roy, S. K., 218 (90), *231*
Rubin, M. B., 187 (62a), *198*, 251 (71, 72), *281*, *282*
Ruccini, J., 13 (63), *31*
Ruchlak, K., 103 (111), 111 (111), *122*
Rufer, C., 27 (130), *33*, 92 (67), 94 (80), *121*
Ruppert, J., *289*
Ruschig, H., 185, 189, *198*
Russey, W. E., 37 (20), *47*
Ruzicka, F. C. J., 96 (88), 103 (88, 115), *122*, *123*
Ruzicka, L., 35 (3, 4, 5), 39 (5), *46*, *47*, 103 (113), *123*, 206, 229
Rydon, H. N., 114 (178), 117 (178), *125*
Rzheznikov, V. M., 87 (10), 90 (10), 91 (10, 45, 46, 48, 52, 56), *119*, *120*, *121*

S

Sabo, E. F., 206 (33), 220 (33), *229*
Saburova, L. A., 110 (159), *124*
Sadovskaya, V. L., 90 (40), *120*, 220 (110), *232*
Sakai, K., 150, *159*, *290*
Sakena, A. K., 144 (4), *158*
Salmond, W. G., 3 (35), *30*
Sannie, C., 241 (20), *280*

Sarett, L. H., 11 (52), 25 (120), *31*, *33*, 65, *82*, 182, 184, 185, 186 (33, 34, 47), *197*, *198*, 213 (74), *231*, 253 (76), *282*
Sarhan, A., *287*
Sarkar, A. K., 70 (153), *83*
Sastry, G. R. N., 73 (173), 74 (173), *83*
Saucy, G., 2, 13 (66), 27 (66), 29, *31*, 39 (27), 47, 267 (127, 128, 129, 130, 131, 132, 133, 134), 268 (131, 134, 135, 136), 270 (133, 135, 137), 271 (131, 136, 138), 272 (132, 133), *283*, *284*, *290*, *292*
Sauer, G., 27 (126), *33*, 98 (95), *122*, 163 (28), *174*, 266 (124), *283*, *289*
Sausen, G. N., 16 (74), *31*, 248 (54), *281*
Sawicki, E., 171 (58), *175*
Sax, M., 277 (168), *285*
Schaap, W. B., 116 (190), *125*
Schäger, H., 158 (64), *160*
Schaffner, K., 63 (79), 69 (152), *81*, *83*, 194 (104, 105), *200*
Scharf, D. J., 56 (41), *80*, 227 (121), *232*
Schick, H., 102 (104), *122*
Schiess, P. W., 251 (68), 257 (68), 258 (68), *281*
Schindler, O., 24 (116), *33*, 186 (54), 187 (62, 63), 189 (62), 191 (63), *198*
Schleigh, W. R., 17 (82), *32*, 77, *84*, 111 (168), *124*
Schlosser, M., 45 (50), *48*
Schlotz, A. G., *289*
Schmidlin, J., 2 (24), 5, 16 (75), 25 (120), 27 (127), 29, *31*, *33*, 187, 189 (61, 67, 69, 70, 74, 75, 76), 192 (64, 66, 67, 68, 69, 70, 79, 80), *198*, *199*, 253 (77, 78, 79), 254 (79), *282*, *291*
Schmiegel, K. K., 45 (45), *48*
Schmitt, J., 129 (6, 8, 9), 131 (19, 20), *132*
Schneider, W. P., 9 (41), 15 (71), *30*, *31*, 58 (54, 55), 61 (54, 55), *80*, 91 (43), *120*, 129 (4), *131*, 135 (1), *140*, 143 (1, 3), 144 (3), *158*, 187 (58a), *198*, 206 (35), *230*, 248 (50), 254 (80), 275 (158), *281*, *282*, *284*
Schneller, J., 169 (40), *174*
Schon, O., 131 (20), *132*
Schramm, G., 49 (3), *79*

Schreiber, J., 234 (7), 235 (7), *238*
Schroder, E., 27 (130), *33*, 92 (67), 94 (80), *121*
Schuck, J. M., 139 (43), *141*
Schütz, S., 2 (25), 5, 19 (25), 25 (25), 29, 151 (35), *159*, 242 (28, 29, 30), *280*
Schumacher, U., *289*
Schwam, H., 105 (130), *123*, 277 (166), *285*
Schwartz, M., 38 (22), *47*
Schwarz, S., 26 (122), *33*
Schwarzel, W. C., 279 (170), *285*
Sciavolino, F. C., 93 (79), *121*
Scott, A. I., 18 (89), *32*, 67 (126), 68 (126), *82*, 115 (180), *125*
Scott, J. W., 3 (34), 13 (66), 27 (66), *30*, *31*, 69 (150), *83*, 251 (69), 267 (134), 268 (134), *281*, *284*
Sears, C. A., 139 (27), *141*
Segal, G. M., 27 (132), *33*
Seiler, M. P., 36 (10), 39 (10), 41 (10), *47*
Semmelhack, M. F., 43 (39), 44 (39), 45 (39), *48*
Sen, B. P., 259 (100), *282*
Sen, H. K., 69, *83*, 178, *197*
Sengupta, P., 136 (8, 9), 138 (18), *140*, *141*
Sengupta, S. C., 103 (109), *122*
Serebryakova, T. A., 90 (41), *120*, *288*
Shafer, P. R., 249 (57), *281*
Shapkina, E. V., 91 (49), *120*
Sharp, J., 107 (141), *123*
Sharpless, K. B., 37 (16), 39 (29), *47*
Shavel, J., 10 (154), *124*
Sheehan, J. C., 58 (61, 67), 61, *80*, *81*
Sheinker, Yu, N., 237 (23), *238*
Shelberg, W. E., 15 (71), *31*, 58 (55), 61 (55), 66 (120), *80*, *82*, 91 (43), *120*, 129 (4), *131*, 143 (3), 144 (3), *158*, 248 (50), 249 (60), *281*
Sherman, A. H. 139 (35), *141*
Shick, H., 241 (21), *280*
Shimizu, R. Z., 56 (36), *80*
Shimojima, H., 27 (129), *33*, 203 (16), 204 (23, 24), *229*
Shmonina, L. I., 20 (97), *32*, 131 (26, 33, 34), *132*

Shner, V. F., 237 (23), *238*, *291*
Shoppe, C. W., 178 (17), *197*
Shoun, E., 139 (27), *141*
Shuck, J. M., 240 (3), *279*
Shulman, S., 68 (144), *82*
Shunk, C. H., 56 (37), *80*, 262 (108), *283*
Shvo, Y., 78 (197), *84*
Siddall, J. B., 22 (107), 25 (117), *32*, *33*, 87 (2, 8), 88 (2), 89 (2), 90 (2), 91 (50), 92 (8), 93 (78a), *118*, *119*, *120*, *121*, 147 (22), 151 (22), *159*, 161 (3, 4, 5, 10), 162 (4), 163 (4, 5), 164 (4, 10), 165 (4, 5), 166 (5), *173*, 220 (111), *232*
Siddall, J. B., 87 (8), 92 (8), 93 (78a), *119*, *121*, 161 (3), *173*
Siegmann, C. M., 187 (63), 191 (63, 77, 78), *198*, *199*
Sieman, H.-J., 26 (122), *33*
Sih, C. J., 27 (128), *33*, 203 (15), *229*
Simak, V., 136 (10) *140*
Simpson, S. A., 187 (59, 60), *198*
Simpson, W. R. J., 64 (103), *81*, 99 (98), *122*
Singh, G., 129 (7, 10), *132*
Singh, I., 118 (207), *125*
Singh, M., 77 (184), *84*, 107 (124), *123*
Sircar, J. C., 109 (153, 155, 156), 110 (157), *124*, 224 (118), *232*
Siuda, J., 25 (117), *33*
Sivanandaiah, K. M., 58 (62), *80*, 136 (13, 14), *140*
Skufca, P., *78*
Skvarchenko, V. R., 116 (191), *125*
Sladkov, V. N., 237 (23), *238*, *291*
Slater, S. N., 17 (86), *32*
Slates, H. L., 241 (8), *279*
Sluyter, M. A. T., 94 (82), *122*
Smith, D. C. C., 147 (17), *159*
Smith, H., 1, 9 (40), 10 (47), 12 (40), 21 (100), 22 (107), 25 (117), 26 (121), *29*, *30*, *32*, *33*, 58 (63), *80*, 87 (2, 4, 8, 15), 88 (2, 4), 89 (2, 4, 24, 25, 31, 32, 33), 90 (2, 4), 91 (50), 92 (8, 32, 68), 93 (15, 73), 94 (86), 95 (86), 98 (86, 96), 103 (107), 104 (116), 105 (128), *118*, *119*, *120*, *121*, *122*, *123*,*125*, 137 (16, 17), *141*, 146 (13, 14), 147, 151 (22), *159*, 161, 162, (4, 9), 163 (4, 5), 164 (2, 4, 10), 165 (4, 5, 23), 166 (5, 7, 16), 171 (52), *173*, *174*, 220 (111), 227 (122), *232*, *233*, 236 (19), *238*, 240 (6), 241 (16), 243 (36), *279*, *280*, *288*
Smith, L. L., 25 (117), 27 (134), *33*, *34*, 91 (50), *120*, 161 (5, 11, 12), 163 (5), 165 (5), 166 (5, 11, 12), *173*
Smith, N. R., 139 (40), *141*
Snatzke, G., 274 (156), *284*
Snozzi, C., 206 (39), *230*
Sobotka, W., 107 (147), 109 (147, 153), *124*, 158 (60), *160*
Solo, A. J., 22 (108), *33*, 66 (117), *82*, 171 (53), *175*
Soloman, D. M., 277 (165), *285*
Solomn, P., 277 (168), *285*
Soloway, A. H., 35 (2), *46*
Sondheimer, F., 2 (19, 28, 27), 3 (27, 28, 31, 35), 4, 6, 12 (19), 18 (19), 27 (19), *29*, *30*, 176 (4, 5), 180 (18), *196*, *197*, 236 (17), *238*, 261 (105, 106), 263 (111, 112, 113), *283*
Sorkina, T. I., 131 (24, 40, 41, 42, 43), *132*, *133*
Sorm, F., 194 (97), *200*
Soto, J. L., 156 (54), *160*
Soto, M., 156 (54), *160*
Soto Martinez, M., 156 (53), *160*
Soulier, J., 78 (200), *84*
Specht, H., 93 (77), *121*
Speckamp, W. N., 10 (48, 49), 21 (48, 101), *30*, *32*, 76 (181), 77, *83*, *84*, 92 (72), 94 (82, 83, 84, 87), 96 (89, 90), 97 (91, 92), 101 (102, 103), 102 (72, 103), 105 (125), *121*, *122*, *123*, *289*
Speziale, A. J., 261 (107), *283*
Sprague, J. W., 117 (193), *125*
Srinivasan, A., 71 (167), *83*
Srinivasan, P. R., *290*
Stache, U., 2 (25), 5, 19 (25), 25 (25), *29*
Stadler, P., 36 (7), *47*
Stalmann, L., 10 (46), 22 (46), *30*, 245 (45), 246 (45), *281*
Stanaback, R. J., 107 (145), *124*, 166 (30, 31, 32, 33, 34, 35), 167 (32, 33), 168 (30, 33, 34, 35), *174*, *288*
Stanovnik, B., *78*
Staskun, B., 43 (41), 44 (41), 45 (41), *48*
Steege, W., 97 (92), *122*

Steelman, S. L., 213 (74), *231*
Stefancich, G., *288*
Stein, R. P., 161 (23), 165 (23), *174*
Stephens, J. A., 261, *283*
Stevens, T. E., 181, *197*
Stevenson, A. C., 13 (60), *31*, 55 (22), 58 (22), 60 (22), 61 (22), 65 (22), *79*
Stewart, R. C., *286*
Stipanovic, R. D., 46 (51), *48*
Stojanac, Z., 129 (11), *132*
Stork, G., 2, 10 (44), *13*, 14 (11, 44, 68), 22 (108), *29*, *30*, *31*, *33*, 35, 36 (6), 39 (6), *47*, 49 (10, 11), 50, 55 (11), 65 (11), 66, 79, 82, 129 (10), *132*, 157 (57), *160*, 171 (53), *175*, 206 (30), 222, 224 (117), 229, 232, 236 (18), *238*, 242 (32), 260, 265 (118), 272 (104), 139), 275, 276, *280*, *283*, *284*, *286*
Stotter, P. L., 14 (68), *31*, 242 (32), *280*
Stournas, S., 13 (64), *31*
Stoutmire, D. W., 258 (99), *282*
Strike, D. P., 87 (15), 93 (15), *119*, 161 (14, 19, 20), *173*, *174*
Stromberg, V. L., 52 (17), 54, 55 (21), *79*
Stubenrauch, G., *289*
Subba Rao, G. S. R., 77, *84*, 93 (78a), 105 (123), 113 (173), 114 (174), *121*, *123*, *125*, 222, 232
Süs, O., 158 (64), *160*
Suffness, M. I., 39 (24), *47*
Sugavanam, B., 53 (19), 79, 219 (102), 221 (102), 222 (102), *232*
Suida, J., 87 (8), 91 (50), 92 (8), *119*, *120*, 161 (3, 5), 163 (5), 165 (5), 166 (5), *173*
Sujeeth, K. P., *288*
Sultanbawa, M. U. S., 43 (39), 44 (39), 45 (39), *48*
Summers, G. H. R., 178 (17), *197*
Sutton, R. E., 21 (102), *32*, 56 (34, 35), *80*, 171 (51), *174*
Suvorov, N. N., 237 (23), *238*, *291*
Swaminathan, S., 68 (141), 82, 241 (10), *280*
Swoboda, J. J., 42 (34), *47*, *287*
Szmuszkovicz J., 10 (46), 22 (46), *30*, 146 (11), *159*, 206 (30), 229, 244 (40, 41), 245 (45), 246 (45), *280*, *281*

Szpilfogel, S. A., 187 (63), 191 (63, 77, 78), *198*, *199*

T

Tadanier, J., 210 (66), *231*
Tait, J. F., 187 (59, 60), *198*
Takasugi, M., 69 (149), *83*
Takeda, K., 23 (111), *33*, 66 (123, 124), 67 (123), 82, 194 (111, 112, 113), *200*, 248 (55), 254 (82, 83), 255 (55, 85, 88), 256 (82), *281*, *282*
Tamm, C., 186 (48), *198*
Tamura, Y., 104 (122), *123*, 139 (41), *141*
Tankart, M. H., *288*
Tarney, R. E., 20 (96), *32*, 136 (11, 12), *140*
Taub, D., 2 (19), 4, 12 (19), 18 (19), 27 (19), *29*, 87, 91 (59, 60), 106 (131), *119*, *121*, *123*, 176 (4), *196*, 236 (17), *238*, 261 (105, 106), 263 (111, 112, 113), 277 (167), *283*, *285*
Taylor, D. A. H., 194 (109, 110), *200*
Taylor, E. C., 17 (83), *32*, 78 (197), *84*, 92 (71), *121*, 169 (45), 170 (45), *174*
Tchen, S.-Y., 43 (43), *48*
Terasawa, T., 10 (43), 23 (111), *30*, *33*, 66 (121, 124), 67 (135), 82, 194 (107, 108, 111, 112, 116), *200*, 248 (55), 254 (82, 83), 255 (55, 85, 88, 89), 256 (82, 89, 94), *281*, *282*
Terekhova, L. N., 131 (26, 36, 37), *132*, *133*
Terrell, R., 206 (30), *229*
Tessier, J., 58 (66), *80*, 201 (8), 203 (8), 204 (8, 20, 21), 206 (29, 36), 207 (41), 208 (53, 54, 55), 209 (59), 213 (41, 59), 214 (59), *229*, *230*
Thill, R. J., 109 (152a), *124*, 169 (47, 48), 170 (47), *174*
Thomas, D. G., 201 (3), 203 (3), *228*, 237 (21), *238*
Thompson, J. M. C., 110 (165), *124*
Thompson, Q. E., 12 (56), *31*, 223 (115), 232, 261 (107), 262 (109), 263 (109),

264 (116, 117), 265 (116, 119, 120), 283
Thornton, R. E., 9 (40), 12 (40), 30, 137 (17), 141, 236 (19), 238
Tien, Y. L., 110 (162, 164), 124
Tikhomirova, O. B., 131 (41), 133
Tikotkar, N. L., 71 (165, 166), 77 (188, 189), 83, 84, 139 (42), 141, 158 (62), 160
Tilak, B. D., 21 (103), 32, 71, 73, 74, (171, 173, 174, 175), 77 (188, 189), 83, 84, 139 (42), 141, 158 (62), 160
Tishler, M., 241 (8), 279
Tisler, M., 78
Titov, Yu. A., 1, 29, 71 (157), 83, 117 (196), 125
Tober, E., 242 (30), 280
Tochtermann, W., 289
Todd, W. M., 112 (172), 125, 235 (9), 238
Tokolics, J., 25 (117), 33, 87 (8), 91 (50), 92 (8), 119, 120, 161 (3, 5), 163 (5), 165 (5), 166 (5), 173
Tomasewski, A. J., 205 (27), 229
Tomasik, G., 291
Tomkins, P. M., 242 (26), 280
Torelti, V., 25 (119), 33, 204 (19), 229, 243 (35, 37), 274 (35, 37), 278 (35), 280, 290
Torgov, I. V., 1, 9 (38), 12 (57, 59), 19 (59), 22 (59), 27 (131, 132), 29, 30, 31, 33, 58 (53, 64), 80, 87 (1, 5, 9, 10, 13), 88 (1, 20, 21), 89 (1, 5, 22, 27, 28, 29, 30), 90 (1, 9, 10, 20, 22, 35, 36, 39, 40, 41, 42), 91 (10, 20, 44, 45, 46, 47, 48, 49, 52, 53, 54, 55, 56, 58), 92 (30, 61, 62), 104 (117, 118), 105, 118, 119, 120, 121, 123, 130 (12), 131 (22, 23, 24, 25, 27, 28, 33, 40, 41, 42, 43), 132, 133, 187, 198, 220, 222, 232, 234 (4, 5, 6), 238, 288
Toromanoff, E., 58 (66), 80, 201 (8), 203 (8), 204 (8), 206 (29, 36), 210 (61), 213 (73), 229, 230, 231
Torupka, E. J., 258 (98), 282
Touet, J., 69 (146, 147), 82, 83
Tourwe, D., 172 (62), 175
Towns, R. L. R., 107 (146), 124, 166 (36), 168 (36), 174

Trefonas, L. M., 107 (146), 124, 166 (36), 168 (36), 174
Trost, B. M., 14, 31
Truett, W. L., 28 (135), 34, 251 (64), 281
Trus, B., 290
Tsai, M., 290
Tsatsos, W. T., 111 (170), 124, 235 (16), 238
Tsuji, J., 157 (57), 160, 276 (163), 284
Tuinman, A., 69 (152), 83, 222, 232, 291
Turchin, K. F., 237 (23), 238
Turner, D. L., 20 (94), 32, 63 (96), 81, 135, 140
Twine, M. E., 218 (95), 232
Tyner, D. A., 2 (21), 4, 29, 56 (36, 40), 80, 181 (26), 197

U

Überwasser, H., 16 (75), 25 (120), 31, 33, 187 (64, 65, 66, 67, 68), 189 (67, 74, 75, 76), 192 (64, 65, 66, 67, 68, 79, 80, 81, 82, 85, 86, 87, 88, 89, 92, 93), 198, 199, 200, 253 (78, 79), 254 (79), 274 (152, 153), 282, 284
Uehlinger, H. P., 186 (48), 198
Uhle, F. C., 3 (30, 32), 30, 180 (19), 197
Urquiza, R., 218 (93), 231
Uskokovic, M., 205 (26), 229, 274 (154), 284

V

Vail, O. R., 218 (88), 231
Valenta, Z., 129 (11), 132
Valls, J., 201 (5), 202 (5), 203 (5), 204 (5), 229, 274 (143), 278 (143), 284
Van Binst, G., 172 (62), 175
van Brynsvoort, J., 92 (72), 102 (72), 121
van Dalen, A. C., 74 (179), 83, 88 (18), 119
Vandegrift, J. M., 182 (28), 197
van der Burg, W. J., 187 (63), 191 (63, 77, 78), 198, 199
van der Gen, A., 42 (34), 47, 287
vander Vlugt, F. A., 74 (179), 83, 88 (18), 119

van Dorp, D. A., 187 (63), 191 (63, 77, 78), *198, 199*
Vanstone, A. E., 104 (121), *123*, 147 (20), 148 (20), *159*, 242 (23), *280*
van Tamelen, E. E., 36, 37 (13, 15, 16, 21), 38 (22), 39 (8, 9, 10, 23, 24, 25, 26, 28, 29), 40 (8), 41 (9, 10), *47*
van Velthyusen, J. A., 96 (89, 90), *122*
Velarde, E., 187 (58c), *198*
Velluz, L., 1, 12 (54), 25 (54), *29, 31*, 58 (66), *80*, 194 (94), *200*, 201, 202 (5, 6), 203 (5, 8, 9), 204 (5, 8, 9, 19), 206 (9, 29, 36), 207 (6, 41), 208 (49), 210 (61), 212 (69), 213 (6, 41, 70, 73), 214 (6), 215 (9), 220 (9, 105), *229, 231, 232*, 243 (35), 274 (35, 143), 278 (35, 143), *280, 284*
Venkiteswaren, 73 (171), 74 (171), *83*
Verkholetova, G. P., 104 (117), *123*, 131 (22, 23, 27, 28), *132*, 234 (5), *238*
Vida, J. A., 206, (31, 32), 220 (31, 32), *229*
Viel, C., 169 (44), *174*
Vignau, M., 58 (66), *80*, 165 (29), *174*, 206 (36, 37), 210 (61), 213 (70, 73), 214 (75), *230, 231, 290*
Villani, F. J., 65 (112), *82*
Vischer, E., 27 (127), *33*, 189 (69, 72), 192 (69, 72), *199*
Volkmann, R. A., *287*
von Eck, R. R., 117 (202, 203), *125*
von Pechmann, H., 156, *160*
von Strandtmann, M., 110 (158), *124*
Voser, W., 35 (3), *46*
Vossing, R., 27 (130), *33*
Vredenburgh, W. A., 23 (112), 27 (112), *33*, 111 (170), *124*, 235 (16), *238*, 251 (62), *281*
Vul'fson, N. S., 90 (40), 91 (47, 53, 54, 55), *120, 121*

W

Waddington-Feather, S. M., 151 (37, 38), *159*
Wakabayashi, T., 256 (92), *282*
Waksmunski, F. S., 100 (101), *122*

Walk, C. R., 10 (47), *30*, 93 (73), 94 (86), 95 (86), 98 (86), *121, 122*, 161 (8, 9, 10, 13, 18), 162 (9), 164 (10), *173, 174*, 220 (111), *232*
Walker, J., 9 (39), 16 (73), *30, 31*, 58 (48, 49, 58), 60 (48, 49), 61 (58), *80*, 127 (2), 129 (2, 5), *131, 132*, 138, *141*, 186 (49), *198*
Wall, M. E., 218 (95), *232*
Walton, E., 187 (55, 56), *198*
Wang, K. C., 27 (128), *33*, 203 (15), *229*
Warnant, J., 214 (77), 215 (77), *231*, 274 (142), *284, 291*
Warnhoff, E. W., *31*, 58 (59), 59 (59), 61 (59), *80*, 144 (5), *159*
Warnock, W. D. C., 258 (98), *282*
Watanabe, H., 139 (44, 45), *141*
Watson, D. H. P., 25 (117), *33*, 87 (8), 91 (50), 92 (8), *119, 120*, 161 (3, 5), 163 (5), 165 (5), 166 (5), *173*
Weber, H. P., 93 (78), *121*
Wehrli, H., 63 (79), *81*, 194 (104, 105), *200*
Weidlich, H. A., 114, 117 (197, 198), *125*
Weill-Raynal, J., 27 (125), *33*, 165 (29), *174*, 274 (144), *284*
Weintraub, A., 28 (138), *34*, 187 (58b), *198*
Welch, S. C., *290*
Wendler, N. L., 87, 89 (26), *119*, 241 (8), *279*
Wendt, G. R., 25 (117), *33*, 87 (8), 89 (32), 91 (50), 92 (8, 32), *119, 120*, 161 (3, 5, 10), 163 (5), 164 (10), 165 (5), 166 (5), *173*, 220 (111), *232*
Weng-Luan, L., 116 (191), *125*
Wenkert, E., 181, *197*
Werthemann, L., 37 (17, 18), 45 (49), *47, 48*
Westra, J. G., 10 (48), 21 (48), *30*, 101 (102, 103), 102 (102), *122*
Wettstein, A., 2 (24), 5, 16 (75), 25 (120), 27 (127), *29, 31, 33*, 187, 189 (61, 67, 69, 70, 72, 74, 75, 76), 192 (64, 65, 66, 67, 68, 69, 70, 72, 79, 80, 81, 82, 83, 84, 85, 86, 87, 88, 89, 90, 91, 92, 93),

194 (99, 100), *198*, *199*, *200*, 253, 254 (79), *282*
Weygand, F., 66 (118), *82*
Wharton, P. S., 224 (117), *232*, 236 (18), *238*
Wheeler, D. M. S., 218 (88, 89, 90), *231*
Whelan, H., 97 (93), *122*
Whetstone, R. R., 22 (104), *32*, 219 (100), *232*
White, M. A., 61 (71), *81*
Whitehurst, J. S., 58 (65), *80*, 87 (2, 6), 88 (6), 104 (121), *118*, *119*, *123*, 147 (18, 19, 20, 23, 24), *159*, 163 (26), *174*, 241 (15, 19), 242 (23, 24), *280*, *288*
Whitlock, H. W., Jr., 3 (34), *30*, 69 (148), *83*, 251 (70), 258 (70), *281*
Wicki, H., 186 (51, 52), *198*
Wiechert, R., 16 (76), 27 (126), *31*, *33*, 163 (28), *174*, 266 (124), *283*, *289*
Wiedhaup, K., 8 (36), *30*, 42 (34, 36), *47*, *286*, *287*
Wieland, P., 16 (75), 25 (120), *31*, *33*, 187 (64, 65, 66, 67, 68), 189 (67, 74, 75, 76), 192 (64, 65, 66, 67, 68, 79, 80, 81, 82, 83, 84, 85, 86, 87, 88, 89, 90, 91, 92, 93), 194 (99, 100), *198*, *199*, *200*, 241 (7), 253 (78, 79), 254 (79), 274 (149, 152, 153), *279*, *282*, *284*
Wierenga, W., 36 (10), 39 (10), 41 (10), *47*
Wijnberg, J. B. P. A., 77 (183), *84*
Wildman, W. C., 56 (39), *80*, 139 (39), *141*, 181 (25), *197*
Wilds, A. L., 2 (17, 21), 4, 8 (17), 13 (60), 16 (17), 17 (17), 19 (17), 21 (102), 22, 26 (17), *29*, *31*, 32, 33, 49, 53, 54, 55 (2), 56, 59 (1), 63, 78 (2), 79, *80*, *81*, 139 (39), *141*, 171 (51), *174*, 181, *197*, 246, 247, 262 (108), *281*, *283*
Wiley, R. H., 139 (40), *141*, 171 (59), *175*
Wilkins, C. K., 75 (180), *83*
Wille, H. J., 171 (60), *175*
Willems, A. G. M., 117 (202, 203), *125*
Willett, J. D., 37 (21), 38 (22), 39 (23), *47*
Williams, D. H., 151 (37, 38), *159*
Williams, T., 274 (154), *284*

Williamson, K. L., 68 (144), *82*, 111 (170), *124*, 235 (16), *238*
Willy, W. E., *287*
Wilson, A. N., 187 (55, 56), *198*
Windholz, T. B., 87 (3, 11), 88 (3), 89 (3), 92 (65), 94 (81), 100 (101), 101 (81), 105 (130), 106 (131), *118*, *119*, *121*, *122*, *123*, 213 (74), *231*, 277 (166, 167), *285*, *289*
Winter, J., 255 (84), 256 (84), *282*
Winternitz, F., 131 (38, 39), *133*
Wolfe, J. F., 67 (131), *82*
Wolff, M. E., 194 (95, 96), *200*
Wollensak, J. C., 227 (124), *233*
Wong, H. N. C., *14*
Woo, E. P., *14*
Wood, J., 77, *84*
Wood, W. D., *31*, 58 (59), 59 (59), 61 (59), *80*, 144 (5), *159*
Woodward, R. B., 2 (19, 22), 3 (22), 4, 5, 12 (19), 18 (19), 27 (19), 29, 35 (1), *46*, 176, *196*, *197*, 236 (17), *238*, 261 (105, 106), 262 (110), 263 (111, 112, 113), *283*
Wulff, G., *287*
Wulfson, N. S., 220 (110), *232*
Wynberg, H., 10 (46), 22 (46), *30*, 244 (41), 245 (45), 246 (45), *280*, *281*

Y

Yamazaki, T., *291*
Yang, R., 267 (131), 268 (131), 271 (131), *283*
Yashin, R., 3 (31), *30*
Yoo, S. C., 277 (168), *285*
Yorka, K. V., 28 (135), *34*, 251 (63, 64), *281*
Yoshikoshi, A., 148 (25), *159*
Yoshioka, K., 92 (70), *121*, *289*
Yoshioka, M., 67 (132, 133, 135), *82*, 194 (114, 116), *200*, 255 (86, 90, 91), 256 (94), *282*
Yoshioko, M., 279 (170), *285*

Z

Zabrocki, K., *287*
Zaikin, V. G., 91 (53, 54, 55), *120*, *121*

Zakharychev, A. V., 87 (17), 88 (21), 89 (22, 27, 28, 29, 30), 90 (22, 41), 91 (56), 92 (30, 61, 62), *119, 120, 121, 288*
Zaretskaya, I. I., 131 (24, 27, 28, 40, 41, 42, 43), *132, 133*
Zaretskii, V. I., 90 (40), 91 (47, 53, 54, 55), *120, 121,* 220 (110), *232*

Zav'yalov, S. I., 10 (45), 20 (97), *30, 32,* 104 (119, 120), *123,* 147 (15), *159,* 234 (1, 2, 3), *238,* 241 (13), *280*
Zderic, J. A., 63 (83), *81*
Zeitschel, R. H., 21 (102), *32,* 56 (35), *80*
Ziffer, H., 117 (192, 193), *125*
Zimmer, H., 78 (198), *84*

Subject Index

A

Absolute configuration, determination of, 28
17β-Acetamido-3,6-dimethoxy-1,3,5(10),6,8-estrapentaene, 58
5-Acetamido-2-naphthol, 177
2-Acetylaminocyclopentane-1,3-dione, 100, 102
2-Acetyl-1,3-cyclohexanedione, 118
1-Acetylcyclohexene, 106, 107
1-Acetylcyclopentene, 106
1-Acetyl-3,4-dihydronaphthalene, 68
17-Acetyl-5α-etiojerva-12,14,16-trien-3β-ol, 69
α-Acetylglutaric esters, acylation of, 110
2-Acetyl-6-methoxynaphthalene, 112
α-Acetylsuccinic ester, acylation of, 110
2-Acetyl-4-thiacyclohexene, 108
Acrylic acid, 12
Acrylonitrile, 111
Acylation, 16, 110
Acyloin condensation, 61, 171
Adrenocortical steroids, 176, 187, 206
Adrenosterone, 206, 220
Aetioallobilianic ester, 179
Agnosterol, 3
Aldol condensation, 58, 111
Aldosterone, 5, 183, 187–194, 251–254
  partial synthesis, 194
  synthesis according to Reichstein, 189
  synthesis according to Szpilfogel, 191
  Wettstein's first synthesis, 27, 187
  Wettstein's second synthesis, 16, 192
13-Alkoxymethyl steroid, 106
Alkylation, 12, 103
  of thallium salts, 92
18-Alkyl-19-norpregnanes, 166
Alkylsteroids, see also Methylsteroids, Methylnorsteroids
13-Alkylsteroids, 86, 163, 288
Allylmagnesium bromide, 208
2-Allyl-2-methyl-1,3-cyclopentanedione, 116
10-Allylsteroids, 207
13-Allylsteroids, 92, 106
Allyltriphenylphosphonium bromide, 19
α and β designation, 24
Aluminum tert-butoxide, 108
1-(2-Aminoethyl)-6-methoxynaphthalene, 105
13β-Amino-17β-hydroxy-4-gonen-3-one, 100
5-Aminoisoquinoline, 17, 110
L-(+)-threo-2-Amino-1-p-nitrophenyl-1,3-propanediol, 203
13-Aminosteroids, 289
11-Amino-12,13,15,16-tetraaza-1,3,5(10),-8,11,14,16-gonaheptaen, 78
Amphetamine, 205
$\Delta^{4,9(11)}$-Androstadiene-3-17-dione, 206
5α-Androstane, 7
4,8,14-Androstatrien-3,17-dione, 289
Androstenedione, 27
8α-Androst-4-ene-3,17-dione, 222
Androsterone, 7, 180
Angular methylation, see also methyl iodide, 14, 16, 240, 251, 252, 255
  via hydrocyanation, 255–256
  with methyl iodide and tert-butoxide, 108, 128, 135, 136, 139, 144, 248, 264, 279
  via Stork's procedure, 276
Angular methyl groups, configuration, 24
p-Anisylcyclohexanes as AC intermediates, 134–137
Anisole, 17
β-m-Anisylethyl bromide, 58
β-m-Anisylethylmalonic ester, 58
Annelation, see also Methyl vinyl ketone, 10, 12, 13, 235, 286
  by cyclization of α-haloketals, 13
  with 1-diethylaminobutan-2-one, 242, 244
  with 1-diethylaminopentan-3-one, 244, 254
  isoxazole method, 13, 275–276
Annelation methods, review, 2
Arndt-Eistert synthesis, 18, 51, 61, 115, 179
Aromatic C-ring steroid, 94
Aromatic D-ring steroid, 10

Aromatization, palladized carbon, 206
Arthrobacter simplex, 207
Asymmetric induction, 266–273
   in the preparation of (+)-estr-4-ene-3,17-dione, 270, 271
   in the preparation of 13β-ethylgon-4-en-3,17-dione, 270, 271
   in the preparation of 17β-hydroxydes-A-androst-9-en-5-one, 267
   in the preparation of 19-norsteroids, 266–273
Axial substituents, 8
13-Aza-8-dehydroestrone, 97
6-Aza-17-dihydroequilenin, 139
6-Azaequilenin, 95, 96
6-Aza-14β-equilenin, 96
15-Azaequilenin, 78
6-Azaestradiol, 95, 140, 290
6-Azaestrogens, 77
8-Azaestrogens, 165–166
6-Azaestrone, 96
8-Azaestrone, 107
8-Aza-D-homoestradiol, 3-methyl ether, 28
9-Aza-D-homosteroids, 158
8-Aza-3-hydroxy-1,3,5(10)-estratrien-12-one, 109
8-Aza-3-hydroxy-D-homo-1,3,5(10),13-gonatetraen-17a-one, 109
8-Aza-3-hydroxy-D-homo-1,3,5(10)-gonatrien-17a-one, 104
8-Aza-17α-hydroxyprogesterone, 167–168
8-Aza-12-ketosteroids, 110
6-Aza-6-methyl-B-norequilenin, 77
10-Aza-19-nor-5α,9α-androst-8(14)-ene-3,17-dione, 172
10-Aza-19-nor-5β,9β-androst-8(14)-ene-3,17-dione, 172
13-Aza-18-norequilenin, 77
6-Aza-B-nor-D-homosteroid, 158
8-Aza-C-nor-D-homosteroids, 172–173
11-Aza-C-nor-D-homosteroids, 118
3-Aza-A-norsteroid, 237
10-Aza-19-nortestosterone, 278–279
Azaoxasteroids, 99
14-Aza-11-oxosteroid, 109
Azasteroids, 74–78, 107
4-Azasteroid, 94
6-Azasteroids, 94, 139, 140

8-Azasteroids, 104, 108, 110, 165–169, 172–173
8-Azasteroids, 11-amino, 288
9-azasteroids, 77, 158
10-Azasteroids, 171–172, 290
11-Azasteroids, 17, 111, 118
12-Azasteroids, 110
13-Azasteroids, 77, 96, 105
14-Azasteroids, 77, 78, 107
15-Azasteroid, 78
Azasteroids, review, 2
13-Aza-15-thia-18-norequilenin methyl ether, 77
13-Aza-6-thiasteroid, 102
18-Aza-6-thiasteroid, 102

B

Baeyer-Villiger degradation, 253
Benedict solution, 225
Benzo[a]fluoren-8-one, 69
trans-Benzohydrindane derivatives, 201, 202, 218
Benzoquinone, 9
Benzyl ether, protecting group, 89
3-Benzyloxy-1-butyl bromide, 13, 222
3-Benzyloxy-1-butyl iodide, 222
13-Benzyl steroids, 92
Bile acids, 7
Biogenetic-like cyclization, 287
3,5-Biscarbethoxy-1,2-cyclopentanedione, 111
2,4-Biscarbomethoxy-2,5-dihydrothiophene, 14
Bischler-Napieralski reaction, 169
Bisdehydrodoisynolic acid, 20, 63–65
cis-Bisdehydrodoisynolic acid, 64, 65, 99
   analogs, 65
$\Delta^{9(11),16}$-Bisdehydro-21-norprogesterone, 260–263
18,19-Bisnor-D-homotestosterone, 146
18,19-Bisnor-14α-hydroxyprogesterone, 65, 66
18,19-Bisnorprogesterone, 65–66, 171
Boron trifluoride, catalyst in Diels-Alder, 129
γ-Bromocrotonic ester, 55
2-Bromo-1,3-cyclohexanedione, 102
5-Bromo-2-ethylenedioxypentane, 67

Subject Index 319

4-Bromo-3-hydroxy-1,3,5(10),8,14-gonapentaen-17-one, 93
2-Bromo-6-methoxynaphthalene, 65
5-Bromo-6-methoxy-1-vinyl-tetralol, 93
α-Bromo-α-methyl succinic ester, 65
5-Bromo-2-pentanone, ethylene ketal, 13, 194
β-Bromo-p-methoxyacetophenone, 235
N-Bromosuccinimide, 210
Bufadienolides, 286
Bufalin, 3
Butenandt's ketone, 49
13-Butylsteroids, 92, 106

C

Calciferol, 25, 144
2-Carbethoxycyclohexanone, 103
2-Carbethoxycyclopentanone, 17, 103, 110
8β-Carbomethoxyestra-3,17-diol, 150
β-Carbomethoxypropionyl chloride, 77
2-Carboxycyclopentanone, 111
Carboxyethyl dimethyl m-methoxyphenyl silane, 102
2-Carboxymethyl-2-methyl-3-oxocyclopentanecarboxylic acid dimethyl ester, 115
2-Carboxymethyl-2-methyl-3-oxocyclopentanecarboxylic anhydride, 110
8-Carboxysteroids, 290
Catalyst hindrance, Linstead's hypothesis, 219
Chiral reagent, 25
4-Chloro-2-butanone, 11
  enol ether with 2,6-dichlorophenol, 14
4-Chloro-2-butyl 2-chlorophenyl thioether, 14
2-Chloro-1,3-cyclopentanedione, 94
2-Chloro-5-iodo-2-pentene, 14
4-Chloro-3-hydroxy-1,3,5(10),8,14-gonapentaen-17-one, 93
5-Chloro-6-methoxy-1-vinyl-1-tetralol, 93
5-Chloro-2-methyl-1-pentene, 224
m-Chloroperbenzoic acid, 215
Cholecalciferol, 5, 19
3β-Cholestanol, 7, 263
Cholesterol, 3, 7, 24, 176, 178, 180, 251
Claisen rearrangement, 116
  of methallyl vinyl ether, 45
Classification of syntheses, 8

Cleve's acid, 49
Conan-4-en-3-one, 195
Conessine, 3, 194, 196, 222, 224–226, 251
Conessine side chain, 13
Configuration
  at C-10, 24
  d-series, 24
  l-series, 24
Conformation, 7
Corticosteroids, 215
Corticosterone acetate, 185
Cortisol, Woodward synthesis, 18
Cortisone, 4, 182–185, 187, 206, 263
  Sarett's synthesis, 11, 182–185
  Woodward's synthesis, 12, 26, 260
Curtius rearrangement, 78
Cyanoacetic acid, t-butyl ester, 236
Cyclization
  with acetic and hydrochloric acids, 150
  with alcoholic hydrogen chloride, 152, 163
  with aluminum chloride, 144, 154, 171, 173
  with aluminum tri-tert-butoxide, 240
  with ethylamine, 235
  with formic acid, 88
  with hydrochloric acid, 88, 104, 118, 237
  with L-phenylalanine and perchloric acid, asymmetric induction, 163
  with phosphorus oxychloride, 97, 167, 168
  with phosphorus pentachloride, 115
  with phosphorus pentoxide, 88, 110
  with piperidine acetate, 191
  with polyphosphoric acid, 105, 111, 113, 116, 118, 146, 148, 150, 154, 155, 163, 173
  by pyrolysis over lead carbonate, 144
  with sulfuric acid, 223
  with p-toluenesulfonic acid, 87, 88, 98, 147, 163, 172, 242, 270, 277
  with triethylammonium benzoate, 163
1,2-Cyclohexanedione, 106
1,3-Cyclohexanedione, 9, 104, 109, 234
  dienol ether, 12
  enol ether, 117
Cyclopentanone, 108

D

Darzens cyclization, 55
11-Dehydroaldosterone, 291
8-Dehydro-13-azaestrone, 77
Dehydrobromination, 212
Dehydrocorticosterone, 185
14-Dehydro-3-deoxy-B-nor-6-thiaequilenin, 73
Dehydro-*cis*-doisynolic acid, 65
Dehydroepiandrosterone acetate, 176
14-Dehydroequilenin, 52, 288
8-Dehydroestrone, 164
9(11)-Dehydroestrone, 148
16-Dehydro-11-ketoprogesterone,3-ethylene ketal, 185
5(10)-Dehydro-9-methyldecalin-1,6-dione, 130
16-Dehydroprogesterone, 42–45
9 (11) - Dehydrotestosterone, 215, 219, 220
Deoxycholic acid, 176, 178
Deoxycorticosterone, 56, 182
3-Deoxyequilenin, pyrrolo analog, 77
3-Deoxyequilenin, thiophene analog, 73
3-Deoxy-14β-equilenin, thiophene analog, 73
3-Deoxy-14β-estradiol, thiophene analog, 74
17-Deoxyestrone, 156
3-Deoxyisoequilenin, *see* 3-Deoxy-14β-equilenin
3-Deoxyisoestradiol, *see* 3-Deoxy-14β-estradiol
3-Deoxy-18-nor-14-azaequilenin, 78
1,5-Diallyl-2,6-dihydroxynaphthalene, 228
Diastereomers, 25
11,14-Diaza-1,3,5(10),6,8-gonapentaene-12,15-dione, 107
6,8-Diaza-D-homosteroids, 107
13,14-Diaza-D-homosteroids, 74
Diaza-A-nor steroid, 97
3,11-Diazasteroids, 17, 111
4,6-Diazasteroids, 289
5,10-Diazasteroids, 157–158
6,7-Diazasteroids, 153–154
  6,7-Diaza-11-oxoequilenin, 154
6,11-Diazasteroids, 118
8,11-Diazasteroids, 291

8,13-Diazasteroids, 17, 169–170
  2-Methoxy-8,13-diaza-18-norestrone, 170
11,12-Diazasteroids, 117
11,13-Diazasteroids, 74, 75
11,14-Diazasteroids, 107, 288
12,14-Diazasteroids, 78
13,14-Diazasteroids, 74, 110
13,16-Diazasteroids, 78
14,16-Diazasteroids, 78
15,16-Diazasteroids, 78, 138
Diazomethane, 16, 104, 236
Di-*t*-butylethoxymagnesiomalonate, 220
1,3-Dichloro-2-butene, 203, 209
2,3-Dichloro-5,6-dicyano-1,4-benzoquinone, 98, 101, 213, 215
1,2-Dichloro-2-propene, 213
Dieckmann cyclization, 23, 51, 56, 59, 60, 65, 67, 135, 145, 237
  with potassium *t*-butoxide, 249, 251
  with sodium hydride, 249
  with sodium methoxide, 249
Diels-Alder reaction, 9, 20, 71, 75, 116, 126, 182
  in preparation of CD-intermediates, 261
  4-vinylindane with quinones, 156
Diethylamine, 105
1-Diethylaminobutane-3-one, 196
4-Diethylamino-2-butanone methiodide, 177, 222
1-(*N*,*N*-Diethylamino)-3-pentanone methiodide, 181
Diethyl α-methyl-α-propionyl succinate, 65
Diethyl (2-naphthylamino)malonate, 107
13-Difluoromethyl steroid, 106
6,6-Difluoronorgestrel, 288
Digitoxigenin, 3, 6
Dihydrobenz[*f*]isoquinoline, 77
Dihydroconessine, 5
Dihydroequilenin, 54, 221
Dihydrolatifoline, 195
24,25-Dihydroparkeol, 40
24,25-Dihydro-$\Delta^{13(17)}$-protolanosterol, 40
Dihydroresorcinol, 181
3,4-Dihydro-5,6,7-trimethoxy-1-vinylnaphthalene, 71
3β,11β-Dihydroxyandrostan-17-one, 250

Subject Index

3β,11β-Dihydroxy-13α-androstan-17-one, 250
γ-(2,4-Dihydroxybenzoyl)butyric acid, 71
11α,17aβ-Dihydroxy-D-homo-4-gonen-3-one, 106
1,6-Dihydroxynaphthalene, 177
2,6-Dihydroxynaphthalene, 227
3β,17aβ-Dihydroxy-13α-18-nor-D-homoandrostane, 247
3β,17aα-Dihydroxy-14β-18-nor-D-homoandrostane, 247
Diisobutylaluminum hydride, 189
3,6-Dimethoxy-17β-acetyl-1,3,5(10),6,8-estrapentaene, 57, 58
1,5-Dimethoxycyclohexa-1,4-diene, 104
1,7-Dimethoxynaphthalene, 110
2,8-Dimethoxynaphthalene, 58
2,3-Dimethoxy steroids, 93, 107
2,4-Dimethoxy steroids, 93
3-Dimethylamino group, 89
N,N-Dimethylaniline, 228
Dimethyl azidodiformate, 117
Dimethyl copper lithium, 222
5,5-Dimethyl-1,3-cyclohexanedione, 92
2,5-Dimethyl-1-cyclopenten-3-one, 130
16,16-Dimethyl-D-homosteroids, 92
Dimethyl marrianolate methyl ether, 61
Dimethyl oxalate, 16, 189
7,7-Dimethylsteroids, 90
Dimethyl succinate, 20
Dimethylsulfoxide, 105
4,8-Dimethyltestosterone, 279
Diosgenin, 3, 180
Diphenylsulfonium cyclopropylide, 14
Doisynolic acid, 65
Double bond isomerization, 22

E

Enamines, 108
of CD-intermediates, 152–154
Enzymic resolutions, 27
with *Arthrobacter simplex*, 28
with *Bacillus thuringiensis*, 27
with *Corynebacterium simplex*, 27
with 17β-hydroxysteroid dehydrogenase, 27
with 3β,17β-hydroxysteroid dehydrogenase from *Pseudomonas testosteroni*, 28
with *Ophiobolus herpotrichus*, 27, 189
with protaminase, 27
with *Pseudomonas testosteroni*, 27, 203
with *Rhizopus arrhizus*, 27
with *Saccharomyces carlsbergensis*, 27
with *Saccharomyces cerevisae*, 27
with *Saccharomyces uvarum*, 27
with steroid 1,2-dehydrogenase, 28
Enzymic resolution attempted
with *Cunninghamella blakeseeana*, 28
with *Flavobacterium dehydrogenans*, 28
with 3α-hydroxysteroid dehydrogenase from *Pseudomonas testosteroni*, 28
Epiandrosterone, 2–4, 16, 176, 178–180, 249
synthesis according to Robinson-Cornforth, 176–180
13α-Epiandrosterone, 249
Epoxidation
with N-bromosuccinimide in glyme, 37
with m-chloroperbenzoic acid, 16, 39
via Corey's procedure, 39
with paracids, 249
Equatorial substituents, 8
Equilenin, 2, 4, 8, 16–18, 49–56, 63, 73, 86, 113, 114, 116, 165, 179, 201, 290
13-aza, 77
14-aza, 77
15-aza, 78
13-aza-18-nor, 77
furano analog, 71
Johnson's first synthesis, 52
Johnson's second synthesis, 54
pyrazole analogs, 72
synthesis according to Bachmann, 49
synthesis according to Banerjee, 53
14β-Equilenin, 51, 53, 114, 147
cis-16-Equilenone methyl ether, 56, 227
Estra-4,9-diene-3,17-dione, 63, 91
5α-Estra-8,14-dien-17-one, 104
Estradiol, 16, 26, 27, 61, 62, 65, 86, 89, 164, 206, 288
3-glyceryl ether, 62
8α-Estradiol, 221
9β-Estradiol, 207
1,3,5(10),6,8,14-Estrahexaen-3-ol, 103

Estrapentaene, 220
1,3,5-Estratrien-3-ol-16-one, 63
Estr-4-ene-3,17-dione, 270
13α-Estr-4-ene-3,17-dione, 63
Estriol, 93
Estrogens, 206
  review, 1
Estrone, 2, 4, 7, 12, 19, 20, 23, 25, 58–63, 86, 88, 91, 92, 114, 115, 127, 129, 136, 137, 142–145, 147, 148, 164, 165, 208, 288
  absolute configuration, 24
  13-aza, 77
  8,9-dehydro-13-aza, 77
  11β-hydroxy, 215
  3-methyl ether, irradiation, 63
  synthesis according to Anner and Miescher, 60
  synthesis according to Bachmann, 58
  synthesis according to Sheehan, 61
d-Estrone, 165
l-Estrone, 24
Estrone-A, 58
Estrone-a, 61
Estrone-b, 61
Estrone-d, 61
Estrone-e, 61
Estrone-f, 61
8α-Estrone, 88, 164, 220
9β-Estrone, 114, 164
14β-Estrone, 136, 137
Ethoxyacetylenemagnesium bromide, 65
2-Ethoxybutadiene, 14
Ethoxyethynyllithium, 190, 191
Ethoxyethynylmagnesium bromide, 183, 188, 190
3-Ethoxy-1,3-pentadiene, 182
Ethyl acrylate, 107
Ethyl bromoacetate, 110
Ethyl cyanoacetate, 111
2-Ethyl-1,3-cyclopentanedione, 98
Ethyl 2-(3,3-ethylenedioxy-2-pyrrolidyl)-acetate, 109
Ethyl formate, 16, 114
13β-Ethylgon-4-en-3,17-dione, 270
Ethylidenetriphenylphosphorane, 19
13-Ethylsteroids, 92
Ethyl vinyl ketone, 12
Ethynylation, 20

17α-ethynyl-17β-hydroxy-19-norandrost-5(10)-en-3-one, 213
Eugenol, 162
Euphol terpenoids, 36, 39

F

Farnesyl acetate, 39
Formal total synthesis, 2
Formic acid, 88
2-Formyl-1,3-cyclohexanedione, 117
8β-Formylestra-3,17-diol, 150
Fractional crystallization, 25
1,4 Free radical transfer reaction, 194
Friedel-Crafts acylation of anisole, 135, 136
Fujimoto-Belleau reaction, 18, 274
A-Furano steroids, 227
Furfural, 112, 235

G

Gitogenin, 180
Glutaric anhydride, 17
Glutarimide, 105
Gonane, 7
Grignard reaction, 18, 212, 220, 234

H

Heterocyclic steroids, 71–78, 94–103, 131, 154–156, 166–170
Hetercyclic steroids, see azadioxasteroids, azanorsteroids, azaoxasteroids, azasteroids, azathiasteroids, diazahomosteroids, diazanorsteroids, diazasteroids, oxadiazasteroids, oxasteroids, pentaazasteroids, phosphasteroids, silasteroids, tetrazasteroids, thiahomosteroids, thiasteroids
Heterocyclic steroids, review, 1, 286
13-Hexadecylsteroids, 92
D-Homo-4,9(11),16-androstatrien-3-one, 223
D-Homo-4-androstene-3,11,16-trione, 223
D-Homobisnorsteroid, 147
D-Homoequilenin, 55, 91
B-Homoestradiol, 93
18-Homoestradiol, 26, 28
D-Homo-4,8(14),9-estratriene-3,17-dione, 105

Subject Index

D-Homoestrone, 90, 150–151
18-Homoestrone, methyl ether, 27
D-Homo-13-gonen-17a-one, 104
Homomarrianolic acid, 129, 137
D-Homo-19-norandrostenedione, 91
D-Homo-18-norestrone, 145
D-Homo-19-nortestosterone, 91
A-Homosteroids, 93
B-Homosteroids, 86, 93
D-Homosteroids, 9, 10, 18, 42, 86, 90, 102, 103, 106, 107, 115, 128, 130, 146, 148, 163, 234, 235, 247, 256–260, 263, 275, 276, 290
18-Homo steroids, 98, 106, 107
D-Homotestosterone, 276
Hunsdiecker decarboxylation, 253
Hydrindane intermediates, 241–243
  via Diels-Alder reactions, 242
  from Hagemann's ester, 242
  via hydroboration, 242
  trans-intermediates, by hydrogenation, 242–243
  from 2-methylcyclopentane-1,3-dione, 268, 269, 277
  5,6,7,8-tetrahydro-8-methylindane-1,5-dione, 241
Hydroboration, 14
Hydrochloric acid, 88
Hydrochrysene approach, 243–254
  modifications, 254–256
  review, 1
Hydrocyanation, 194
Hydrogenation
  palladium-on-barium carbonate, 185
  palladium-on-calcium carbonate, 89, 94
  palladium-on-charcoal, 88, 91, 94, 96, 114
  palladium-on-strontium carbonate, 94, 107, 222
  platinum-on-carbon, 114
  platinum oxide, 115
  Raney nickel, 221
  rhodium-on-charcoal, 97, 224
  ruthenium, 187
  ruthenium-on-carbon, 218
  ruthenium oxide, 224, 236
  stereochemistry of, 21
Hydronaphthalene intermediates, 239–241
  from 1,5-dihydroxynaphthalene, 239

  from 2,5-dimethoxynaphthalene, 240
  5-methoxy-2-tetralone, 243
  6-methoxy-2-tetralone, 254
  from 4-methoxytoluquinone, 260
  10-methyl-$\Delta^{1,9}$-octalin-2,5-dione, 276
3β-Hydroxy-5β-androstan-17-one, 3
3β-Hydroxyandrost-5-en-16-one, 180
3-Hydroxy-16-equilenin, 63
3-Hydroxy-16-equilenone, 56
3-Hydroxy-1,3,5(10),8,14-estrapentaen-17-one, 87
3-Hydroxy-1,3,5(10),8(14)-estratetraen-17-one, 106
11β-Hydroxyestrone, 215
3-Hydroxyetiochola-1,3,5(10),6,8-pentaenoic acid, 112
3-Hydroxy-1,3,5(10),8,11,13-gonahexaen-17-one, 101
3-Hydroxy-14β-gona-1,3,5(10),6,8-pentaen-17-one, 114
3-Hydroxy-1,3,5(10),13-gonatetraen-17-one, 110
16-Hydroxy-D-homo-4,9(11)-androstadien-3-one, 223
3-Hydroxy-D-homo-1,3,5(10),6,8,13-estrahexaen-17a-one, 104
3-Hydroxy-D-homo-1,3,5(10),13-gonatetraen-17a-one, 110
3-Hydroxy-D-homo-14β-gona-1,3,5(10)-trien-17a-one, 129
3-Hydroxy-D-homo-$9\beta,14\beta$-gona-1,3,5(10)-trien-17a-one, 129
4-Hydroxy-2-methylcyclopentane,-1,3-dione, 93
3-Hydroxy-17-methyl-D-homo-14β-estra-1,3,5(10),9(11),16-pentaene-15,17a-dione, 129
5-Hydroxy-10-methyl-$\Delta^{1(9)}$-2-octalone, 26
4-Hydroxy-2-methyl-3-oxopentanoic acid lactone, 98
17β-Hydroxy-19-nor-8α-androst-5(10)-en-3-one, 221
17β-Hydroxy-B-nor-$9\beta,10\alpha$-estra-4-en-3-one, 94
3-Hydroxy-19-norpregna-1,3,5(10)-trien-20-one, 68
17β-Hydroxy-3-oxo-4-estren-16β-ylacetic acid lactone, 63

3β-Hydroxy-5α-pregnan-20-one, 194, 195
17-Hydroxy-5β-pregnan-20-one, 287
3-Hydroxy-1,3,5(10),16-pregnatetraen-20-one, 129
3β-Hydroxy-5α-pregn-16-en-20-one, 256
11α-Hydroxyprogesterone, 187
11β-Hydroxyprogesterone, 3-ketal, 185
11-Hydroxysteroids, 208
14-Hydroxysteroid, 88
3-Hydroxy-17-thia-D-homo-1,3,5(10)-estratrien-12-one, 108
Hypobromous acid, 131

I

A,B-Indolosteroids, 237, 291
Intramolecular condensation, 19, 235
1-Iodo-6-methoxynaphthalene, 49
Iodosobenzene diacetate, 206
13-Isobutylsteroids, 92
Isoeuphenol, 40
Isomerization, palladized charcoal, 50
2-Isopropylcyclopentanone, 106
Isopropyl 1-methyl-2-oxo-1-cyclopentanecarboxylate, 115
13-Isopropylsteroids, 92, 106
Isosteroids, *see also* specific compounds, 218–222
8-Isosteroids, 220
Isothiouronium salts, 12, 88, 97, 98, 100, 102
(−)-Isotirucallenol, 40

J

Johnson's isoxazole method, 52
Julia olefin synthesis, 43

K

$\Delta^{5(10)}$-3-Ketosteroids, 213
Köster-Logemann ketone, 176, 178, 181

L

Lanosterol, 3, 5, 36, 39
Latifoline, 194, 195–196
Lead tetraacetate, 129
(S)-(−)-Limonene, 39
Liquid ammonia, solvent for Michael addition, 107

Lithium acetylide, 98
Lumiestrone, 142–145

M

Maleic anhydride, 116, 117
Malonic ester, 12
Mannich base, 11
Methallyl iodide, 16, 187, 192
p-Methoxyacetophenone, 235
Methoxyacetylene, lithium salt, 20
3-Methoxy-17-acetyl-18-nor-1,3,5(10)-estratriene-14-ol, 66
8-Methoxybenz[h]chroman-4-one, 72
3-Methoxybenzosuberone, 93
8-Methoxy-3-cyano-3-methylbenz[h]-chroman-4-one, 72
6-Methoxy-3,4-dihydronaphthalene, 67
3-Methoxy-1,3,5(10),8,14-estrapentaen-17-one, 27, 147
3-Methoxy-1,3,5,9(11)-estratetraen-17-one, 152
2-Methoxymethylcyclopentanone, 117
6-Methoxy-9-methyl-Δ^6-1-decalone, 129
3-(3-Methoxy-2-methylphenyl)propanoic acid, 236
6-Methoxy-1-methyl-2-tetralone, 194
β-Methoxynaphthalene, 50
4-(6-Methoxy-1-naphthalene)butyric acid, 49
6-Methoxy-1-naphthol, 71
6-Methoxy-2-naphthyl lithium, 115
3-Methoxy-B-nor-8α-estra-1,3,5(10)-trien-17-one, 94
3-Methoxy-B-nor-9β-estra-1,3,5(10)-trien-17-one, 94
7-Methoxy-1-oxo-1,2,3,4,9,10-hexahydrophenanthrene, 65
7-Methoxy-4-oxo-1,2,3,4-tetrahydrodibenzofuran, 71
7-Methoxy-4-oxo-1,2,3,4-tetrahydrodibenzothiophene, 74
2-(m-Methoxyphenyl)ethylamine, 17
6-(m-Methoxyphenyl)-1-hexen-3-one, 92
6-Methoxy-2-propionylnaphthalene, 63
4-Methoxysteroids, 93
6-Methoxysteroids, 103
5-Methoxy-2-tetralone, 177

Subject Index 325

6-Methoxy-1-tetralone, 50, 67, 87, 106, 108, 118, 194, 202, 222
6-Methoxy-2-tetralone, 66, 74, 109
6-Methoxy-1-vinyl-3,4-dihydronaphthalene, 129
4-Methoxy-2-vinylpyridine, 171–172
6-Methoxy-1-vinyl-1-tetralol, 87
Methyl acrylate, 58, 205
Methyl bromoacetate, 17, 50, 182
Methyl γ-bromocrotonate, 50
Methyl 2-chloroacrylate, 12
2-Methyl-1,3-cyclohexanedione, 12, 90, 91, 104, 105, 234
2-Methyl-1,3-cyclopentanedione, 12, 27, 87, 92, 93, 96–99, 101, 102, 104, 105, 107, 129, 163, 165
  iso-butyl enol ether, 19
  t-butyl enol ether, 116
3′-Methyl-1,2-cyclopentanophenanthrene, Diels hydrocarbon, 181
Methyl 3,4-dihydro-1-naphthoate, 110
2-Methyldihydroresorcinol isobutyl ether, 235
5,10-Methylenesteroid, 16
Methylenetriphenylphosphorane, 99
19-Methyl group, stereospecific introduction, 217
8β-Methyl-D-homoestranes, 148–150
Methyl iodide, 14, 108, 113, 128
Methyl 3-oxoetio-5β-cholanoate, 56
Methylmagnesium bromide, 16, 18, 215, 217
5-Methyl-6-methoxy-1-tetralone, 224
8β-Methyl-B-nor-D-homoestranes, 148–150
6-Methyl-B-norsteroid, 288
2-Methyl-3-oxa-A-nor-1,5(10),8,14-estratetraen-17β-ol, 98
Methyl 5-oxo-6-heptenoate, 12
4-Methyl-4-penten-1-ylmagnesium bromide, 19
6-Methylsteroids, 103
7-Methylsteroids, 98, 213
9-Methylsteroids, 291
12-Methylsteroids, 92
14-Methylsteroids, 130
15-Methylsteroids, 103
16-Methylsteroids, 103, 290
17-Methylsteroids, 103

17α-Methyltestosterone, 212
6-Methyl-1-tetralone, 227
β-Methyltricarballylic acid anhydride, 56, 58
Methyltriphenylphosphonium bromide, 19
Methyl vinyl ketone, 11, 12, 203, 210, 220, 234, 240, 241, 244, 270, 274
Michael addition, 11, 58, 68, 105, 181, 234
Microbiological aromatization with *Arthrobacter simplex*, 207
Microbiological hydroxylation, 189
Morpholide, 194
Morpholine, 88
3-Morpholino group, 89

N

1,5-Naphthalenediol, 111
1,6-Naphthalenediol, 111
1,7-Naphthalenediol, 111
1-Naphthylamine, 110
1-Naphthylamine-6-sulfonic acid, 49
2-Naphthyl lithium, 117
Neotigogenin, 3, 180
Nerolin, 50
Nomenclature for computer use, 7
19-Nor-4-androstene-3,17-dione, 63
19-Nor-8α-anthratestosterone, 219
18-Norestrone, 145
Norethynodrel, 213
9β-Norethynodrel, 207
Norgestrel, 270, 271
18-Nor-D-homoepiandrosterone, 248
18-Nor-D-homoequilenin, 117
18-Nor-16-homomarrianolic acid, 145
C-Nor-D-homosteroids, 28, 256–260, 287
  from fluorene derivatives, 259–260
B-Nor-6-oxaequilenin, 71
A-Nor-3-oxasteroids, 97
19-Norpregnanes, 65–68
19-Norprogesterone, 63, 208
B-Norsteroids, 86, 93, 154
19-Norsteroids, 213, 288, 290, 292
18-Nortestosterone, 251
19-Nortestosterone, 62, 90, 165–166, 201, 203–206
19-Nor-8α-testosterone, 222
19-Nor-8α,10α-testosterone, 221

19-Nor-9β,10α-testosterone, 166, 207
B-Nor-6-thiaequilenin, 74
B-Nor-6-thia-14β-equilenin, 74

O

$\Delta^{1(9)}$-Octalone, 14
Osmium tetroxide, 129
Ovral, see Norgestrel
11-Oxa-15,16-diaza-14-dehydroequilenin methyl ether, 72
11-Oxa-15,16-diaza steroids, 72
6-Oxaestranes, 156–157
Oxasteroids, 71–72, 107
  11-oxa-15,16-diaza, 72
4-Oxasteroids, 91, 98, 289
6-Oxasteroids, 98
11-Oxasteroids, 17, 111
12-Oxasteroids, 110
16-Oxasteroids, 99
Oxidation
  with Benedict solution, 225
  with chromium trioxide, 98, 195, 210
  with chromium trioxide–pyridine complex, 184–186, 210
  with Collins reagent (pyridine–chromic oxide), 45, 150
  with Jones reagent (chromic acid and sulfuric acid), 16, 114
  Oppenauer, 95, 98, 101, 128, 178, 183, 190
  with osmium tetroxide, 210
  with *Pseudomonas testosteroni*, 203
3-Oxo-1-butyl p-bromobenzenesulfonate, dithioketal, 13
3-Oxo-1-butyl tosylate, ethylene ketal, 13
16-Oxoestradiol, 61
3-Oxoetio-5β-cholanoic acid, 4, 182, 263
3-Oxoetio-5β-chola-4,9(11),16-trienoic acid, 263
3′-Oxo-7-methoxy-3,4-dihydro-1,2-cyclopentenophenanthrene, 52
β-Oxopimelic acid, diethyl ester, 59
11-Oxoprogesterone, 187
$\Delta^{5(10)}$-3-Oxosteroids, 213
11-Oxosteroid, 13, 115
12-Oxosteroids, 107, 109
15-Oxosteroids, 111, 130
4-Oxo-1,2,3,4-tetrahydrodibenzothiophene, 73

11-Oxygenated steroids, 103, 222
Organometalic coupling, 115

P

Pentaazasteroids, 78
4-Penten-1-ylmagnesium bromide, 19
13-Pentylsteroids, 106
Perphthalic acid, 208
Phenethylcyclohexanes as AC intermediates, 137–140
Phenylhydrazine, 97
13-Phenylsteroids, 92
Phosphasteroids, 289
11-Phosphasteroid, 117
Phosphorus oxychloride, 97
Phosphorus pentachloride, 102
Phosphorus pentoxide, 88, 104
Phosphorus tribromide, 189, 206
Photochemical transformation, 194
Piperidine, 88
Polyolefin, cyclization of, 8
Polyolefinic cyclization, 35–46
  of acetals, 42
  of allylic alcohols, 42, 44
  with asymmetric induction, 46
  of demethylfarnesic acid, 36
  with stannic chloride in benzene, 38, 42
  with stannic chloride in nitromethane, 40, 42, 45
  of sulfonate esters, 41
  with trifluoroacetic acid, 45
Polyphosphoric acid, 92, 104, 105, 114, 116, 118
Potassium t-butoxide, 87, 128
Potassium hydroxide, 87
Pregnenolone, 3
Progesterone, 45–46, 56, 182, 251, 276
  microbiological oxidation of, 187
17α-Progesterone, 46
α-Propionylpropionic ester, 63
13-Propylnorestradiol, 208
10β-Propylsteroids, 207
13-Propylsteroids, 92, 106
Protective group, trimethylsilyl ether, 218
Pseudomonas testosteroni, 203
Pyrazole A-ring steroids, 97
Pyridine hydrochloride, 91, 96
Pyrrolidine, 88

Subject Index

1-Pyrrolidinocyclopentene, 110
Pyruvic acid, ethyl ester, 237

R

Reduction
  Birch, 16, 22, 62, 91, 92, 94, 96, 98, 100, 106, 146, 147, 150, 171, 194, 203, 208, 219, 222, 235, 236, 240, 247, 258
  calcium and ammonia, 94
  Clemmensen, 235
  diisobutylaluminum hydride, 189
  Huang-Minlon, 68, 195, 256
  lithium aluminum hydride, 89, 95, 96, 104, 178, 189, 194, 195, 218
  lithium and ammonia, 14, 22, 89, 93–95, 113, 130, 220, 222, 226, 246
  lithium tri-t-butoxyaluminum hydride, 129, 227
  Potassium and ammonia, 62, 87, 92, 100
  potassium–ammonia–isopropyl alcohol, 183
  Raney nickel, 16
  sodium and ammonia, 89, 94, 114
  sodium and amyl alcohol, 108
  sodium borohydride, 73, 89, 90, 95, 97, 98, 100, 183, 190, 196, 202, 208, 218, 220, 222, 223
  Wolff-Kishner, 65, 128, 210
Reformatsky reaction, 17, 50, 55, 59, 61, 136, 179
Reich diketone, 176–178
Relay compound, 2
Resibufogenin, 3
Resolution, *see also* enzymic resolutions, 23, 287
  with 2-amino-1-butanol, 26, 27
  via amphetamine salt, 205
  with brucine, 25, 26, 178, 186
  with chloramphenicol, 25
  with dehydroabietylamine, 26
  with diacetyltartaric acid, methyl ester, 26
  with digitonin, 27
  with ephedrine, 26
  with menthol, 26
  with menthoxyacetic acid, 26, 178
  with phenylalanine, 27
  with strychnine, 25, 185
  with tartaric acid, monohydrazide, 27
Retrosteroids, 270, 274–275
  retroandrostenedione, 274
  retroprogesterone, 19, 274
  retrotestosterone, 270, 274
Reviews, 1, 286
Ring-C aromatic steroids, 196
Robinson annelation, 11, 177, 182, 194
Robinson-Rapson synthesis, 112, 146

S

Samanine, 180
Sandmeyer reaction, 49
Sapogenins, 180
Sarett's ketone, 16, 183, 189, 191, 192
Schiff's bases, 118
Sheehan acyloin condensation, 137
6-Silasteroid, 102
Simmons-Smith reagent, 16
Smilagenin, 180
Sodium hydride, 95
Solasodine, 3
Squalene oxide, 36, 37
Steric control, 23
Steroid antibiotics (fusidane series), 279
Steroidal alkaloids, 180
  C-nor-D-homo, 69
8α-Steroids, 98
14β-Steroids, 96
Stobbe condensation, 20, 53, 63, 70, 72, 74, 135, 235
Stork-Eschenmoser hypothesis, 35, 42
Succinic acid, di-t-butyl ester, 235
Succinic acid methyl ester acid chloride, 105
Succinimide, 96, 102, 105
Sulfone analogs of steroids, 101
Sulfuric acid, 103

T

Testosterone, 5, 7, 16, 180, 219, 220, 251, 287
8α-Testosterone, 218, 219, 222
13α-Testosterone, 251
$9\beta,10\alpha$-Testosterone, 270
$19(10\rightarrow9\beta)abeo$-$10\alpha$-Testosterone, 222
Tetrazasteroids, 78

Tetrahydropyranyl ether, protecting group, 89
2-Tetralone, 107
6-Thiaestrone, 101
17-Thia-D-homo steroid, 108
17a-Thia-D-homo steroids, 131
6-Thia-B-nor-1,3,5(10),8-estratetraen-17β-ol, 73
Thiasteroids, 73–4, 101–102, 131, 154–156, 228
6-Thiasteroids, 102, 131, 155
Thiazole A-ring steroids, 102
Thorpe cyclization, 236
Thorpe-type addition, 53
Tigogenin, 3, 6, 180
p-T.oluenesulfonic acid, 88, 97, 98, 100, 105
Tomatidine, 3, 180
Torgov synthesis, 9, 12, 19, 86–103
Total synthesis, definition, 2
Trans-anti-trans-anti-trans backbone, 7
Trienes, 211
Triethylamine, 87
13-Trifluoromethyl steroid, 106
2,3,4-Trimethoxy steriods, 93
trans-5,13,17-Trimethyloctadeca-5,9,13,17-tetraenal ethyleneketal, 8
Trimethylsilyl ether, protective group, 218

Triphenylmethylsodium, 16, 58, 179, 181
Triton B, 87, 93, 105, 220

U

Urinary steroids, 54

V

Veratramine, 3, 69, 170, 251
1-Vinylcyclohexene, 14
Vinylmagnesium bromide, 19, 87, 93, 234
6-Vinyl-2-picoline, 12
2-Vinylpyridine, 107
Vitamin D, 6, 286
von Pechmann coumarin synthesis, 156–157

W

Wittig reaction, 19, 39, 151, 208, 210, 289
  synthesis of calciferol, 151
  synthesis of estrone, 151–152
  synthesis of 18-homoestrone, 151–152
  with triphenylphosphonium isopropylide, 39

X

X-Ray diffraction studies, 90, 107, 207
2,6-Xyloquinone, 129

ORGANIC CHEMISTRY
A SERIES OF MONOGRAPHS

EDITORS

ALFRED T. BLOMQUIST
Department of Chemistry
Cornell University
Ithaca, New York

HARRY WASSERMAN
Department of Chemistry
Yale University
New Haven, Connecticut

1. Wolfgang Kirmse. CARBENE CHEMISTRY, 1964; 2nd Edition, 1971
2. Brandes H. Smith. BRIDGED AROMATIC COMPOUNDS, 1964
3. Michael Hanack. CONFORMATION THEORY, 1965
4. Donald J. Cram. FUNDAMENTALS OF CARBANION CHEMISTRY, 1965
5. Kenneth B. Wiberg (Editor). OXIDATION IN ORGANIC CHEMISTRY, PART A, 1965; Walter S. Trahanovsky (Editor). OXIDATION IN ORGANIC CHEMISTRY, PART B, 1973
6. R. F. Hudson. STRUCTURE AND MECHANISM IN ORGANO-PHOSPHORUS CHEMISTRY, 1965
7. A. William Johnson. YLID CHEMISTRY, 1966
8. Jan Hamer (Editor). 1,4-CYCLOADDITION REACTIONS, 1967
9. Henri Ulrich. CYCLOADDITION REACTIONS OF HETEROCUMULENES, 1967
10. M. P. Cava and M. J. Mitchell. CYCLOBUTADIENE AND RELATED COMPOUNDS, 1967
11. Reinhard W. Hoffman. DEHYDROBENZENE AND CYCLOALKYNES, 1967
12. Stanley R. Sandler and Wolf Karo. ORGANIC FUNCTIONAL GROUP PREPARATIONS, VOLUME I, 1968; VOLUME II, 1971; VOLUME III, 1972
13. Robert J. Cotter and Markus Matzner. RING-FORMING POLYMERIZATIONS, PART A, 1969; PART B, 1; B, 2, 1972
14. R. H. DeWolfe. CARBOXYLIC ORTHO ACID DERIVATIVES, 1970
15. R. Foster. ORGANIC CHARGE-TRANSFER COMPLEXES, 1969
16. James P. Snyder (Editor). NONBENZENOID AROMATICS, VOLUME I, 1969; VOLUME II, 1971

17. C. H. Rochester. ACIDITY FUNCTIONS, 1970
18. Richard J. Sundberg. THE CHEMISTRY OF INDOLES, 1970
19. A. R. Katritzky and J. M. Lagowski. CHEMISTRY OF THE HETEROCYCLIC N-OXIDES, 1970
20. Ivar Ugi (Editor). ISONITRILE CHEMISTRY, 1971
21. G. Chiurdoglu (Editor). CONFORMATIONAL ANALYSIS, 1971
22. Gottfried Schill. CATENANES, ROTAXANES, AND KNOTS, 1971
23. M. Liler. REACTION MECHANISMS IN SULPHURIC ACID AND OTHER STRONG ACID SOLUTIONS, 1971
24. J. B. Stothers. CARBON-13 NMR SPECTROSCOPY, 1972
25. Maurice Shamma. THE ISOQUINOLINE ALKALOIDS: CHEMISTRY AND PHARMACOLOGY, 1972
26. Samuel P. McManus (Editor). ORGANIC REACTIVE INTERMEDIATES, 1973
27. H.C. Van der Plas. RING TRANSFORMATIONS OF HETEROCYCLES, VOLUMES 1 AND 2, 1973
28. Paul N. Rylander. ORGANIC SYNTHESIS WITH NOBLE METAL CATALYSTS, 1973
29. Stanley R. Sandler and Wolf Karo. POLYMER SYNTHESES, VOLUME I, 1974
30. Robert T. Blickenstaff, Anil C. Ghosh, and Gordon C. Wolf. TOTAL SYNTHESIS OF STEROIDS, 1974

In preparation

Barry M. Trost and Lawrence S. Melvin, Jr. SULFUR YLIDES: EMERGING SYNTHETIC INTERMEDIATES